Heavy Metal Contamination in the Environment

This reference book explores the multifaceted problem of heavy metal contamination in the environment. Through its in-depth analysis, the book provides a thorough overview of the sources and pathways of heavy metals, their persistence in ecosystems, and the resulting health impacts on individuals and ecosystems. The chapters explore the diverse sources of contamination, including industrial activities, mining, agriculture, and urbanization, while examining the types of heavy metals found in the environment and their toxicological properties. The book further reviews the profound health effects associated with heavy metal exposure, such as neurological disorders, developmental abnormalities, carcinogenicity, and organ damage. Furthermore, the book provides insights into risk assessment methodologies, regulatory frameworks, and guidelines aimed at controlling and minimizing heavy metal exposure. It highlights the challenges and gaps in current regulations, identifies potential areas for improvement, and presents analytical techniques for heavy metal analysis and removal. This book is an important source for researchers and professionals working in the fields of environmental science, toxicology, and public health.

Heavy Metal Contamination in the Environment

Health Impacts and Potential Remediation Approaches

Edited by
Veer Singh, Ashish Kumar, Vishal Mishra
and Sachchida Nand Rai

CRC Press
Taylor & Francis Group
Boca Raton London New York

CRC Press is an imprint of the
Taylor & Francis Group, an **informa** business

Designed cover image: Shutterstock

First edition published 2024
by CRC Press
2385 NW Executive Center Drive, Suite 320, Boca Raton FL 33431

and by CRC Press
4 Park Square, Milton Park, Abingdon, Oxon, OX14 4RN

CRC Press is an imprint of Taylor & Francis Group, LLC

ISBN: 9781032685762 (hbk)
ISBN: 9781032685786 (pbk)
ISBN: 9781032685793 (ebk)

DOI: 10.1201/9781032685793

Typeset in Times
by KnowledgeWorks Global Ltd.

Contents

Preface

Environmental pollution, particularly through heavy metals, has emerged as a critical concern worldwide. The pervasive distribution of heavy metals in soil, water, and air results from various anthropogenic activities such as industrial operations, mining, agriculture, and improper waste disposal. This widespread contamination poses significant risks to environmental health and human wellbeing, as heavy metals are known for their toxicity, persistence, and bioaccumulation potential. Heavy metals, including lead (Pb), cadmium (Cd), mercury (Hg), arsenic (As), and chromium (Cr), among others, are naturally occurring elements that have been extensively utilized in industrial processes. However, their non-degradable nature leads to persistent environmental pollution. When released into the environment, these metals can undergo complex chemical transformations, influencing their mobility and toxicity. Soil and water, as primary repositories of these contaminants, act as conduits for their entry into the food chain, ultimately impacting human health.

The health impacts of heavy metals are profound and multifaceted. Acute exposure can lead to immediate health issues such as gastrointestinal distress, respiratory problems, and skin lesions. Chronic exposure, on the other hand, is associated with long-term health conditions, including neurological disorders, cardiovascular diseases, renal dysfunction, and carcinogenic effects. Vulnerable populations, particularly children and pregnant women, are at heightened risk, with developmental and reproductive health being significantly compromised. Understanding the pathways of heavy metal contamination and their impacts on health is crucial for developing effective remediation strategies. Traditional methods of heavy metal remediation, such as excavation, chemical treatment, and stabilization, often come with high costs and secondary environmental impacts. Hence, there is an increasing shift toward sustainable and eco-friendly remediation approaches that leverage biological systems and innovative technologies.

In this context, this book, *Heavy Metal Contamination in the Environment: Health Impacts and Potential Remediation Approaches*, aims to provide a comprehensive overview of the current state of knowledge on heavy metal contamination. It covers the sources, distribution, and environmental fate of heavy metals, elucidating their pathways from release to bioaccumulation in living organisms. The book also delves into the toxicological profiles of various heavy metals, detailing their mechanisms of action and health implications. A significant portion of the book is dedicated to exploring potential remediation approaches. Emphasis is placed on bio-based solutions, such as phytoremediation, microbial bioremediation, and the use of biopolymers and biosorbents. These methods harness the natural detoxifying capabilities of plants, microorganisms, and other biological materials, offering sustainable alternatives to conventional techniques. Additionally, the book discusses the role of advanced technologies, including nanotechnology and artificial intelligence, in enhancing the efficiency and effectiveness of remediation efforts.

The interdisciplinary nature of this book makes it a valuable resource for a wide range of readers, including researchers, environmental scientists, policymakers, and students. By integrating perspectives from environmental science, toxicology, and bioremediation, the book provides a holistic understanding of heavy metal contamination and its remediation. It aims to foster a deeper appreciation of the complexities involved in managing environmental pollutants and the need for innovative, sustainable solutions. In conclusion, addressing the challenge of heavy metal contamination requires a concerted effort from all sectors of society. This book serves as a call to action for researchers and practitioners to collaborate in developing and implementing effective remediation strategies. By advancing our understanding and application of sustainable remediation technologies, we can mitigate the health impacts of heavy metal pollution and protect our environment for future generations.

I hope that this book will inspire further research and innovation in the field of environmental remediation and contribute to the global effort to achieve a cleaner, healthier environment.

AIM AND SCOPE

This book, *Heavy Metal Contamination in the Environment: Health Impacts and Potential Remediation Approaches*, aims to provide a comprehensive examination of the sources, distribution, and environmental fate of heavy metals, highlighting their pathways from release to bioaccumulation in living organisms. It explores the toxicological profiles of various heavy metals, detailing their mechanisms of action and associated health risks. The book also delves into potential remediation approaches, with a focus on sustainable, bio-based solutions such as phytoremediation, microbial bioremediation, and biosorbents, alongside advanced technologies like nanotechnology and artificial intelligence. By integrating perspectives from environmental science, toxicology, and bioremediation, this book serves as a valuable resource for researchers, environmental scientists, policymakers, and students, fostering a holistic understanding of heavy metal contamination and promoting innovative strategies for effective remediation and environmental protection.

Acknowledgments

It gives us immense pleasure to acknowledge ICMR – Rajendra Memorial Research Institute of Medical Sciences, Agamkuan, Patna. Veer Singh would like to acknowledge the Department of Health Research (DHR), Ministry of Health and Family Welfare, Govt. of India, for support through the Young Scientist Fellowship Grant (File No.R.12014/37/2022-HR). Dr. Ashish Kumar would like to acknowledge ICMR-Rajendra Memorial Research Institute of Medical Sciences, Agamkuan, Patna, for providing workplace and necessary research funding. Dr. Vishal Mishra would like to acknowledge the Indian Institute of Technology (BHU), Varanasi, for necessary research support. Dr. Sachchida Nand Rai would like to acknowledge Banaras Hindu University, Varanasi – 221005, India. It is also our great pleasure to acknowledge and express our enormous debt to all the contributors who have provided their quality material to prepare this book. We are grateful to our beloved family members, who joyfully supported and stood with us in many hours of our absence to finish this book project. We are also grateful to our friends and colleagues who offered their support and encouragement throughout the writing process. I would also like to acknowledge the publishing team at CRC Press for their professionalism, enthusiasm, and belief in this project. Their expertise in design, production, and marketing has been essential in bringing this book to life.

Veer Singh
Ashish Kumar
Vishal Mishra
Sachchida Nand Rai

Contributors

Maaz Abdullah
Civil Engineering Department
Aligarh Muslim University
Aligarh, Uttar Pradesh, India

Syed Fayez Ali
Civil Engineering Department
Aligarh Muslim University
Aligarh, Uttar Pradesh, India

Nasreen Ghazi Ansari
Analytical Chemistry Laboratory, Regulatory Toxicology Group
CSIR-Indian Institute of Toxicology Research (CSIR-IITR)
Lucknow, Uttar Pradesh, India
Academy of Scientific and Innovative Research (AcSIR)
Ghaziabad, Uttar Pradesh, India

Shashi Lata Bharati
Department of Chemistry
North-Eastern Regional Institute of Science and Technology
Nirjuli, Arunachal Pradesh, India

Pankaj Kumar Chaurasia
Department of Chemistry (P.G.)
L.S. College
Muzaffarpur, Bihar, India

Asmita De
Department of Chemistry
Birla Institute of Technology
Mesra, Ranchi, Jharkhand, India

Sapna Desai
Department of Pharmacology
Pioneer Pharmacy College
Vadodara, Gujarat, India

Akanksha Dwivedi
Laboratory of Ecotoxicology
Department of Botany
Institute of Science
Banaras Hindu University
Varanasi, Uttar Pradesh, India

Shivani Gupta
Laboratory of Ecotoxicology
Department of Botany
Institute of Science
Banaras Hindu University
Varanasi, Uttar Pradesh, India

Jyoti
Department of Environmental Science
Babasaheb Bhimrao Ambedkar Central University
Lucknow, Uttar Pradesh, India

Saif Ullah Khan
Civil Engineering Department
Aligarh Muslim University
Aligarh, Uttar Pradesh, India

Asheesh Kumar
Department of Civil Engineering
Zakir Husain College of Engineering and Technology
Aligarh Muslim University
Aligarh, Uttar Pradesh, India

Indrajeet Kumar
Laboratory of Ecotoxicology
Department of Botany
Institute of Science
Banaras Hindu University
Varanasi, Uttar Pradesh, India

Umesh Kumar
Laboratory of Ecotoxicology
Department of Botany
Institute of Science
Banaras Hindu University
Varanasi, Uttar Pradesh, India

Sneh Lata
Analytical Chemistry Laboratory, Regulatory Toxicology Group
CSIR-Indian Institute of Toxicology Research (CSIR-IITR)
Lucknow, Uttar Pradesh, India
Academy of Scientific and Innovative Research (AcSIR)
Ghaziabad, Uttar Pradesh, India

Parvati Madheshiya
Laboratory of Ecotoxicology
Department of Botany
Institute of Science
Banaras Hindu University
Varanasi, Uttar Pradesh, India

Ashish Kumar Mishra
Laboratory of Ecotoxicology
Department of Botany
Institute of Science
Banaras Hindu University
Varanasi, Uttar Pradesh, India

Saumya Mishra
Laboratory of Ecotoxicology
Department of Botany
Institute of Science
Banaras Hindu University
Varanasi, Uttar Pradesh, India

Soumitra Mishra
T. S. Misra Medical College and Hospital
Lucknow, Uttar Pradesh, India

Sumit Mishra
Department of Chemistry
Birla Institute of Technology
Mesra, Ranchi, Jharkhand, India

Vishal Mishra
School of Biochemical Engineering
Indian Institute of Technology
Banaras Hindu University
Varanasi, Uttar Pradesh, India

Nagraj
Department of Chemistry
L.S. College
Muzaffarpur, Bihar, India

Himanshu N Panchal
Department of Pharmacology
PIP, Parul University
Vadodara, Gujarat, India

Manish Pant
Land, Air and Water – Centre of Excellence
Department of Chemistry
School of Physical Sciences
DIT University
Dehradun, Uttarakhand, India

Sushant Pathak
Department of Civil Engineering
Zakir Husain College of Engineering and Technology
Aligarh Muslim University
Aligarh, Uttar Pradesh, India

Avanish Prajapati
Laboratory of Ecotoxicology
Department of Botany
Institute of Science
Banaras Hindu University
Varanasi, Uttar Pradesh, India

Bhavani R
Department of Biotechnology and Medical Engineering
National Institute of Technology
Rourkela, Odisha, India

Nagendra Sharma
P.G. Department of Chemistry
L.S. College
Muzaffarpur, Bihar, India

Rajesh Kumar Sharma
Laboratory of Ecotoxicology
Department of Botany
Institute of Science
Banaras Hindu University
Varanasi, Uttar Pradesh, India

Swati Sharma
Viral Research and Diagnostic Laboratory
Department of Virology
Rajendra Memorial Research Institute of Medical Sciences (RMRIMS)
Patna, Bihar, India

Priya Shukla
School of Biochemical Engineering
Indian Institute of Technology
Banaras Hindu University
Varanasi, Uttar Pradesh, India

Anjali Singh
School of Biochemical Engineering
Indian Institute of Technology
Banaras Hindu University
Varanasi, Uttar Pradesh, India

Bhanu Pratap Singh
School of Biochemical Engineering
Indian Institute of Technology
Banaras Hindu University
Varanasi, Uttar Pradesh, India

Jabrinder Singh
Land, Air and Water – Centre of Excellence
Department of Chemistry
School of Physical Sciences
DIT University
Dehradun, Uttarakhand, India

Kamal Singh
Viral Research and Diagnostic Laboratory
Department of Virology
Rajendra Memorial Research Institute of Medical Sciences (RMRIMS)
Patna, Bihar, India

Prince Kumar Singh
Laboratory of Ecotoxicology
Department of Botany
Institute of Science
Banaras Hindu University
Varanasi, Uttar Pradesh, India

Priyanka Singh
Laboratory of Ecotoxicology
Department of Botany
Institute of Science
Banaras Hindu University
Varanasi, Uttar Pradesh, India

Pradeep Srivastava
School of Biochemical Engineering
Indian Institute of Technology
Banaras Hindu University
Varanasi, Uttar Pradesh, India

Divya J Tandel
Department of Pharmacology
PIP, Parul University
Vadodara, Gujarat, India

Supriya Tiwari
Laboratory of Ecotoxicology
Department of Botany
Institute of Science
Banaras Hindu University
Varanasi, Uttar Pradesh, India

Surabhi Zainith
Babasaheb Bhimrao Ambedkar Central University
Lucknow, Uttar Pradesh, India

The Editors

Dr. Veer Singh is currently working as a Young Scientist (DHR) at the Department of Biochemistry, Rajendra Memorial Research Institute of Medical Sciences, Patna, India. He got his PhD from the Indian Institute of Technology (Banaras Hindu University), Varanasi. He has more than 45 publications and his area of work involved environmental contamination and toxicity, the role of drinking-water heavy metal contamination in visceral leishmaniasis (VL), and post-kala-azar dermal leishmaniasis (PKDL) development.

Dr. Ashish Kumar is currently working as Scientist-C and Head of Department at the Department of Biochemistry, Rajendra Memorial Research Institute of Medical Sciences, Patna, India. He has published more than 40 publications in international/national journals. He has received several awards and received several funding (sponsored projects) from many national and international agencies. Currently, four doctoral and two post-doctoral students are working under the supervision of Dr. Ashish Kumar. Dr. Kumar is working in the interdisciplinary area of life science. His work is mainly focused on visceral leishmaniasis (VL) and post kala-azar dermal leishmaniasis (PKDL). He is also working on VL and PKDL diagnosis, treatment, and environmental impacts on disease development.

Dr. Vishal Mishra is currently working as Assistant Professor in the School of Biochemical Engineering, Indian Institute of Technology (BHU), Varanasi, India. He got his PhD degree from IIT Roorkee, India. He has more than 100 potential publications and his area of work involved wastewater treatment, heavy metal contamination, life-cycle assessment, agro-waste biomass, and carbon materials/green synthesized nanomaterials for enhanced capacitance.

Dr. Sachchida Nand Rai is currently working as a Postdoctoral Fellow at the Centre of Experimental Medicine & Surgery, Banaras Hindu University, Varanasi, India. He got his PhD degree from Banaras Hindu University, Varanasi, India. He has more than 60 potential publications and his area of work involved neurological diseases, neurotoxicity, mushroom biotechnology, heavy metal toxicity, environmental contaminations, and their remediation.

1 The Overview of the Problem of Heavy Metals Contamination in the Environment

*Maaz Abdullah, Syed Fayez Ali, and Saif Ullah Khan**
Civil Engineering Department, Aligarh Muslim University, Aligarh, India

1.1 INTRODUCTION

Pollution of the environment through the inclusion of harmful substances is an existing concern among environmentalists on a global scale. The formulation of potentially effective solutions to combat these issues requires a thorough examination of the constituents of contaminants, along with their impacts and treatment methodologies (Khan et al. 2023). Recent studies have noted that geological activities, including volcanic eruptions and rock weathering, coupled with human activities in the vicinity like industrial processes, pharmaceuticals, household waste, and fertilizers are significant contributors to soil and water pollution (Onakpa et al. 2018; Kumar et al. 2020). The necessity of water for survival adds to the pre-existing concern for addressing the deteriorating quality of ground and surface water. Numerous researchers, including Moss (2008), Dwivedi (2017), and Walker et al. (2019). Soil pollution, have examined the contamination of these water sources, along with regulatory and remediation strategies. Soil pollution, with its implications for agriculture and urban settings, has similarly drawn considerable attention from researchers (He et al. 2015; Li et al. 2019).

Heavy metals are of particular concern among the pollutants responsible for soil and water contamination and are usually defined as metals with atomic weights between 63.5 and 200.6 g/mol with densities greater than 5 g/mL (Sadak 2023). These metals naturally exist in infinitesimal quantities within the environment. However, the accumulation of these metals in the environment due to their non-degradable and persistent nature in the ecosystem is known as heavy metal contamination. This contamination poses serious risks to both ecosystems and public health, arising from natural sources as well as human activities (Gautam et al. 2016). Recently, the increase in human intervention in the environment has also led to the build-up of heavy metals in urban road sediments (Hanfi et al. 2020). Metals such as arsenic (As), cadmium (Cd), lead (Pb), and mercury (Hg) are particularly harmful to plants and animals, even at low levels (Gumpu et al. 2015; Zaidi et al. 2021a). These four are exclusively referenced for the representation of heavy metals as they are also included in the identified group of ten chemicals of major public health concern by WHO (Brathwaite and Rabone 1985; Ojo et al. 2017).

* **Corresponding Author:** saifkhan@zhcet.ac.in

DOI: 10.1201/9781032685793-1

Heavy metals pose more potential threats on a long-time scale. This is because their concentration is continuously increasing as they are not degradable (Salomons and Förstner 2012). A rise in the flux of metallic compounds in the aquatic and terrestrial environment is a natural consequence of the massive increase in the use of heavy metal ions during the past several decades (Knopf and König 2010; Gautam et al. 2016). The global level of industrialization and urbanization is considered a primary contributor to the problem of heavy metals contamination (Khan et al. 2015; Kumar et al. 2020). This study intends to provide an overview of the problem of heavy metal contamination by discussing its potential sources in the environment and the pathways through which it is transported in agricultural lands, sediments, and water. Furthermore, it also sheds light on the consequential effect of heavy metals contamination on human lives and ecosystems, its potential mitigation and remedial measures, along with regulatory frameworks and international agreements concerning heavy metal poisoning.

1.2 SOURCES OF HEAVY METALS IN THE ENVIRONMENT

The problem of heavy metals contamination in surface and groundwater, soil, and atmosphere can be attributed to numerous sources of natural or anthropogenic origin. The sources responsible for the contamination of different parts of the ecosystem vary accordingly along with the distribution of heavy metals. Romic and Romic (2003) conducted a case study and noted the strong skewness of Cd, Cu, Ni, Pb, and Zn in their study area. Few studies reporting heavy metal contamination around the globe are enlisted in Table 1.1.

TABLE 1.1
Few Instances of Reported Heavy Metal Contamination around the Globe

S. No.	Heavy Metals	Country/Region	Media Type	Source of Pollution	Reference
1.	As, Cd, Co, Cr, Cu, Fe, Mn, Ni, Pb, Zn	India	Aquatic sediments	Anthropogenic activities, lithogenic factors, and sand intrusion	Kumar et al. (2020)
2.	Cu, Hg, Pb, Zn	New Zealand	Sulphide deposits	Rock and mineral deposits	Brathwaite and Rabone (1985)
3.	Cd, Cr, Cu, Hg, Ni, Pb, Zn	Jamaica	Stream sediments	Anthropogenic pollution	Knight et al. (1997)
4.	Cd, Co, Cr, Cu, Hg, Mn, Ni, Pb, Zn	Pakistan	Indus drainage system	Waste water discharges and runoff	Khan et al. (2022)
5.	Cd, Cu, Pb, Zn	Croatia	Krka river	Organic matter, carbonates, and hydrous oxides	Šurija and Branica (1995)
6.	As, Cd, Cr, Cu, Hg, Ni, Zn	Dalian, China	Drinking water	Ore mining and industrial influents	Dong et al. (2020)
7.	Cd, Cr, Co, Ni, Pb, Zn	Ranipet town, India	Surface water	Leather tanneries	Srinivasa Gowd and Govil (2008)
8.	Cu, Ni, Zn	Turkey	Küçük Menderes River	Pesticides	Turgut (2003)
9.	As, Cd, Pb, Zn	Egypt	Tap water	Corrosion and impurity in pipe fittings	El-Harouny et al. (2008)
10.	Cd, Pb, Zn	Oklahoma, USA	Soil	Mining and smelting activities	Basta and Gradwohl (2000)
11.	As, Cd, Cu, Pb, Zn	Cabezo Rajao, Spain	Surface water and soil	Abandoned mine sites	Navarro et al. (2008)

(Continued)

TABLE 1.1 *(Continued)*
Few Instances of Reported Heavy Metal Contamination around the Globe

S. No.	Heavy Metals	Country/Region	Media Type	Source of Pollution	Reference
12.	Cd, Cu, Fe, Ni, Pb, Zn	Iraq	Tigris and Euphrates rivers	Industry and domestic sewage, weathering of minerals	Abdullah (2013)
13.	Cd, Cr, Cu, Fe, Mn, Ni, Pb, Zn	Iran	Drinking water and irrigation water from Tembi River	Industrial and domestic waste	Shanbehzadeh et al. (2014)
14.	As, Cd	China	Changjiang River	Anthropogenic sources	Wang et al. (2011)

1.2.1 Natural Sources of Heavy Metal Contamination

Heavy metals are distributed into the environment through several natural processes. Magmatic rocks or igneous rocks are transported to the surface of the Earth by plate tectonics or volcanic eruptions. These primary rocks are solidified magma and contain a variety of chemicals including some trace amounts of heavy metals in their crystal lattice (Bradl 2005). Another form of rock-induced sources of contamination is physical and chemical weathering of rocks. Carbonate rocks are well-known to possess traces of heavy metals and have been studied comprehensively by numerous researchers (Šurija and Branica 1995; Talluri et al. 2000). Ni et al. (2009) noted the potential migration of heavy metals from the weathered carbonate rocks to smaller carbonate rocks and the surrounding soil.

Acid rain is also considered an environmental concern for an extended duration of time. The H^+ ion in the acidic water displaces the cations from their binding sites and allows the dispersion of cations in the surrounding area including Na^+, Mg^{2+}, K^+, etc. Similarly, toxic cations like Zn^{2+}, Cr^{3+}, and Cd^{2+} are also displaced from the negatively charged particles and result in heavy metals contamination (Wu et al. 2010; Zheng et al. 2012). Flash floods can also be considered as one of the sources of heavy metals in the environment as they are known to play an important role in creating and reshaping the pollutant pattern in river systems (Grygar et al. 2014). The sediments transported with flood water are carriers of harmful microorganisms and heavy metals that are gradually dispersed in the flood plain when the flood water enters a locality. Ciszewski and Grygar (2016) noted that extreme precipitation events cause failures of settling ponds and accumulate washes from stockpiles. The sediments carrying the harmful products remain in the surrounding area even after the flood water sweeps away and causes heavy metal contamination in the urban soils. The distribution of sediment in floodwater events requires a much deeper study of the riverbed morphology and flow, as noted by Paul and Biswas (2019), and Liu et al. (2022).

1.2.2 Anthropogenic Sources of Heavy Metal Contamination

Anthropogenic sources of heavy metal in soil and water are mainly hazardous solid waste disposal and combustion processes in industry, mining activities, and transportation (Mateen et al. 2020). Effluents from the industrial sector are the primary source of heavy metals contamination in the environment. The untreated water used for irrigation in agricultural lands contains effluents released from paper mills and factories that increase the concentration of alkalis, cyanide, and heavy metals in the soil (Basta and Gradwohl 2000; Gautam et al. 2016). Hordes of pollutants are potentially released into the atmosphere, nearby soils, and water as a result of mining activities including the extraction and manufacture of metal products (Parth et al. 2011). The concentration of heavy metals in the area surrounding the mine alters based on the weather and the geochemical properties of its sediments and surroundings (Sherameti and Varma 2015). Vareda et al. (2019) performed a case study analysing the accumulation of heavy metals in the environment under mining and industrial activities and inferred that the sites in the neighbourhood of places where heavy metals are extracted through mining or processed are contaminated in water, sediments, and soil.

Developing countries supposedly require more production of agricultural goods to sustain the lives of their civilians. The Green Revolution came into existence to solve the problem of adequate food supply by encouraging the use of fertilizers and pesticides (Sharma and Singhvi 2017). Agriculture watersheds are greatly influenced by the usage of fertilizers and pesticides as an appreciable amount of chemicals are supplied to the agricultural soils through them (Atafar et al. 2010). Surface water and aquifers in the vicinity of the agricultural lands are also affected by their usage when the irrigation or flood water washes away the particles containing traces of heavy metals. Nicholson et al. (2003) also described how long-term usage of agricultural land with frequent use of pesticides results in the accumulation of heavy metals in the topsoil. Anthropogenic activities are also considered a primary reason for the increase in the frequency of floods, forest fires, acid rain, etc. These sources of heavy metals are of natural origin; however, human activities interfere with the intensity of their impact by altering their magnitude and frequency.

1.3 COMMON HEAVY METALS AND THEIR PROPERTIES

Several heavy metals are deemed to be a source of heavy metals contamination inflicting harm on the lives of flora and fauna. Heavy metals along with their toxic limits in drinking water are shown in Table 1.2. Moreover, specifically prominent heavy metals are briefly discussed below:

TABLE 1.2
Heavy Metals and Their Toxic Limits in Drinking Water

S. No	Heavy Metals	Atomic Number	Atomic Weight (g/mol)	Drinking Water Limit (ppm)
1.	Arsenic (As)	33	74.92	0.010
2.	Cadmium (Cd)	48	112.41	0.005
3.	Chromium (Cr)	24	52.00	0.100
4.	Copper (Cu)	29	63.54	1.300
5.	Lead (Pb)	82	207.21	0.015
6.	Mercury (Hg)	80	200.59	0.002
7.	Nickel (Ni)	28	58.69	0.020
8.	Zinc (Zn)	30	65.41	5.000

1.3.1 Arsenic (As)

Arsenic naturally occurs in minuscule quantities on the surface of the Earth and is mainly transported through dust particles accompanied by wind and runoff from water sources. Recently, arsenic has been considered a primary cause of health problems for numerous thousand people because of arsenic-contaminated groundwater (Khan et al. 2021). Minerals bearing arsenic were initially mined in Chinese, Indian, Greek, and Egyptian civilizations as noted by Matschullat (2000). Arsenic is mainly used in the preparation of special types of glass and for the preservation of wood. It is quite terrifying that even a low concentration of arsenic in drinking water supplies is believed to be a cause of skin and other types of cancer inside the human body (National Research Council 1999).

1.3.2 Cadmium (Cd)

Cadmium is known to be a hazardous non-essential transition metal and has been a prevalent contaminant since the 20th century (Byrne et al. 2009) for the lives of humans and animals alike as faces threats regarding health problems due to its presence. It is a contaminant that comes from industrial and agricultural sources and can also be found to exist naturally in the environment (Genchi et al. 2020). It is also displaced by H^+ ions during acid rain and the Cd^{2+} ion becomes a cause of heavy metals contamination. It usually occurs in zinc ores and is used as electricity conductors, for the process of galvanizations and electroplating and in the manufacture of several goods including pigments and alloys (Byrne et al. 2009). Cadmium is also known as a primary pollutant in the arctic regions (Rahman and Singh 2019).

1.3.3 Chromium (Cr)

Chromium is an element that does not occur in the natural environment in its elemental form but is found in ores, predominantly in the form of chromite ($FeOCr_2O_3$) (Ray and Ray 2009). The two stable and predominately present oxidation states of chromium are Cr(III) and Cr(VI). Despite being a necessary element for both humans and animals and playing a significant part in the metabolism of glucose and cholesterol, chromium is hazardous at large concentrations (Rahman and Singh 2019). Chromium is used in metal corrosion inhibition, wood preservation, and tanning of leather. In particular, the wastewater control of tanneries is not adequate to remove the unprocessed and becomes a major source of contamination of chromium in the environment (Dhal et al. 2013; Zaidi et al. 2021b).

1.3.4 Copper (Cu)

Copper generally exists in both Cu(I) and Cu(II) oxidation states. Conversely, the property of copper to cycle between these two oxidation states also results in the potential toxicity of this metal by the release of superoxide and hydroxyl radicals (Stern 2010). The extraction of copper from ores is a major source of contamination in the environment. This is attributed to the extraction from low-grade ores and the role of smelter plants, which overburdens the environment with copper as well as other heavy metal contaminants (Punia 2021).

1.3.5 Lead (Pb)

According to the report provided by USGS (2016), China is the prime supplier of lead and is followed by Australia. Anthropogenic sources of lead in the environment include ammunition, pesticides, pipes (Roman civilizations also used lead pipes due to their corrosion resistance), and dyes (Rahman and Singh 2019). Richardson et al. (2001) noted the natural sources of lead contamination in the environment including volcanic eruptions, natural deposits, and fires.

1.3.6 Mercury (Hg)

Mercury is one of the most known and worked upon heavy metal contaminants due to its toxic nature. Lindqvist (1991) noted how mercury received special attention for several years due to its toxicological relevance after the outbreak of Minamata disease in 1956. Major sources of anthropogenic origin that result in contamination of mercury in the environment are small-scale gold mining, combustion of coal, production of cement, ferrous and non-ferrous metals, and use of fertilizers and pesticides (Wang et al. 2004; Rahman and Singh 2019). While the natural sources of mercury contamination include wildfire, weathering of rocks, and volcanic eruptions (Wang et al. 2004). Xu et al. (2015) also noted the re-emission of mercury from pre-existing deposition in land and ocean.

1.3.7 Nickel (Ni)

It was noted by Bartzas et al. (2021) that the employment of pyro-metallurgical and hydrometallurgical processes is the primary methodology of the nickel industry for creating nickel metal or nickel alloys, which leads to the generation of significant amounts of solid and liquid wastes, especially slags and leaching residues. Additionally, the manufacturing of nickel through pyro-metallurgy leads to the release of large amounts of greenhouse gases (GHGs). Nickel is also transported in the soil through the efforts of industrial waste materials, lime, fertilizers, and sewage sludge (Iyaka 2011) and is known to exist in several forms.

1.3.8 Zinc (Zn)

Galvanization of iron to prevent corrosion is considered to be a primary source of zinc metal and a major cause of heavy metal contamination in the environment due to its mining and smelting processes (Xu et al. 2023). Zinc is known to be an essential element for the functioning of human and animal metabolisms. However, an increased concentration of zinc leads to several problems and disadvantages in the life of flora and fauna. This is mainly attributed to the use of irrigation water containing effluents from smelting processes without proper treatment, which eventually gives rise to severe health problems in arid regions (Xu et al. 2023).

1.4 PATHWAYS OF HEAVY METALS CONTAMINATION

The distribution of heavy metals is influenced by the medium being studied as well as the proximity to emission sources. Contaminants can end up in the soil even when emissions are aerial. The adsorption and retaining of heavy metals in the soil, water, and air depend upon several factors including the pH of the transport medium and destination, the number of electrons in the valence shell of the atom, ionic strength, and temperature of the surrounding (Sherene 2010). Heavy metals are known for their reputation as bioaccumulative elements in the human body and environment. A variety of pathways are responsible for the bioaccumulation of heavy metals in the human body; however, food intake is considered to be the main route for the majority of the population (Hu et al. 2019). It has also been noted several times about how the higher concentration of heavy metals is more prominent in the animals at the culmination of a food chain. It has been a common gesture for researchers to use the term bio-magnification to explain the concentration pattern of heavy metals in terrestrial animals (Janssen et al. 1993).

1.4.1 Air

The majority of heavy metals that settle down on the surface of the Earth after precipitation are airborne. This includes metals that exist in the air due to natural Earth processes as well as those vented out by chimneys and manufacturing units. The weathered and fine-grained particles of overburdened rocks are also potential sources of heavy metal contamination in the environment. These fine-grained particles often travel across distant regions through wind, primarily in the arid and semi-arid regions (Punia 2021). A similar study by Qin et al. (2023) noted the role of winds in grassland and the consequent transportation of heavy metals to long distances through winter winds.

Large amounts of fine and coarse particulate matter (PM) are produced when the Earth's upper layer is removed during mining and excavation activities of ores. The finer percentage moves to remote locations, with the wind being the primary carrier, while the coarser PM accumulates close to the proximity of mines (Punia 2021). PM, accompanied by toxic metals also results in occupational health hazards due to poor ventilation among the mine workers as described by Widiatmojo et al. (2015) by entering through their respiratory organs.

1.4.2 Water

The transport of heavy metals often occurs when fine or coarse-grained sediment is accompanied by river water through a variety of natural and artificial means. The development of industries and increased urbanization are thought to be the main contributors to the rise in these trace metal concentrations, particularly heavy metals, through water runoff. This runoff combines with river water, which is already carrying untreated surface runoff from agricultural lands, including fertilizers and pesticides (Dwivedi 2017). It was also noted by Bradl (2005) that the physical, chemical, and biological parameters of water are influenced by the rocks in contact with the flow of river water.

Some rivers are deemed prime carriers of heavy metals when their trajectories are artificially modified by engineering structures to pass through densely populated localities and industrial areas (Ciszewski and Grygar 2016). The same outcome can also be expected if one of its major tributaries follows the aforementioned course of flow. The intermixing of sewage with river water during flood events is also deemed as a potential threat to exposure to heavy metals. This river water may further interact with agricultural soils and aquifers to further degrade the quality of food and water consumed by humans (Gall et al. 2015).

1.4.3 Soil

More challenging to regulate and control than air and water contamination, soil contamination has a wide spectrum of distribution. The environmental authorities in developed countries have recently been promoting soil pollution in agricultural and urban watersheds as a crucial aspect of environmental problems (Atafar et al. 2010). The magnitude of metal concentration of the same concentration can be one to several times higher in the case of fine-grained sediments when compared to the dissolved phase (Ciszewski and Grygar 2016). Water bodies near mining operations are at a significant risk of being contaminated by the effluent from mining and smelting. In most cases, the metals are either stored or absorbed in the sediments. The metals accumulated in soils can also be inherited through their respective parent rocks (He et al. 2015). Rural soils are generally affected by the use of fertilizers and pesticides to increase productivity and crop yield. However, the amount of heavy metal contamination varies with the type of fertilizers used as reported by Su (2014) and Hanfi et al. (2020) noted the various sources of sediments in the urban surface including roofs, driveways, and roads, with an emphasis on the deposition and removal of heavy metals in the road sediments.

1.5 HEALTH IMPLICATIONS OF HEAVY METAL EXPOSURE

The awareness of health implications due to heavy metal contamination has resulted in several regulations, and restrictions being enforced to constrain the use of toxic metals. Government agencies have placed limits on the concentration of heavy metals in various media, particularly water, to reduce the exposure of ecosystems to these hazardous substances. However not all governments are equally concerned about this problem, and as a result, some nations have more legislation than others. However, the mandated limitations for contaminant concentrations in natural media have tightened.

There are various mechanistic components of heavy metal-induced toxicity and carcinogenicity, some of which are not well-described and understood. Toxicological methods of action are known to be particular to each metal, each of which is known to have distinct physical and chemical characteristics (Tchounwou et al. 2012). Therefore, it requires separate discussion before elucidating potential solutions. Depending upon the toxicity and persistence of the heavy metals, direct ingestion or skin contact exposure to riverine water pollution can cause several health issues (Khan et al. 2022). Metals including cadmium, chromium, lead, mercury, and nickel are highly toxic and are known to be metals of high concern due to their direct impacts on human and aquatic health, whereas metals like cobalt, copper, manganese, and zinc are essentially required for the human metabolism but could be harmful if their level exceeds the safe limit (Abdullah 2013; Khan et al. 2013).

TABLE 1.3
Impact of Some Heavy Metals on the Human Body

S. No.	Heavy Metals	Impacts on the Human Body
1.	Arsenic (As)	Skin lesions, Increased risk of skin, lung, bladder cancer, and cardiovascular issues
2.	Cadmium (Cd)	Kidney and lung damage. Increased risk of lung cancer in smokers
3.	Chromium (Cr)	Hexavalent chromium is a known carcinogen and increases the risk of lung cancer when inhaled over long periods
4.	Copper (Cu)	Gastrointestinal problems such as nausea and vomiting in high concentrations
5.	Lead (Pb)	Impairs neurological development causing behavioural and learning problems in Children and high blood pressure, fertility, and kidney problems in adults
6.	Mercury (Hg)	Neurological symptoms such as tremors and memory problems. Severe mercury poisoning can lead to organ damage and even death
7.	Nickel (Ni)	Skin irritations, allergies, and respiratory problems
8.	Zinc (Zn)	Long-term exposure can lead to immune system suppression

Population exposed to these heavy metals may have adverse effects on their nervous system, gastrointestinal tract, and vascular systems of kidneys. It can also interfere with the circulation of phosphate bio-compounds. Table 1.3 showcases the impacts of some heavy metals on the human body. It was noted by Vellingiri et al. (2022) that there is a reasonable cause-and-effect relationship between nutritional deficiencies and metabolic abnormalities in many harmful illnesses. Even though many metal ions are necessary, their excessive buildup can be exceedingly poisonous and most likely lethal.

The neurological system is harmed and the effects on mental health are negative due to the presence of heavy metals in drinking water. Heavy metal toxicity also results in abnormalities in blood composition and damage to the body's important organs, such as the liver and kidneys (Kumar et al. 2020). One of the primary reasons for instigating the concern for controlling heavy metals contamination before their entrance into the human body is their replacement of proteins and enzymes, thereby creating disturbances in the enzymatic pathways (Witkowska et al. 2021). They are also reported to cellular organelles including mitochondria and cell membranes, and enzymes responsible for metabolism and damage repair (Tchounwou et al. 2012). Malfunctioning caused by disruption of normal activities and body processes including high blood pressure, type 2 diabetes, and later life obesity eventually end up risking individuals to cardiovascular diseases (Yang et al. 2020).

Studies have shown how the consumption of vegetables and fruits, aquatic organisms, and other poultry animals in the vicinity of contaminated sites have resulted in the accumulation of heavy metals inside the neurological and cardiovascular systems of the human body (Zhou et al. 2018). Gall et al. (2015) conducted a thorough study on the transfer of heavy metals in the ecosystem, analysed metal accumulation in the food web by both non-destructive (hair, milk, and blood) and destructive samples (muscle, kidney, and lever), and devised the methodology of tackling this particular problem by survey, discovery, and conservation of species that can cope with heavy metal stress.

1.6 ENVIRONMENTAL CONSEQUENCES

Heavy metals are naturally occurring elements that are essential for the development of plants and organisms in limited concentrations, but as these metals start to accumulate, they have a toxic effect on the flora and fauna, as well as human health.

The release of heavy metals in soil, air, and water is very common due to increasing industrialization and several anthropogenic activities such as mining, chemical and metal processing industries, and the application of fertilizers, pesticides, and insecticides in agriculture (Sandeep et al. 2019).

The runoff from the polluted lands drains into nearby waterbodies thus polluting them and threatening the biodiversity of the area.

1.6.1 Effects on Soil Quality

The main contaminants of the soil in heavy metals are chromium, manganese, iron, nickel, copper, zinc, cadmium, lead, and mercury. Soil poisoning by the compound of these elements poses major concerns in this industrialized era. Characteristics of such contamination in urban soils are found to be different than in agricultural soils according to some studies in the past years (Babula et al. 2010). The heavy metals present in the agricultural soil are taken up and accumulated in plants and consumed by the biota of the region whereas in urban soils they reach the body directly through inhalation of particulates or skin contact thus directly damaging human health (Su 2014). Heavy metal contamination not only impacts numerous parameters associated with plant yield and development but also affects the microbial activities of the region. The quality of the soil can be sensitively assessed by the microbial activity and the enzymatic activity i.e. soil heavily contaminated with heavy metals will have a significantly reduced level of microbial activity (Lee et al. 1996), and soil's microbial biomass is a crucial indicator of the amount of heavy metals present in the soil. The level of contamination of soil and harmful effects by heavy metals depends on various chemical and physical properties of the soil i.e. pH, organic matter, clay/silt content, etc., as well as metal properties i.e. specific metal, its chemical form, and concentration (Singh and Kalamdhad 2011).

1.6.2 On Aquatic Life

Heavy metals have detrimental effects on aquatic life like fishes, molluscs, and plants. Organisms absorb these metals that accumulate in the water bodies through various sources like industrial and residential discharges, including runoff from urban areas. The presence of metals like mercury, cadmium, lead, and nickel, in high concentrations, can lead to harmful impacts such as reproductive problems, impaired growth, and damage to the internal organs of organisms (Ali et al. 2014). Moreover, they disrupt the entire aquatic ecosystem by affecting the food chain and water quality. Some heavy metals are more soluble and toxic in acidic environments and the presence of any form of acidification in the water body increases the bioavailability of these metals, making them more harmful to aquatic life.

These metals tend to accumulate in the tissues of the organisms over time. When smaller species take up these metals from water or sediments and are later eaten by larger predators, the metals manifest into the food chain. This process is called bioaccumulation which results in larger metal concentrations in organisms further up the food chain (Borrill et al. 2019). Heavy metals interfere with enzyme activities, cellular functions, DNA alterations, and other vital physiological processes within aquatic organisms which affects the growth rates and immune system of the species, rendering them more vulnerable to diseases. The ability of aquatic organisms to reproduce can be significantly impacted by a wide variety of heavy metals. They may result in reduced fertility, abnormal egg and embryo development, and even population reductions in the affected species (Jezierska et al. 2009).

The effects of heavy metal exposure might last for very long periods even after the source of pollution is eradicated. Sediments act as reservoirs for the toxins and gradually release them back into the water, prolonging the impacts on aquatic life (Woo et al. 2009). Table 1.4 illustrates the adverse impacts of some heavy metals on aquatic life (if found in large quantities).

1.6.3 On Biodiversity

Plants require certain essential metals for their development namely cobalt, copper, iron, manganese, molybdenum, nickel, and zinc, but the ability of plants to uptake essential metals equally

TABLE 1.4
Impact of Some Heavy Metals on Aquatic Life

S. No.	Heavy Metal	Impacts on Aquatic Life
1.	Arsenic (As)	Damages internal organs, impairs reproduction, and alters behaviour
2.	Cadmium (Cd)	Causes kidney damage and growth abnormalities, disrupts the food chain, and impacts predator-prey relationships
3.	Chromium (Cr)	Affects the respiratory functions of fish Accumulates in sediments
4.	Copper (Cu)	Excessive levels can affect the ability of fish to detect predators, affect their sense of smell, and lead to gill damage
5.	Lead (Pb)	Disrupts neurological development, impairs growth, and damages internal organs
6.	Mercury (Hg)	Neurological and reproductive problems
7.	Nickel (Ni)	Affects fish gills and causes stress responses. Bioaccumulates in organisms
8.	Zinc (Zn)	Excessive levels can affect fish behaviour, reproduction, and growth

regulates the uptake of non-essential metals. The concentration of heavy metals to a certain tolerance limit does not affect the growth of a plant; however, a high concentration of these metals becomes toxic and might also result in the withering away of the plants (Sandeep et al. 2019). Heavy metal pollution in the soil affects crucial microbial processes and decreases the quantity and activity of microorganisms in the soil (Singh and Kalamdhad 2011). The discharge of heavy metals through anthropogenic activities such as mining activities, agricultural activities, and industrial activities, and wastewater disposal from various industries, along with natural phenomena such as weathering of rocks and volcanic eruptions, is absorbed by plants, fish, and is consumed by terrestrial animals, which not only affects the ecology of the area but also follows their way up the food chain eventually impacting human health as well (Singh and Kalamdhad 2011).

Human exposure to heavy metals is a result of anthropogenic activities and can ingress through multiple channels such as inhalation, ingestion, and physical contact and can have severe and potentially life-threatening effects (Abd Elnabi et al. 2023). Heavy metals pose threats to human metabolism due to their presence in the environment. An acute amount of intoxication can damage the central nervous functions, cardiovascular system, lungs, kidneys, endocrine glands, and bones (Sankhla et al. 2016). The accumulation of heavy metals in aquatic ecosystems can lead to the decline or extinction of certain species that are more sensitive to these contaminants and the loss of biodiversity can hamper the overall balance of the ecosystem.

1.7 METHODS AND TOOLS USED FOR MONITORING HEAVY METALS IN THE ENVIRONMENT

Looking into the harmful environmental impacts of heavy metals, it becomes an unsaid necessity to control and minimize if not eradicate the pollution by these toxins. Monitoring the presence of these metals and assessing their traces in the sediments gives us a clear path towards understanding and minimizing the harmful effects of these and protecting ecosystems and human health. There are various methods and tools for monitoring heavy metals in the environment. The methods and tools used for assessment are outlined as follows.

1.7.1 Analytical Techniques

The following section states a few of the analytical methods for heavy metal detection and monitoring:

1.7.1.1 Atomic Absorption Spectrometry (AAS)

Atomic absorption spectrometry (AAS) is the most commonly employed analytical technique for determining the amount of metals and metalloids present in environmental samples (Soodan et al. 2014). It works on the principle that when the atoms in the sample are vaporized and are exposed to a light source of a specific wavelength, they absorb light at that specific wavelength and the concentration of the element sample is directly proportional to the degree of absorption (Melendez-Pastor et al. 2011). Due to its precision and sensitivity in identifying even low quantities of metals, AAS is widely utilized in environmental, clinical, culinary, and materials analysis (Soodan et al. 2014).

1.7.1.2 Inductively Coupled Plasma-Mass Spectrometry (ICP-MS)

Inductively coupled plasma-mass spectrometry (ICP-MS) is a powerful analytical technique used for the precise and sensitive determination of trace and ultra-trace elements in various samples (Soodan et al. 2014). The multi-elemental ultra-trace detection capability of ICP-MS makes it the most powerful technique (Beauchemin 2008). It combines two essential components: ICP and MS. This method can simultaneously analyse a wide range of elements and can even distinguish between isotopes of elements. This method is employed in the industrial examination of metals, chemicals, environmental specimens, clinical and biological products, foods, and drinks (Soodan et al. 2014).

1.7.1.3 Inductively Coupled Plasma-Atomic Emission Spectrometry (ICP-AES)

This method is used for simultaneous qualitative (based on emitted wavelength) and quantitative (based on the intensity of light) analysis of chemical elements. Due to the instrument's ability to determine multiple elements across a wide range in the periodic table at once, it was adopted for rock sampling in place of commonly used methods including AAS and X-ray fluorescence spectrometry (Soodan et al. 2014). Although this method is not as sensitive as ICP-MS, it can still detect elements in parts per million to parts per billion ranges.

ICP-AES is an important tool for laboratories and industries that require rapid and accurate elemental analysis.

1.7.1.4 X-ray Fluorescence (XRF) Spectrometry

This is an analytical technique used for quantitative and qualitative determination of the elemental composition of a wide range of materials. Atoms in the sample are excited using X-rays and the emitted characteristic X-ray radiation is detected and analysed (Soodan et al. 2014). It is a highly sensitive analytical tool with a wide range of applications such as geology, metallurgy, environmental monitoring, and material science.

Types of XRF:

a. *Energy Dispersive XRF* (*EDXRF*): The detector measures the energies of X-rays emitted simultaneously from the sample. These are used for qualitative and semi-quantitative analysis.
b. *Wavelength-Dispersive XRF* (*WDXRF*): WDXRF separates X-rays by their wavelengths before detection and offers higher resolution and accuracy. These are used for quantitative analysis.

1.7.2 Voltammetry Techniques

Voltammetry is a method that involves a number of electroanalytical techniques used to measure the current as a function of an applied potential (voltage) in an electrochemical cell. These are highly valuable for monitoring heavy metals in the environment due to their sensitivity, selectivity, and being able to deliver real-time or close to real-time data. Voltammetry techniques provide less complex, lower cost, and smaller footprint instrumentation with similar detection at the source (Borrill et al. 2019). Some voltammetry techniques are given as follows:

- Differential pulse voltammetry (DPV)
- Anodic stripping voltammetry (ASV)
- Cathodic stripping voltammetry (CSV)
- Field-portable voltammetry

1.7.3 Remote Sensing Technique

The remote sensing technique involves the use of satellite-mounted or airborne sensors for ground mapping and data collection without any physical contact with the ground. Remote sensing techniques are valuable tools for monitoring heavy metals in the environment, especially when assessing large and inaccessible areas (Melendez-Pastor et al. 2011). This technique can cover extensive areas and since there is no direct contact with the ground, this technique is non-invasive and reduces the need for physical sample collection. However, in complex environments with mixed vegetation and soil types, this method can face challenges, moreover, the resolution (spatial or spectral) may affect the ability to detect and differentiate heavy metals.

1.7.4 Biomonitoring

Biomonitoring is a method of using biological indicators such as mosses, algae, fishes, or other organisms and analysing their response in a specific region, to assess and monitor the quality and health of the environment.

Despite having less precision and less standardized methodology, biomonitoring is chosen over the direct monitoring method because it is cost-effective and does not contribute to other forms of pollution (Ştefănuţ et al. 2021).

Bioindicators are chosen based on their sensitivity to specific environmental pollutants.

Biomonitoring has two divisions (Wacławek et al. 2022).

1. *Active biomonitoring*: Monitoring at the site.
2. *Passive biomonitoring*: Shifting the indicators to a different environment and observing their response to the contamination.

1.8 REGULATORY FRAMEWORKS AND INTERNATIONAL AGREEMENTS

Regulatory standards and guidelines hold critical importance in minimizing the fatal impacts of heavy metals on the environment and biodiversity. Heavy metals even in trace amounts can have profound and lasting effects on human health, ecosystems, and industries. Compliance with standards and guidelines is pivotal for safeguarding human health, preserving the natural environment, and promoting responsible and sustainable industry practices. Some heavy metals such as lead, mercury, cadmium, and arsenic are naturally occurring elements that are hazardous even in small amounts and it becomes necessary to keep them under standard limits.

These regulations empower consumers to make informed choices and provide a scientific basis for monitoring and responding to heavy metal contamination incidents. Table 1.5 illustrates few regulatory measures globally adopted so as to regulate the issue pertaining to heavy metals. Moreover, they give society a strong basis for risk management, sustainable development, and well-informed decision-making.

TABLE 1.5
Regulatory Measures Globally Adopted to Regulate the Issue Pertaining to Heavy Metals

S. No.	Name	Aim
1.	Long-range transboundary air pollutants (LRTAP) 1979	The Convention's goal is for parties to make an effort to inhibit air pollution, particularly LRTAP, and to limit and gradually reduce it to the extent attainable. Mercury, lead, and cadmium are the three specific hazardous elements that are the focus of the protocol.
2.	Rotterdam Convention 2004	The Rotterdam Convention is an international agreement to regulate the trade of dangerous chemicals and pesticides. It encompasses the objective of making it easier for exporting and importing countries to share information about these substances and implement a prior informed consent (PIC) process that enables nations to decide on chemical imports with information and lower trade barriers.
3.	The Basel Convention 1992	The Basel Convention is an international agreement that regulates the management and transportation of hazardous waste across international borders. Its main objectives are to stop the environmental and health concerns connected to the transboundary transfer and disposal of hazardous waste. In addition to requiring prior informed consent from receiving nations, the treaty fosters environmentally sound waste management, promotes waste reduction and cleaner production methods, and streamlines information exchange.
4.	Aarhus Protocol on Heavy Metals 1998 Amended 2012	The aim of the protocol is to reduce emissions from ferrous and non-ferrous industries, processes involving combustion, waste disposal, and incineration. It was amended in 2012 to include more strict controls of heavy metals emissions and to add flexibility making it easier for new parties to join, particularly those in Eastern Europe, South-Eastern Europe, the Caucasus, and Central Asia. The protocol mainly targets three heavy metals namely Hg, Pb, and Cd.
5.	The Minamata Convention 2013	The Minamata Convention on mercury is a progression of the Aarhus Protocol 1998, which was adopted in 2013. The United Nations Environment Programme (UNEP) served as the negotiating body for the treaty. The convention elevated mercury's profile to a worldwide level.
6.	Lead National Ambient Air Quality Standards (NAAQS) Estd.1978 Revised 2008	National Ambient Air Quality Standards (NAAQS) for lead are regulations set by the U.S. Environmental Protection Agency (EPA) to supervise the amount of lead in the country's ambient air. Lead is a hazardous metal that can have major negative health effects, especially in children. These standards are intended to protect human health and the environment by decreasing exposure to lead.
7.	Mercury and Air Toxics Standards (MATS) for Power Plants 2012	Mercury and Air Toxics Standards (MATS), also known as the Mercury Air Toxics Standards, was established under the U.S. Environmental Protection Agency (EPA), as regulations to control emissions of hazardous air pollutants, including mercury, from power plants in the country. Power plants subject to MATS must adhere to strict emission limits for mercury and other toxic pollutants.
8.	Food Safety and Standards Regulations 2011	According to the (FSSR 2011), Food Safety and Standards (Contaminants, Toxins and Residues) Regulations, metal contaminants are allowed in quantities no higher than the specified limits. The regulations set maximum limits (MLs) for metal pollutants in a variety of foods.
9.	Jal Shakti Abhiyan 2019	To conserve water and manage water resources, the Indian government initiated this project in July 2019. It is a thorough campaign that aims to address the water-related issues faced by the country, such as water scarcity and deteriorating groundwater quality and conditions.

(Continued)

TABLE 1.5 (Continued)
Regulatory Measures Globally Adopted to Regulate the Issue Pertaining to Heavy Metals

S. No.	Name	Aim
10.	Drugs and Cosmetics Rule 1945	The Drugs and Cosmetics Rules are a set of regulations in India that govern the manufacture, distribution, sale, and import of drugs and cosmetics. These rules are framed under the Drugs and Cosmetics Act, 1940 by BIS (Bureau of Indian Standards) and are implemented by the Central Drugs Standard Control Organization (CDSCO) in India. The primary purpose of these rules is to ensure the safety, quality, and efficacy of drugs and cosmetics available in the Indian market. It prohibits cosmetics containing mercury compounds and the use of lead and arsenic in coloured cosmetics
11.	Central Pollution Control Board 1974	The Central Pollution Control Board (CPCB) is a regulatory organization in India that functions under the Ministry of Environment, Forests, and Climate Change. The CPCB was established under the Water (Prevention and Control of Pollution) Act, 1974, and later the Air (Prevention and Control of Pollution) Act, 1981, and it plays a crucial role in environmental protection and pollution control in the country. It has specified limits for effluents discharged containing some of these metals, but not limited to, arsenic, mercury, lead, cadmium, chromium, etc., in inland surface water.

REFERENCES

Abd Elnabi, Manar K., Nehal E. Elkaliny, Maha M. Elyazied, Shimaa H. Azab, Shawky A. Elkhalifa, Sohaila Elmasry, and Moustafa S. Mouhamed et al. "Toxicity of heavy metals and recent advances in their removal: A review." Toxics 11, no. 7 (2023): 580.

Abdullah, Enaam J. "Quality assessment for Shatt Al-Arab River using heavy metal pollution index and metal index." Journal of Environmental & Earth Sciences 3, no. 5 (2013): 114–120.

Afshan, Sehar, Shafaqat Ali, Uzma Shaista Ameen, Mujahid Farid, Salma Aslam Bharwana, Fakhir Hannan, and Rehan Ahmad. "Effect of different heavy metal pollution on fish." Research Journal of Chemistry and Environment 2, no. 1 (2014): 74–79.

Atafar, Zahra, Alireza Mesdaghinia, Jafar Nouri, Mehdi Homaee, Masoud Yunesian, Mehdi Ahmadimoghaddam, and Amir Hossein Mahvi. "Effect of fertilizer application on soil heavy metal concentration." Environmental Monitoring and Assessment 160 (2010): 83–89.

Babula, Petr, Vojtech Adam, Radka Opatrilova, Josef Zehnalek, Ladislav Havel, and Rene Kizek. "Uncommon heavy metals, metalloids and their plant toxicity: A review." In Organic Farming, Pest Control and Remediation of Soil Pollutants: Organic Farming, Pest Control and Remediation of Soil Pollutants, pp. 275–317, Springer, 2010.

Bartzas, Georgios, Petros E. Tsakiridis, and Kostas Komnitsas. "Nickel industry: Heavy metal (loid)s contamination-sources, environmental impacts and recent advances on waste valorization." Current Opinion in Environmental Science & Health 21 (2021): 100253.

Basta, Nicholas, and R. Gradwohl. "Estimation of Cd, Pb, and Zn bioavailability in smelter-contaminated soils by a sequential extraction procedure." Journal of Soil Contamination 9, no. 2 (2000): 149–164.

Beauchemin, Diane. "Inductively coupled plasma mass spectrometry." Analytical Chemistry 80, no. 12 (2008): 4455–4486.

Borrill, Alexandra J., Nicole E. Reily, and Julie V. Macpherson. "Addressing the practicalities of anodic stripping voltammetry for heavy metal detection: A tutorial review." Analyst 144, no. 23 (2019): 6834–6849.

Bradl, H. B. "Sources and origins of heavy metals." Heavy metals in the environment: Origin. Interaction and remediation, 1–27, Elsevier, London, 2005.

Brathwaite, R. L., and S. D. C. Rabone. "Heavy metal sulphide deposits and geochemical surveys for heavy metals in New Zealand." Journal of the Royal Society of New Zealand 15, no. 4 (1985): 363–370.

Byrne, Celia, Shailaja D. Divekar, Geoffrey B. Storchan, Daniela A. Parodi, and Mary Beth Martin. "Cadmium—A metallohormone?" Toxicology and Applied Pharmacology 238, no. 3 (2009): 266–271.

Ciszewski, Dariusz, and Tomáš Matys Grygar. "A review of flood-related storage and remobilization of heavy metal pollutants in river systems." Water, Air, & Soil Pollution 227 (2016): 1–19.

Dhal, B., H. N. Thatoi, N. N. Das, and B. D. Pandey. "Chemical and microbial remediation of hexavalent chromium from contaminated soil and mining/metallurgical solid waste: A review." Journal of Hazardous Materials 250 (2013): 272–291.

Dwivedi, Anil K. "Researches in water pollution: A review." International Research Journal of Natural and Applied Sciences 4, no. 1 (2017): 118–142.

Dong, Wanwan, Ying Zhang, and Xie Quan. "Health risk assessment of heavy metals and pesticides: A case study in the main drinking water source in Dalian, China." Chemosphere 242 (2020): 125113.

El-Harouny, Mona A., Sahar A. El-Dakroory, Sohayla M. Attalla, Nermin A. Hasan, and Ragia Hegazy. "Chemical quality of tap water versus bottled water: Evaluation of some heavy metals and elements content of drinking water in Dakahlia governorate-Egypt." Mansoura Journal of Forensic Medicine and Clinical Toxicology 16, no. 2 (2008): 1–15.

Gall, Jillian E., Robert S. Boyd, and Nishanta Rajakaruna. "Transfer of heavy metals through terrestrial food webs: A review." Environmental Monitoring and Assessment 187 (2015): 1–21.

Gautam, Pavan Kumar, Ravindra Kumar Gautam, Sushmita Banerjee, M. C. Chattopadhyaya, and J. D. Pandey. "Heavy metals in the environment: Fate, transport, toxicity and remediation technologies." Nova Science Publishers 60 (2016): 101–130.

Genchi, Giuseppe, Maria Stefania Sinicropi, Graziantonio Lauria, Alessia Carocci, and Alessia Catalano. "The effects of cadmium toxicity." International Journal of Environmental Research and Public Health 17, no. 11 (2020): 3782.

Grygar, T. Matys, J. Elznicová, O. Bábek, M. Hošek, Z. Engel, and T. Kiss. "Obtaining isochrones from pollution signals in a fluvial sediment record: A case study in a uranium-polluted floodplain of the Ploučnice River, Czech Republic." Applied Geochemistry 48 (2014): 1–15.

Gumpu, Manju Bhargavi, Swaminathan Sethuraman, Uma Maheswari Krishnan, and John Bosco Balaguru Rayappan. "A review on detection of heavy metal ions in water—An electrochemical approach." Sensors and Actuators B: Chemical 213 (2015): 515–533.

Hanfi, Mohamed Y., Mostafa YA Mostafa, and Michael V. Zhukovsky. "Heavy metal contamination in urban surface sediments: Sources, distribution, contamination control, and remediation." Environmental Monitoring and Assessment 192 (2020): 1–21.

He, Zhenli, Jiali Shentu, Xiaoe Yang, Virupax C. Baligar, Tiequan Zhang, and Peter J Stoffella. "Heavy metal contamination of soils: Sources, indicators and assessment." Journal of Environmental Indicators 9 (2015): 17–18.

Hu, Yuanmei, Jun Zhou, Buyun Du, Hailong Liu, Wantong Zhang, Jiani Liang, Wenhui Zhang, Laiyong You, and Jing Zhou. "Health risks to local residents from the exposure of heavy metals around the largest copper smelter in China." Ecotoxicology and Environmental Safety 171 (2019): 329–336.

Iyaka, Yahaya Ahmed. "Nickel in soils: A review of its distribution and impacts." Scientific Research and Essays 6, no. 33 (2011): 6774–6777.

Janssen, M. P. M., W. C. Ma, and N. M. Van Straalen. "Biomagnification of metals in terrestrial ecosystems." Science of the Total Environment 134 (1993): 511–524.

Jezierska, Barbara, Katarzyna Ługowska, and Małgorzata Witeska. "The effects of heavy metals on embryonic development of fish (a review)." Fish Physiology and Biochemistry 35 (2009): 625–640.

Khan, Kifayatullah, Yonglong Lu, Hizbullah Khan, Shahida Zakir, Sardar Khan, Akbar Ali Khan, Luo Wei, and Tieyu Wang. "Health risks associated with heavy metals in the drinking water of Swat, northern Pakistan." Journal of Environmental Sciences 25, no. 10 (2013): 2003–2013.

Khan, Kifayatullah, Muhammad Younas, Hafiz Muhammad Adeel Sharif, Chenchen Wang, Muhammad Yaseen, Xianghui Cao, Yunqiao Zhou, Sobhy M. Ibrahim, Baninla Yvette, and Yonglong Lu. "Heavy metals contamination, potential pathways and risks along the Indus Drainage System of Pakistan." Science of the Total Environment 809 (2022): 151994.

Khan, Saif Ullah. Azmatullah Noor, and Izharul Haq Farooqi. "GIS application for groundwater management and quality mapping in rural areas of district Agra, India." International Journal of Water Resources and Arid Environments 4, no. 1 (2015): 89–96.

Khan, Saif Ullah, Mohammad Khalid, Khalid Hashim, Mehdi Hassanvand Jamadi, Milad Mousazadeh, Farrukh Basheer, and Izharul Haq Farooqi. "Efficacy of electrocoagulation treatment for the abatement of heavy metals: An overview of critical processing factors, kinetic models and cost analysis." Sustainability 15, no. 2 (2023): 1708.

Khan, Saif Ullah, Rumman Zaidi, Feroz Shaik, Izharul Haq Farooqi, Ameer Azam, Hatem Abuhimd, and Faheem Ahmed. "Evaluation of Fe-Mg binary oxide for As (III) adsorption—synthesis, characterization and kinetic modelling." Nanomaterials 11, no. 3 (2021): 805.

Knight, C., J. Kaiser, G. C. Lalor, H. Robotham, and J. V. Witter. "Heavy metals in surface water and stream sediments in Jamaica." Environmental Geochemistry and Health 19 (1997): 63–66.

Knopf, Burkhard, and Helmut König. "Biomethylation of heavy metals in soil and terrestrial invertebrates." In Soil Heavy Metals, pp. 315–328, Springer, Berlin, 2010.

Kumar, Vinod, Anket Sharma, and Artemi Cerda, eds. Heavy Metals in the Environment: Impact, Assessment, and Remediation. Elsevier, Chennai, 2020.

Lee, Dong-Hun, Young-Gun Zo, and Sang-Jong Kim. "Nonradioactive method to study genetic profiles of natural bacterial communities by PCR-single-strand-conformation polymorphism." Applied and Environmental Microbiology 62, no. 9 (1996): 3112–3120.

Li, Changfeng, Kehai Zhou, Wenqiang Qin, Changjiu Tian, Miao Qi, Xiaoming Yan, and Wenbing Han. "A review on heavy metals contamination in soil: Effects, sources, and remediation techniques." Soil and Sediment Contamination: An International Journal 28, no. 4 (2019): 380–394.

Lindqvist, O. Mercury as an Environmental Pollutant. Water, Air and Soil Pollution 56, Kluwer Academic Publisher, Dordrecht, 1991.

Liu, Hongxi, Jizeng Du, and Yujun Yi. "Reconceptualising flood risk assessment by incorporating sediment supply." Catena 217 (2022): 106503.

Mateen, Qazi Shabihul, Saif Ullah Khan, Dar Tafazul Islam, Nadeem Ahmad Khan, and Izharul Haq Farooqi. "Copper (II) removal in a column reactor using electrocoagulation: Parametric optimization by response surface methodology using central composite design." Water Environment Research 92, no. 9 (2020): 1350–1362.

Matschullat, Jörg. "Arsenic in the geosphere—A review." Science of the Total Environment 249, no. 1–3 (2000): 297–312.

Melendez-Pastor, I., J. Navarro-Pedreño, I. Gómez, and M. Almendro-Candel. "The use of remote sensing to locate heavy metal as source of pollution." Advances in Environmental Research 7 (2011): 225–233.

Moss, Brian. "Water pollution by agriculture." Philosophical Transactions of the Royal Society B: Biological Sciences 363, no. 1491 (2008): 659–666.

National Research Council. Arsenic in Drinking Water. National Academies of Sciences: Washington, 1999.

Navarro, M. C., C. Pérez-Sirvent, M. J. Martínez-Sánchez, J. Vidal, P. J. Tovar, and J. Bech. "Abandoned mine sites as a source of contamination by heavy metals: A case study in a semi-arid zone." Journal of Geochemical Exploration 96, no. 2–3 (2008): 183–193.

Ni, Shanqin, Yiwen Ju, Quanlin Hou, Shijie Wang, Qing Liu, Yudong Wu, and Lingling Xiao. "Enrichment of heavy metal elements and their adsorption on iron oxides during carbonate rock weathering process." Progress in Natural Science 19, no. 9 (2009): 1133–1139.

Nicholson, F. A, S. R. Smith, B. J. Alloway, C. Carlton-Smith, and B. J. Chambers. "An inventory of heavy metals inputs to agricultural soils in England and Wales." Science of the Total Environment 311, no. 1–3 (2003): 205–219.

Ojo, Adeola A., I. B. Osho, Samuel O. Adewole, and Lawrence K. Olofintoye. "Review on heavy metals contamination in the environment." European Journal of Earth and Environment 4, no. 1 (2017): 1–6.

Onakpa, Michael Monday, Anoka Ayembe Njan, and Ogbureke Chidiebere Kalu. "A review of heavy metal contamination of food crops in Nigeria." Annals of Global Health 84, no. 3 (2018): 488.

Parth, Vandana, N. N. Murthy, and Praveen Raj Saxena. "Assessment of heavy metal contamination in soil around hazardous waste disposal sites in Hyderabad city (India): Natural and anthropogenic implications." Journal of Environmental Research and Management 2, no. 2 (2011): 027–034.

Paul, Ankita, and Mery Biswas. "Changes in river bed terrain and its impact on flood propagation–a case study of River Jayanti, West Bengal, India." Geomatics, Natural Hazards and Risk 10, no. 1 (2019): 1928–1947.

Punia, Anita. "Role of temperature, wind, and precipitation in heavy metal contamination at copper mines: A review." Environmental Science and Pollution Research 28, no. 4 (2021): 4056–4072.

Qin, Muhan, Yuanliang Jin, Tianyue Peng, Bin Zhao, and Deyi Hou. "Heavy metal pollution in Mongolian-Manchurian grassland soil and effect of long-range dust transport by wind." Environment International 177 (2023): 108019.

Rahman, Zeeshanur, and Ved Pal Singh. "The relative impact of toxic heavy metals (THMs)(arsenic (As), cadmium (Cd), chromium (Cr)(VI), mercury (Hg), and lead (Pb)) on the total environment: An overview." Environmental Monitoring and Assessment 191 (2019): 1–21.

Ray, Suranjana, and Manas Kanti Ray. "Bioremediation of heavy metal toxicity-with special reference to chromium." Al Ameen Journal of Medical Sciences 2, no. 2 (2009): 57–63.

Richardson, G. Mark, R. Garrett, I. Mitchell, M. Mah-Poulson, and T. Hackbarth. "Critical review on natural global and regional emissions of six trace metals to the atmosphere." Prepared for the International Lead Zinc Research Organisation, the International Copper Association, and the Nickel Producers Environmental Research Association (2001).

Romic, Marija, and Davor Romic. "Heavy metals distribution in agricultural topsoils in urban area." Environmental Geology 43 (2003): 795–805.

Sadak, O. "Chemical sensing of heavy metals in water." In Barhoum, A., Altintas, Z. (eds), Advanced Sensor Technology, pp. 565–591. Elsevier, Amsterdam, 2023.

Salomons, Wim, and Ulrich Förstner. Metals in the Hydrocycle. Springer Science & Business Media, Munich, 2012.

Sandeep, G., K. R. Vijayalatha, and T. Anitha. "Heavy metals and its impact in vegetable crops." International Journal of Chemical Studies 7, no. 1 (2019): 1612–1621.

Sankhla, Mahipal Singh, Mayuri Kumari, Manisha Nandan, Rajeev Kumar, and Prashant Agrawal. "Heavy metals contamination in water and their hazardous effect on human health—A review." International Journal of Current Microbiology and Applied Sciences 5, no. 10 (2016): 759–766.

Sharma, Nayana, and Ritu Singhvi. "Effects of chemical fertilizers and pesticides on human health and environment: A review." International Journal of Agriculture, Environment and Biotechnology 10, no. 6 (2017): 675–680.

Sherameti, Irena, and Ajit Varma. Heavy metal Contamination of Soils. Soil biology, 44, Springer, Berlin/Heidelberg, 2015.

Sherene, T. "Mobility and transport of heavy metals in polluted soil environment." Biological Forum—An International Journal 2, no. 2 (2010): 112–121.

Srinivasa Gowd, S., and Pradip K. Govil. "Distribution of heavy metals in surface water of Ranipet industrial area in Tamil Nadu, India." Environmental Monitoring and Assessment 136 (2008): 197–207.

Shanbehzadeh, Saeed, Marzieh Vahid Dastjerdi, Akbar Hassanzadeh, and Toba Kiyanizadeh. "Heavy metals in water and sediment: A case study of Tembi River." Journal of Environmental and Public Health 2014 (2014): 1–5

Singh, Jiwan, and Ajay S. Kalamdhad. "Effects of heavy metals on soil, plants, human health and aquatic life." International Journal of Research in Chemistry and Environment 1, no. 2 (2011): 15–21.

Soodan, Rajneet Kour, Yogesh B. Pakade, Avinash Nagpal, and Jatinder Kaur Katnoria. "Analytical techniques for estimation of heavy metals in soil ecosystem: A tabulated review." Talanta 125 (2014): 405–410.

Ştefănuţ, Sorin, Kinga Öllerer, Mihaela C. Ion, Larisa I. Florescu, Marian Constantin, Cristian Banciu, Marilena Onete et al. "Country-scale complementary passive and active biomonitoring of airborne trace elements for environmental risk assessment." Ecological Indicators 126 (2021): 107357.

Stern, Bonnie Ransom. "Essentiality and toxicity in copper health risk assessment: Overview, update and regulatory considerations." Journal of Toxicology and Environmental Health, Part A 73, no. 2–3 (2010): 114–127.

Su, Chao. "A review on heavy metal contamination in the soil worldwide: Situation, impact and remediation techniques." Environmental Skeptics and Critics 3, no. 2 (2014): 24.

Šurija, Budimir, and Marko Branica. "Distribution of Cd, Pb, Cu and Zn in carbonate sediments from the Krka river estuary obtained by sequential extraction." Science of the Total Environment 170, no. 1–2 (1995): 101–118.

Turgut, C. "The contamination with organochlorine pesticides and heavy metals in surface water in Küçük Menderes River in Turkey, 2000–2002." Environment International 29, no. 1 (2003): 29–32.

Talluri, Jawahar Kumar, H. S. Pandalai, and G. N. Jadhav. "Fluid chemistry and Depositional mechanism of the epigenetic, discordant ores of the proterozoic, carbonate-hosted, Zawarmala Pb-Zn Deposit, Udaipur District, India." Economic Geology 95, no. 7 (2000): 1505–1525.

Paul, B Tchounwou., Clement G. Yedjou, Anita K. Patlolla, and Dwayne J. Sutton. "Heavy metal toxicity and the environment." In Molecular, Clinical and Environmental Toxicology, pp. 133–164, Springer, Berlin, 2012.

USGS. Mineral commodity summaries 2016, U.S. Geological Survey (2016). http://minerals.usgs.gov/minerals/pubs/mcs/2016/mcs2016.pdf

Vareda, João P., Artur J. M. Valente, and Luisa Durães. "Assessment of heavy metal pollution from anthropogenic activities and remediation strategies: A review." Journal of Environmental Management 246 (2019): 101–118.

Vellingiri, Balachandar, Atchaya Suriyanarayanan, Priyanka Selvaraj, Kripa Susan Abraham, Md Younus Pasha, Harysh Winster, Abilash Valsala Gopalakrishnan et al. "Role of heavy metals (copper (Cu), arsenic (As), cadmium (Cd), iron (Fe) and Lithium (Li)) induced neurotoxicity." Chemosphere 301 (2022): 134625.

Wacławek, Maria, Paweł Świsłowski, and Małgorzata Rajfur. "The biological monitoring as a source of information on environmental pollution with heavy metals." Chemistry-Didactics-Ecology-Metrology 27 (2022): 53–78.

Walker, D. B., D. J. Baumgartner, C. P. Gerba, and K. Fitzsimmons. "Surface water pollution." In Brusseau, M. L. Pepper, I. L., Gerba, C. P. (eds), Environmental and Pollution Science, Academic Press (2019): 261–292.

Wang, Qianrui, Daekeun Kim, Dionysios D. Dionysiou, George A. Sorial, and Dennis Timberlake. "Sources and remediation for mercury contamination in aquatic systems—A literature review." Environmental Pollution 131, no. 2 (2004): 323–336.

Wang, Lan, Yaping Wang, Chunxue Xu, Ziyi An, and Suming Wang. "Analysis and evaluation of the source of heavy metals in water of the River Changjiang." Environmental Monitoring and Assessment 173 (2011): 301–313.

Widiatmojo, Arif, Kyuro Sasaki, Yuichi Sugai, Yoshiaki Suzuki, Hiroyuki Tanaka, Kagemi Uchida, and Hiroyuki Matsumoto. "Assessment of air dispersion characteristic in underground mine ventilation: Field measurement and numerical evaluation." Process Safety and Environmental Protection 93 (2015): 173–181.

Witkowska, Danuta, Joanna Słowik, and Karolina Chilicka. "Heavy metals and human health: Possible exposure pathways and the competition for protein binding sites." Molecules 26, no. 19 (2021): 6060.

Woo, Seonock, Seungshic Yum, Hong-Seog Park, Taek-Kyun Lee, and Jae-Chun Ryu. "Effects of heavy metals on antioxidants and stress-responsive gene expression in Javanese medaka (*Oryzias javanicus*)." Comparative Biochemistry and Physiology Part C: Toxicology & Pharmacology 149, no. 3 (2009): 289–299.

Wu, Fuzhong, Wanqin Yang, Jian Zhang, and Liqiang Zhou. "Cadmium accumulation and growth responses of a poplar (*Populus deltoids* × *Populus nigra*) in cadmium contaminated purple soil and alluvial soil." Journal of Hazardous Materials 177, no. 1–3 (2010): 268–273.

Xu, Jingying, Andrea Garcia Bravo, Anders Lagerkvist, Stefan Bertilsson, Rolf Sjöblom, and Jurate Kumpiene. "Sources and remediation techniques for mercury contaminated soil." Environment International 74 (2015): 42–53.

Xu, Jun, Yueyue Li, Shengli Wang, Song Long, Yining Wu, and Zhaoming Chen. "Sources, transfers and the fate of heavy metals in soil-wheat systems: The case of lead (Pb)/zinc (Zn) smelting region." Journal of Hazardous Materials 441 (2023): 129863.

Yang, Ai-Min, Kenneth Lo, Tong-Zhang Zheng, Jing-Li Yang, Ya-Na Bai, Ying-Qing Feng, Ning Cheng, and Si-Min Liu. "Environmental heavy metals and cardiovascular diseases: Status and future direction." Chronic Diseases and Translational Medicine 6, no. 04 (2020): 251–259.

Zaidi, Rumman, Saif Ullah Khan, Ameer Azam, and Izharul Haq Farooqi. "A study on effective adsorption of lead from an aqueous solution using copper oxide nanoparticles." IOP Conference Series: Materials Science and Engineering, 1058, no. 1 (2021a): 012074.

Zaidi, Rumman, Saif Ullah Khan, I. H. Farooqi, and Ameer Azam. "Rapid adsorption of Pb (II) and Cr (VI) from aqueous solution by Aluminum hydroxide nanoparticles: Equilibrium and kinetic evaluation." Materials Today: Proceedings 47 (2021b): 1430–1437.

Zheng, Shun-an, Xiangqun Zheng, and Chun Chen. "Leaching behavior of heavy metals and transformation of their speciation in polluted soil receiving simulated acid rain." PLoS One 7, no. 11 (2012): e49664.

Zhou, Jun, Jiani Liang, Yuanmei Hu, Wantong Zhang, Hailong Liu, Laiyong You, Wenhui Zhang, Min Gao, and Jing Zhou. "Exposure risk of local residents to copper near the largest flash copper smelter in China." Science of the Total Environment 630 (2018): 453–461.

2 Sources and Pathways of Heavy Metals and Their Persistence in Ecosystems

*Jyoti and Surabhi Zainith**
Babasaheb Bhimrao Ambedkar Central University, Lucknow, India

2.1 INTRODUCTION

The environment consists of both biotic and abiotic components. Abiotic elements include air, water, soil, land, mountains, clouds, and temperature, while biotic components encompass all living organisms and microorganisms. The environment is made up of various spheres, with the biosphere being crucial as it supports all living organisms that interact with each other and their non-living surroundings, such as soil, air, and water (Masindi and Muedi 2018).

Heavy metals naturally exist in the Earth's crust, with 35 metals of concern, 23 of which are considered hazardous, including arsenic, cadmium, chromium, lead, and mercury. These metals enter the food chain through both natural processes, such as volcanic eruptions and soil erosion, and anthropogenic activities like mining, smelting, and industrial operations (Herawati et al. 2000; He et al. 2005; Walker et al. 2012; Gautam et al. 2016; Masindi and Muedi 2018).

Heavy metals are defined by their atomic number greater than 20 and a density exceeding 4 g/cm^3. Some, like iron, cobalt, copper, manganese, molybdenum, and zinc, are essential for growth, while others, such as arsenic, cadmium, chromium, lead, and mercury, are highly toxic even in small amounts. These metals are non-biodegradable and tend to accumulate in living organisms, leading to various health issues, including organ dysfunction and cancer (Jarup 2003). Their bioaccumulation and biomagnification through the food chain pose serious health risks, prompting prioritization as pollutants by regulatory authorities (Babel and Kurniawan 2003; Fu and Wang 2011; Zwolak et al. 2019).

Recent studies highlight heavy metal contamination in Southeast Asian countries, including India, Bangladesh, Pakistan, Thailand, and Indonesia (Shaji et al. 2021). In India, the Central Pollution Control Board (CPCB 2011) reported that states like Maharashtra, Andhra Pradesh, and Gujarat generate 80% of hazardous waste, including heavy metals, due to high levels of industrialization (Kubier et al. 2019). Industrial effluents degrade air, soil, and water quality, leading to contamination of vegetation. Vegetables such as parsley, beet leaves, coriander, radish leaves, and basil have been found to contain hazardous levels of metals like Cr, Pb, Zn, As, and Cd, making them unsafe for consumption (Kumar et al. 2020). Manzoor et al. (2018) discussed the impact of heavy metal accumulation on vegetable quality and the associated health risks.

This chapter focuses on the sources and pathways of toxic heavy metals like lead, arsenic, chromium, and cadmium due to their persistence, bioaccumulation, and non-biodegradable nature.

* Corresponding Author: Jyoti16december@gmail.com

DOI: 10.1201/9781032685793-2

2.2 SOURCES OF HEAVY METALS

The toxicity, persistence, and bioaccumulative properties of heavy metals make them significant environmental contaminants. These metals originate from both natural (geological/lithogenic) and anthropogenic (man-made) sources. Natural emissions of heavy metals occur under specific environmental conditions through volcanic eruptions, weathering of metal-bearing rocks, sea-salt sprays, and forest fires. Various forms of heavy metals, such as oxides, hydroxides, sulfates, phosphates, sulfides, silicates, and organic compounds, are present in the environment (Masindi and Muedi 2018) (Figure 2.1).

Anthropogenic activities have increased heavy metal pollution in the environment. Industries, agriculture, wastewater treatment, foundries, mining, and metallurgical processes are primary sources of pollutants. Additionally, the use of chemical fertilizers, herbicides, and insecticides in agriculture and the combustion of fossil fuels contribute to heavy metal pollution (Ali et al. 2019). In some cases, anthropogenic emissions of heavy metals exceed natural fluxes. Transportation, home cooking, and industrial processes release heavy metals, with automobiles emitting lead (Pb), chromium (Cr), zinc (Zn), and cadmium (Cd); smelting releasing arsenic (As), copper (Cu), and Zn; and the combustion of fossil fuels emitting nickel (Ni), vanadium (V), mercury (Hg), selenium (Se), and tin (Sn) (Briffa et al. 2020).

Phosphate fertilizers, rich in toxic heavy metals, contribute to the global distribution of these contaminants (Rashmi et al. 2020). Groundwater can become contaminated by heavy metals from inorganic fertilizers used in agriculture (Khan et al. 2018). Elevated concentrations of heavy metals have been reported in soils and plants near roads in urban areas.

Heavy metals enter the ecosystem through the hydrosphere, lithosphere, and atmosphere. They pollute natural water bodies, sediments, and soils from both natural and anthropogenic sources. Volcanic eruptions and industrial emissions release heavy metals into the atmosphere, which eventually deposit on land and contaminate water, air, and soil. These metals pose significant threats to

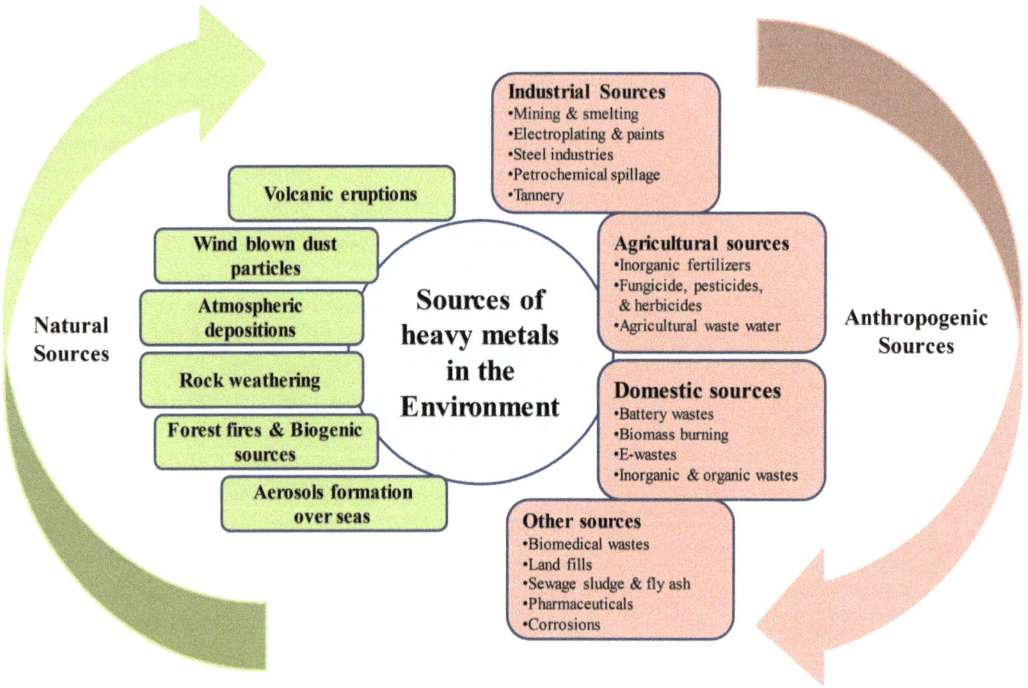

FIGURE 2.1 Natural and anthropogenic sources of heavy metals in the environment.

both terrestrial and aquatic ecosystems due to their persistence, causing atmospheric pollution, soil accumulation, and groundwater contamination. This results in severe environmental and human health impacts (Islam et al. 2018). Common heavy metals that pollute the environment include mercury (Hg), cadmium (Cd), arsenic (As), nickel (Ni), copper (Cu), lead (Pb), and chromium (Cr).

The toxic level of heavy metals in soil leads to the bioaccumulation of metals in plants, animals, and human bodies. Several studies reported that some plants play as an indicator of environmental pollution and grown in the vicinity of contaminated sites or sewage drains can accumulate the heavy metals (Nouchi 2002; Parmar et al. 2016; Bayouli et al. 2021). Plants are the pathways to move these heavy metals from contaminated water to get accumulated in animals and human beings. The process of metals accumulation in a plant or animal tissue is known as bioaccumulation which adversely affects the growth and metabolism of the individual. Stunted growth, chlorosis, necrosis, and discoloration of parts of plants are few symptoms of phytotoxicity due to heavy metals accumulation (CPCB Bhopal 2012). Pine, mosses, and lichens are widely used as specific environmental pollution indicators (Abas 2021).

2.3 SOURCES AND PATHWAYS OF LEAD (Pb) CONTAMINATION

Lead is an element, and its atomic number is 82 and symbol Pb. It is categorized as a heavy metal due to its high density, specific gravity, and high atomic weight. WHO has confirmed that lead (Pb) is one of the major toxic chemicals of public health concern and needed strict actions to control. It is present in combination with other elements i.e., lead sulfide, lead (II) sulfate, and lead (II) carbonate. Now, the generation of high amount of Pb waste due to various anthropogenic activities such as various industries, Pb-based paints, Pb dust, street runoff, agricultural fields, and municipal wastewater are leading to adverse effects on animals, human beings, aquatic life, and the environment (Sepe et al. 2003). The Pb is produced by different sources such as metal mining, coal and oil combustion, battery fabrication, pesticides containing lead-arsenate, lead-based food cans, pigments, and paints (Abadin et al. 2007). The physiochemical properties of Pb, for instance, low melting point, anti-corrosive, and malleability makes it amenable to wide range of uses (UNEP and Pure Earth 2019). The solubility of lead in water is depending upon pH, salinity, hardness, etc. Highest solubility of lead is observed in soft and acidic water.

Pb is highly toxic to humans especially children and pregnant women. In humans, the common exposure route of Pb is respiration, unknowingly consumption, and skin contact (United Nations Environment Programme [UNEP] 2022). Inhalation of fumes and dust is also an extensive route of exposure for people working around especially the workers, children, and women of reproductive age. Lead exposure to young children while playing in contaminated soil and airborne household dust is a major concern because it can cause various disabilities in them such as learning disabilities, antisocial behavior, infertility, and heart-related diseases. There is no safe level of Pb has been fixed by government authorities because once exposed and taken in the body, it leads to cause permanent damage to the neuro and cardiovascular systems (UNEP and Pure Earth 2019). Although, WHO has developed some guidelines on clinical management of lead exposure and suggested that ≥5 µg/dl blood Pb concentration in a person should be identified and appropriate action should be taken to reduce and stop its exposure.

2.3.1 Lead in Soil

Soil and dust are the main routes of lead (Pb) contamination from various anthropogenic sources for instance, Pb-based paints, leaded-gasoline combustion, automobile exhaust, and various industrial processes. Perhaps, it is reported that Pb has been released in a large amount by gasoline (4–5 million metric tons) and persist in the environment in the form of dust and soil. Since Pb doesn't decay or degrade, it becomes a long-term source of Pb exposure for children (Agency for Toxic Substances and Disease Registry [ATSDR] 1988). Pb gets immobilized by the organic constituents of soil, so it generally found in the upper 2–5 cm of undisturbed soil (U.S. EPA 1986). On the other hand, the urban and other soils mostly get disturbed and possibly get contaminated to more deeper depths. The high level of Pb has been reported in soil within the range of 20–25 m of roadways and highways and higher than the natural soil levels. Some roadside soils reported to have high concentrations of Pb up to 10,000 ppm. Houses get painted with lead-based paints and smelters are reported to contaminate the adjacent soil to the very high Pb concentrations (60,000 ppm) (U.S. EPA 1986). Children exposure to Pb contamination soil can leads to the serious health issues, as young children may breathe in or ingest contaminated soil while playing (U.S. EPA 1986). Various studies have reported about varying relationships between levels of lead in soil and dust and children's blood lead levels (U.S. EPA 1986; ATSDR 1988; NRS 1993; Gandolff 2023). The particle size and chemical forms of Pb may affect its bioavailability in soil and dust (Barltrop and Meek 1979).

2.3.2 Lead in Water

Lead levels are naturally low in ground and surface water but can increase when water enters the distribution system. Due to its rust-resistant properties, lead piping was widely used until the twentieth century. Lead contamination of drinking water has been reported in various places, including residential buildings, schools, public areas, and offices, primarily due to plumbing using lead connectors (goose necks or pigtails), service pipelines, lead-soldered joints, lead-carrying water fountains and coolers, brass faucets, and other fixtures. The Safe Drinking Water Act of 1986, amended by the EPA, banned the use of lead in public drinking water distribution systems and restricted the lead content in brass plumbing to a maximum of 8%. The EPA regulates permissible lead levels in water and introduced the Lead and Copper Rule (LCR) in 1991 to control lead and copper in drinking water. The LCR has been amended multiple times since its inception. According to the regulations, water systems must monitor lead and copper concentrations at customer taps. If more than 10% of sampled taps exceed lead levels of 15 ppb or copper levels of 1.3 ppm, action must be taken to control corrosion.

2.3.3 Lead in Air

Sources of Pb contamination in the air vary with reference to area. The most important Pb sources in the air are metal and ore processing and Piston-engine aircraft working on leaded aviation fuel. Various other sources include Pb-based battery manufacturers, waste incinerators, smelters, etc., and the highest pb concentration in air is typically found near lead smelters. In India, coal is the largest source of Pb pollution in the environment, after leaded gasoline was banned. Most of the countries successfully withdrawn the use of leaded gasoline with other lead controlling measures, leading to substantial public health benefits such as reduction in Pb level concentrations in human blood (UN News 2021). Since July 2021, leaded fuel for automobiles has been banned throughout the world (Angrand et al. 2022). Moreover, according to WHO there is a lot of work to be done to complete the removal of Pb-based paint from the environment, only 48% of countries have introduced it after March 2023. WHO has united with the United Nations Environment Programme to form the Global Alliance for complete removal of Pb-based paint and to encourage countries all over the world to have legally binding laws to control the use of Pb in paint.

2.4 SOURCES AND PATHWAYS OF ARSENIC (As) CONTAMINATION

Arsenic is an element also called as metalloid, its atomic number is 33 and symbol As. Various organic and inorganic forms of As is present in the environment in combination with more than 200 minerals, the common is arsenopyrite (WHO 2001) and sometimes as a pure elemental crystal.

As is known to cause several diseases including arsenicosis, keratosis, neurodegeneration, diabetes, cardiovascular, dermatological, and carcinogenic effects. As pollution is a worldwide issue and 21 countries have reported in their environment (Mandal and Suzuki 2002; Shrivastava et al. 2017; Bhattacharyya and Alam 2020; Shahid et al. 2020). There are various geogenic and anthropogenic sources, through which As enters into the environment and food chain; affect the human and animal health. As is widely present in the Earth's crust with an average concentration of 5 mg kg^{-1} and is found in more than 200 different minerals such as arsenopyrite, realgar, and orpiment (De and Roy 2023). Geogenic sources include volcanic emissions, weathering of rocks, and the breakdown of natural organic matter. Anthropogenic sources of As include various types of mining and smelting, the use of As-containing pesticides, herbicides, wood preservatives, paints, dyes, and semiconductors (Bhattacharyya and Alam 2020). The concentrations of As varies in air, soil, and water in rural and urban areas (Figure 2.2).

There are three important modes of As biotransformation in the environment such as redox transformation of As(III) to As(V), reduction, methylation, and the biosynthesis of organoarsenic forms. There is biogeochemical cycle for all above processes. Various toxic forms of arsenic such as arsenite (As(III)) and arsenate (As(V)), monomethylarsonic acid (MMA), dimethylarsinic acid (DMA), trimethylarsine oxide (TMAO), arsenobetaine (AsB), etc., are present in the environment (Akter et al. 2005). As(III) is extremely toxic than As(V), and due to high mobility it is generally found in reducing environments viz., groundwater and sediments (Sanyal et al. 2020). Contrariwise, As(V) is less toxic and is generally found in oxidizing environments such as surface water and soils

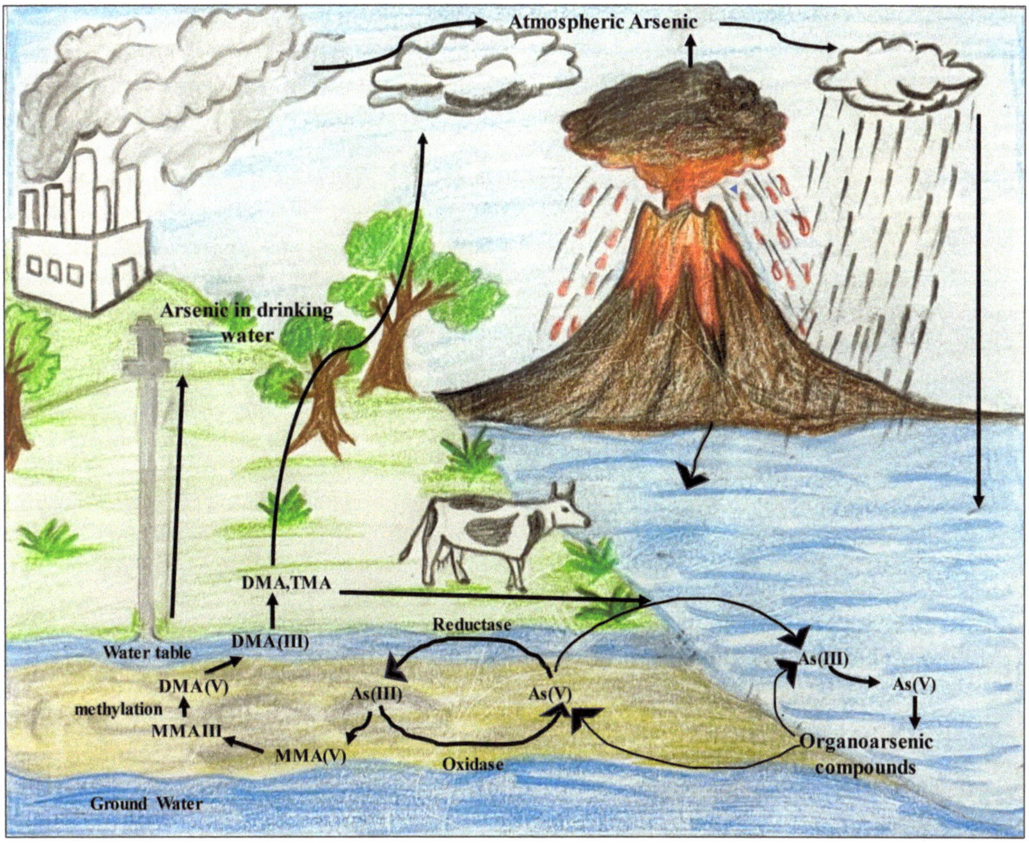

FIGURE 2.2 Arsenic geochemical cycle.

(Mandal and Suzuki 2002). The toxicity, mobility, and solubility of arsenic varies with its forms, oxidation state, and pH of the environment (Sharma and Sohn 2009; Alka et al. 2021). In acidic pH, arsenic likely to be more mobile and bioavailable, while in alkaline, it is comparatively less toxic and mobile. The presence of minerals such as iron and manganese oxides also impacts the bioavailability and mobility of arsenic in the surroundings (Shahid et al. 2020).

2.4.1 Arsenic in Soil

The concentration of arsenic in soil depends upon the type of soils, ranging from 5 to 10 mg kg^{-1} (Abbas et al. 2018). Arsenic is present naturally in soil by natural and anthropogenic activities. Mostly arsenic originated in soil by burning of fossil fuels and biomass, various mining activities, industrial toxic wastes, use of chemical pesticides in agricultural activities (e.g., monosodium methyl arsenate [herbicide], cacodylic acid, and disodium methyl arsonate[DSMA]), usage of wood preservatives (e.g., chromated copper arsenate [CCA] and ammonical copper zinc arsenate [ACZA]), glass manufacturing (arsenic trioxide), alloys (brass and bronze), optics and electronic materials (light emitting or laser diodes, and gallium arsenide microchips), leather preservative, various pigments, and paints (Shankar et al. 2014; Baker et al. 2018).

2.4.2 Arsenic in Water

Arsenic in water originates from natural processes such as atmospheric deposition, rock weathering, mining activities, and volcanic eruptions, as well as from anthropogenic sources like agricultural runoff (from pesticides, herbicides, livestock feed additives, and wood preservatives) and industrial wastewater discharge (Smedley and Kinniburgh 2002). Soil arsenic gradually leaches into groundwater through rainwater, surface runoff, rock weathering, and the irrigation of arsenic-contaminated soils (Nriagu et al. 2007; Qaiser et al. 2023). In well-oxygenated water and sediments, arsenic predominantly exists in the As(V) form (Chen et al. 2019; Hull et al. 2021). However, the most common forms of arsenic in natural waters are arsenite [As(III)] and arsenate [As(V)], along with organic forms such as monomethyl arsenic acid (MMA) and dimethyl arsenic acid (DMA) (Smedley and Kinniburgh 2002; Raju 2022). Arsenic mobility and solubility increase with higher alkalinity and salinity in the environment. Various chemical and physical methods, including adsorption/desorption and precipitation/dissolution reactions, are employed to reduce arsenic concentrations in water (Brusseau and Chorover 2020). Naturally occurring biotransformation and mobilization of arsenic are facilitated by various bacterial species (Darma et al. 2022). Key modes of arsenic biotransformation include redox transformations between As(III) and As(V), reduction and methylation of arsenic, and the bio-production of organic compounds (Shankar et al. 2014; Alonso et al. 2020; Hussain et al. 2021). The permissible limits for arsenic in drinking and irrigation water are 10 and 100 µg/L, respectively. In many developing countries, arsenic-contaminated groundwater is used for drinking and other purposes. High levels of arsenic have been reported in wells, tube wells, and surface waters in several Asian countries, including Bangladesh, Cambodia, China, India, Iran, Nepal, Pakistan, and Vietnam, due to agricultural and industrial activities (Shahid et al. 2020).

2.4.3 Arsenic in Air

Arsenic enters the air through various pathways. It is non-reactive in air and is primarily emitted as arsenic trioxide (As_2O_3) and arsines, which then transform into arsenite [As(III)], arsenate [As(V)], and organoarsenicals. Sources include volcanic emissions, high-temperature processes like coal-fired power plants, burning vegetation, wind erosion, low-temperature soil volatilization, aquatic aerosols, and pollution (WHO 2022). In polluted rural areas, the total average arsenic concentration in the air ranges from 0.02 to 4 ng/m^3, while in urban areas, it ranges from 3 to approximately 200 ng/m^3.

Significantly higher concentrations (>1000 ng/m³) have been reported near industrial areas (WHO 2001). Arsenic returns to the ground, particularly into water bodies, through atmospheric deposition. It then moves via surface runoff and groundwater discharge, eventually binding with suspended particles and sinking to the sediments (Masuda 2018; Bhadauria 2019).

2.5 SOURCES AND PATHWAYS OF CHROMIUM (Cr) CONTAMINATION

In the Earth's mantle, Cr is the 17th most prevalent element (Avudainayagam et al. 2003). At different concentrations, chromium is found in water, air, and soil environments. Chromium is released into the environment (air, soil, and water) by geogenic (Lilli et al. 2015) and natural sources such as dust from rocks and volcanic activity. Around 43,000 tonnes of Cr are thought to be released annually from natural sources worldwide (Nriagu and Nieboer 1988). Numerous small- and large-scale industries make substantial use of Cr and its compounds (Xia et al. 2019). Anthropogenic activities like leather tanning as well as other applications like metallurgy, alloying, electroplating, ceramic glazes, inhibition of water corrosion, wood preservation, refractory bricks, textile dyes, and mordants, production of paints and pigments, pulp and paper production, and wood preservation release significant concentrations of Cr into the environment (Lian et al. 2019). This extensive anthropogenic usage of Cr led to increased environmental contamination, which has grown to be a growing problem in recent years (Oliveira 2012; Saha et al. 2022). The United States Environmental Protection Agency (USEPA) has listed Cr as one of the top pollutants because of its high toxicity, and the Comprehensive Environmental Response, Compensation, and Liability Act classifies Cr as the fifth potentially toxic element (Xia et al. 2019).

2.5.1 Chromium in Soil

Chromium is one of the most common heavy metal pollutants in soil. Its concentration in soil can increase due to anthropogenic activities such as the disposal of chromium-bearing liquid and solid wastes, mining byproducts, and atmospheric deposition. Improper handling of chromium-bearing solid waste from mining and illegal dumping from these sectors cause significant environmental pollution (Gao and Xia 2011). Naturally occurring chromium levels in soil typically range from 10 to 100 mg/kg, depending on the parent rock material, but can rise to 125 g/kg in serpentine mineral-based soils (Prasad et al. 2021).

Chromium exists in various valence states from 0 to VI, with trivalent chromium (CrIII) and hexavalent chromium (CrVI), being the most stable and prevalent in the natural environment (Tumolo et al. 2020). Compared to CrIII, CrVI is more mobile in soil, highly toxic (10–100 times more), and has the potential to be mutagenic, carcinogenic, and teratogenic. CrVI toxicity has been reported even at very low concentrations (<1 mg/kg) in soils (Shahid et al. 2017). The leather tanning industry is a major source of soil and groundwater contamination, prompting significant concern from the European Legislation regarding Cr discharge (Vaiopoulou and Gikas 2020). Balasubramanian and Pugalenthi (1999) reported Cr levels ranging from 217 to 2375 mg/L in the effluents of the leather industry. Additionally, a substantial amount of chromium is present in organic fertilizers (biosolids and phosphorus fertilizers), contributing significantly to soil pollution in agricultural areas (Vogel et al. 2015).

2.5.1.1 Chromium Uptake, Translocation, and Accumulation in Plants

The uptake of chromium (Cr) by plants begins when Cr first contacts the plant. Although the specific pathway of Cr uptake by plants is not fully understood, Cr, being a nonessential element, lacks a dedicated mechanism for its uptake. Consequently, Cr is taken up through carriers that typically transport essential metals for plant metabolism. The toxic effects of Cr are primarily influenced by its speciation, which determines its uptake, transport, and accumulation. Carriers of essential nutrients, such as sulfate and phosphate, play active roles in the transport of Cr(VI)

(Cervantes et al. 2001). For upward translocation, Cr utilizes iron (Fe), sulfur (S), and phosphate (P) channels, leading to competition among these elements.

Members of the Brassicaceae family, known as sulfur accumulators, have been found to accumulate significant amounts of Cr, suggesting that Cr is transported from roots to shoots via sulfur uptake and translocation mechanisms, or sulfate carriers (Singh et al. 2013). Cr predominantly accumulates in roots and shoots, with roots often containing the majority, and only a small portion transferred to the shoots (Sundaramoorthy et al. 2010). The accumulation of Cr in different parts of the plant follows this order: roots > stem > leaves > seeds (Tiwari et al. 2009).

These findings are supported by other studies, such as Liu et al. (2008), who found that *Amaranthus viridis* L. grown hydroponically under varying concentrations of Cr(VI) primarily accumulated Cr in the roots. Additionally, López-Luna et al. discovered that wheat, oat, and sorghum exhibited Cr translocation from roots to shoots, although the roots accumulated more Cr than the shoots.

2.5.2 Chromium in Water

Surface and groundwater can frequently become contaminated with chromium from both natural and anthropogenic sources. Chromium may enter natural streams through soil leaching, weathering of chromium-containing rocks, and direct discharge from industrial activities (Tumolo et al. 2020). Industries such as chromate manufacturing, metal finishing, textile dyeing, and leather tanning are major sources of Cr(VI) contamination, significantly contributing toxic residues to freshwater streams and soil, thereby causing severe environmental impacts (He and Li 2020).

Potential causes of groundwater pollution include seepage from chromite mines, improper disposal of mining waste, and industrial equipment finishing. In aquatic environments, chromium can undergo various transformations, including sorption, desorption, reduction, oxidation, dissolution, and precipitation (Gorny et al. 2016). Cr(III) compounds are generally soluble in water at normal pH ranges but tend to precipitate at neutral to basic pH levels and solubilize at acidic pH levels. Cr(VI), primarily in the form of chromate and dichromate, remains soluble across all pH values but may precipitate with divalent cations.

The recommended concentration limits for Cr(III) and Cr(VI) in drinking water are 8.0 μg/L and 1.0 μg/L, respectively. Chromium levels in wastewater from industries can range from 2 to 5 g/L (Oliveira 2012).

2.5.3 Chromium in Air

According to Shanker and Venkateswarlu (2011), there is limited information on chromium levels in the air. The largest natural source of atmospheric chromium is continental dust flux, with minor contributions from volcanic dust and gas flux (Fishbein 1981). Most chromium is released into the atmosphere through anthropogenic activities, such as the combustion of natural gas, oil, and coal in industrial, commercial, and residential settings. Metal industries, including chrome plating and steel production, are significant stationary point sources of anthropogenic chromium emissions.

Additional sources include cement manufacturing, the wearing down of chromium-containing asbestos brake linings, incineration of municipal and sewage sludge, and emissions from chromium-based automotive catalytic converters (Pacyna and Pacyna 2001). The final processing stages of chromium are likely the most significant airborne sources of chromium in mining locations. Approximately one-third of atmospheric chromium releases are estimated to be in the hexavalent form, Cr(VI) (Johnson et al. 2006). Annually, around 1,723 metric tonnes of chromium are released into the air from burning fossil fuels, though only 0.2% of this is Cr(VI). The chrome-plating process releases about 700 metric tonnes of chromium into the air each year, nearly all of which is Cr(VI) (ATSDR 2000). Workers may be exposed to Cr(VI) during activities such as grinding, abrasive blasting, sanding, and spray painting, particularly in the context of airplane painting and repair (Rosaaen 2017).

2.6 SOURCES AND PATHWAYS OF CADMIUM (Cd) CONTAMINATION

Cadmium (Cd) is one of the most significant toxic heavy metals in the environment. It is abundant in the Earth's crust, typically found with copper, zinc, and lead ores (Cheng et al. 2014). Natural processes that contribute to environmental cadmium contamination include the weathering of rocks, volcanic eruptions, wind dust, marine aerosols, forest fires, and erosion (Liu et al. 2015). However, its prevalence is largely due to anthropogenic activities such as mining, alloy production, cement manufacturing, metallurgy, pigment and plastic stabilizer production, fossil fuel combustion, high phosphate fertilizers, municipal and sewage sludge incineration, and nickel-cadmium battery production (Zeng et al. 2010). According to the United Nations Environment Programme (UNEP), between 150 and 2,600 tonnes of cadmium are emitted into the environment annually (UNEP 2008).

2.6.1 Uptake and Transportation of Cd in Plants

Cadmium (Cd), a nonessential element for plants, enters them through pathways meant for essential elements like calcium (Ca), iron (Fe), manganese (Mn), and zinc (Zn) (Clemens 2006). Cd can enter root cells either as a free ion (Cd^{2+}) or as chelated compounds such as Cd-phytochelatin (PC) and Cd-glutathione (GSH). This process requires ion transporters that are often element-specific, transporting one or several types of metal ions (Lux et al. 2011). Additionally, non-specific cation channel proteins, including voltage-sensing channels (VICC), depolarization-activated calcium channels (DACC), and hyperpolarization-activated calcium channels (HACC), can transport free Cd^{2+} into cells. The yellow stripe 1-like (YSL) transporter is known to transfer Cd into root cells as chelates (Curie et al. 2009; Lux et al. 2011).

The process of Cd absorption and accumulation in plants involves several steps: absorption by roots, loading and transport into the xylem, transfer from xylem to phloem, redistribution to aboveground parts, and accumulation in grains (Uraguchi et al. 2009). Once in the soil, Cd reaches the root xylem, where it is loaded and transported throughout the plant. The xylem unloads Cd, which is then transported to various tissues, including stems, leaves, flowers, and other organs. At the stem node, the scattered vascular bundle facilitates the transfer of Cd from the xylem to the phloem, completing its journey to the plant's aboveground parts (Tanaka et al. 2007).

2.6.2 Uptake and Distribution of Cd in Human Body

The human body is exposed to Cd toxicity both directly and indirectly. Direct exposure occurs through skin contact and inhalation, whereas indirect exposure occurs through the ingestion of vegetables and crops. The Cd travels from the soil to the roots of plants, then transforms into edible plant components before entering the human body (Alloway and Steinnes 1999). About 90% of a person's exposure to Cd comes from cereals and vegetables, including lettuce, cabbage, beets, coriander, radishes, carrots, spinach, parsley, and potatoes. The route through which cadmium enters the body depends on factors such as age, sex, nutritional levels, diet composition, exposure time, and exposure dose (Gustin et al. 2018). People with low hemoglobin levels have been found to be more susceptible to Cd exposure, and their Cd accumulation ratio is 6% greater than that of healthy individuals. Therefore, those who have anemic are at risk, particularly girls and women who are menstruating. Ion carrier proteins that transport Cd into the gastrointestinal (GI) track are activated by low iron levels. DCT-1, a metal ion transporter that acts as a gate for Cd re-absorption, is expressed more frequently in the GI tract in response to low blood iron levels (Godt et al. 2006). Seafood like fish and prawns are also sources of cadmium through food, in addition to vegetables and cereals. Fish is the main source of Cd consumption among sea creatures for two reasons: (a) fish is the primary source of food for individuals who live in coastal areas and (b) fish has a high Cd content (Fatima et al. 2019). Smoke from cigarettes is another way of Cd exposure to the lungs as it contains two times more Cd through air. Because fertilizers are applied to tobacco plants, cigarette smoke contains significant levels of Cd. Cigarette smoke has a direct correlation with the amount of toxicity of cadmium. When a person smokes, their lungs absorb 50–70% of the Cd (Kim et al. 2010).

2.6.3 Cadmium in Water and Soil

Anthropogenic activities are the primary sources of Cd, which is highly mobile, toxic, and has the ability to bioaccumulate in the ecosystem. It may contaminate groundwater and river water, which would have a negative effect on agricultural and water supplies. Anthropogenic activities like mining and smelting, agriculture, sewage irrigation, and traffic are the main causes of Cd poisoning of groundwater. Cd in the nearby soil and water has often been primarily derived from acidic drainage, mine wastewater, atmospheric deposition, and toxic slag (Du et al. 2020). Electronic waste (e-waste) has emerged as one of industrialization's fastest-growing waste products. Since these e-wastes comprise a mixture of plastic and metals, irresponsible recycling practices such as open-air burning and disposal in fields and waterways contribute to the heavy metal pollution (Wong et al. 2007). Guiyu, a town in Southeast China, has been a center for recycling e-waste since 1995, resulting in extremely worrying cadmium concentration of up to 42.9 mg/kg. Cd has been found in the paddy soils and river sediments in the nearby areas as a result of these operations (Wong et al. 2007). Agriculture is another significant factor in the Cd contamination of groundwater. Higher plants absorb cadmium (Cd) through their roots from soil and water, depending on its availability and concentration in the external medium (Clemens 2006). Several factors influence the availability of Cd, as well as its absorption and transport by plants. These include soil type, pH, temperature, water availability, microbial community composition in the rhizosphere, and genotypic variations among plants (Eriksson 1989). For example, plants growing in Cd-contaminated soils tend to have higher Cd levels at pH 4.0 compared to pH 5.0. The increased acidity of soils, especially in southern China, has enhanced Cd activity in the soil, making it easier for crops to absorb (Zhao et al. 2015). Low-molecular-weight organic acids released by plant roots significantly influence the solubility and availability of Cd. These acids may also play a role in mediating Cd uptake and transport within plants (Eriksson 1989). Additionally, plants can absorb higher levels of Cd in sandy soils than in clayey soils because Cd tends to remain in a soluble or exchangeable form in sandy soils (Mann & Ritchie 1995).

2.7 CONCLUSION AND PROSPECTS

Heavy metal contamination is one of the most pressing environmental issues today. The entry of toxic heavy metals into the ecosystem can lead to bioaccumulation and biomagnification, posing significant threats to living organisms and plants. These metals are indestructible and can become toxic at specific concentrations, making their pollution a global concern. Naturally, heavy metals exist in soil due to the pedogenetic processes of parent material degradation. However, anthropogenic activities are major contributors to heavy metal pollution, including transportation emissions, industrial production, vehicle exhaust, waste disposal, and the combustion of coal and fuel. High levels of heavy metals degrade soil quality, reduce crop yields, and compromise the quality of agricultural products, posing serious risks to human, animal, and ecosystem health. Continuous assessment and monitoring of heavy metal levels in the environment are crucial to understanding human exposure and maintaining a sustainable environment.

ACKNOWLEDGMENT

Authors of this chapter are thankful to their University for providing necessary support during this research.

REFERENCES

Abadin, H., A. Ashizawa, Y.W. Stevens et al. 2007. Toxicological profile for lead. DOI:10.1201/9781420061888_ch106

Abas, A. 2021. A systematic review on biomonitoring using lichen as the biological indicator: A decade of practices, progress and challenges. Ecol. Indic., 121: 107197. DOI: 10.1016/j.ecolind.2020.107197

Abbas, G., B. Murtaza, I. Bibi et al. 2018. Arsenic uptake, toxicity, detoxification, and speciation in plants: Physiological, biochemical, and molecular aspects. Int. J. Environ. Res. Public Health, 15: 59, DOI: 10.3390/ijerph15010059

Akter, K. F., G. Owens, D. E. Davey et al. 2005. Arsenic speciation and toxicity in biological systems. Rev. Environ. Contam. Toxicol., 184: 97–149. DOI: 10.1007/0-387-27565-7_3

Ali, H., E. Khan and I. Ilahi. 2019. Environmental chemistry and ecotoxicology of hazardous heavy metals: Environmental persistence, toxicity, and bioaccumulation. J. Chem., 4: 1–14. DOI: 10.1155/2019/6730305

Alka, S., S. Shahir, N. Ibrahim et al. 2021. Arsenic removal technologies and future trends: A mini review. J. Clean. Prod., 278: 123805. DOI: 10.1016/j.jclepro.2020.123805

Alloway, B.J. and E. Steinnes. 1999. Anthropogenic additions of cadmium to soils. In: McLaughlin, M.J., Singh, B.R. (eds). Cadmium in Soils and Plants, Springer, Dordrecht, 97–123.

Alonso, D.L., R. Pérez, C.K.Y.A. Okio et al. 2020. Assessment of mining activity on arsenic contamination in surface water and sediments in southwestern area of Santurban Paramo, Columbia. J. Environ. Manage., 264: 110478. DOI: 10.1016/j.jenvman.2020.110478

Angrand, R.C., G. Collins, P.J. Landrigan et al. 2022. Relation of blood lead levels and lead in gasoline: an updated systematic review. Environ. Health, 21: 138. DOI: 10.1186/s12940-022-00936-x

ATSDR (Agency for Toxic Substances and Disease Registry). 1988. The Nature and Extent of Lead Poisoning in Children in the United States: A Report to Congress. Agency for Toxic Substances and Disease Registry, Atlanta.

ATSDR (Agency for Toxic Substances and Disease Registry). 2000. Toxicological Profile for Chromium, U.S. Department of Health and Human Services, Public Health Service.

Avudainayagam, S., M. Megharaj, G. Owens et al. 2003. Chemistry of chromium in soils with emphasis on tannery waste sites. Rev. Environ. Contam. Toxicol., 178: 53–91.

Babel, S. and T.A. Kurniawan. 2003. Low-cost adsorbents for heavy metals uptake from contaminated water: A review. J. Hazard. Mater., B97: 219–243. DOI: 10.1016/S0304-3894(02)00263-7

Baker, B.A., V.A. Cassano and C. Murray. 2018. Arsenic exposure, assessment, toxicity, diagnosis, and management: Guidance for occupational and environmental physicians. J. Occup. Environ. Med., 60: 634–639. DOI: 10.1097/JOM.0000000000001485

Balasubramanian, S. and V Pugalenthi. 1999. Determination of total chromium in tannery waste water by inductively coupled plasma-atomic emission spectrometry, flame atomic absorption spectrometry and UV–visible spectrophotometric methods. Talanta, 50(3): 457–467.

Barltrop, D. and F. Meek. 1979. Effect of particle size on lead absorption from the gut. Arch. Environ. Health, 34: 280–285.

Bayouli, I.T., H.T. Bayouli, A. Dell'Oca et al. 2021. Ecological indicators and bioindicator plant species for biomonitoring industrial pollution: Eco-based environmental assessment. Ecol. Indic., 125(6): 107508. DOI: 10.1016/j.ecolind.2021.107508.

Bhadauria, R. 2019. Arsenic toxicity: An overview. Horticult. Int. J., 3: 20–22. DOI: 10.15406/hij.2019.03.00106.

Bhattacharyya, P. and M.M. Alam. 2020. Arsenic-contaminated soil toxicity and its mitigation through monocot crops. In: Naeem, M., Ansari, A., Gill, S. (eds). Contaminants in Agriculture. Springer, Cham, 327–337. DOI: 10.1007/978-3-030-41552-5_16

Briffa, J., E. Sinagra and R. Blundell. 2020. Heavy metal pollution in the environment and their toxicological effects on humans. Heliyon, 6(9): e04691.

Brusseau, M.L. and J. Chorover. 2020. Chapter 8-Chemical processes affecting contaminant transport and fate. In: Brusseau, M.L., Pepper, I.L., Gerba, C.P. (eds). Environmental and Pollution Science (Third Edition). Academic Press, 113–130. DOI: 10.1016/B978-0-12-814719-1.00008-2

Cervantes, C., J. Campos-García, S. Devars et al. 2001. Interactions of chromium with microorganisms and plants. FEMS Microb. Rev., 25(3): 335–347.

Chen, Y. W., X. Yu and N. Belzile. 2019. Arsenic speciation in surface waters and lake sediments in an abandoned mine site and field observations of arsenic eco-toxicity. J. Geochem. Explor., 205: 106349. DOI: 10.1016/j.gexplo.2019.106349

Cheng, K., H.Z. Tian, D. Zhao et al. 2014. Atmospheric emission inventory of cadmium from anthropogenic sources. Int. J. Environ. Sci. Technol., 11: 605–616.

Clemens, S. 2006. Toxic metal accumulation, responses to exposure and mechanisms of tolerance in plants. Biochimie, 88(11): 1707–1719.

CPCB. 2011. Performance Evaluation and Monitoring of the Common Hazardous Waste Treatment Storage and Disposal Facilities including Common Hazardous Waste Incinerators. Hazardous Waste Management Series: HAZWAMS/(cpcb.nic.in)

CPCB Bhopal. 2012. Presence of Heavy Metals in Vegetation & Their Effects. CPCB_Bhopal_presence of Heavy Metal in Vegetation at Mandideep.pdf

Curie, C., G. Cassin, D. Couch et al. 2009. Metal movement within the plant: Contribution of nicotianamine and yellow stripe 1-like transporters. Ann. Bot., 103(1): 1–11.

Darma, A, J. Yang, E. Bloem et al. 2022. Arsenic biotransformation and mobilization: The role of bacterial strains and other environmental variables. Environ. Sci. Pollut. Res. Int., 29(2): 1763–1787. DOI: 10.1007/s11356-021-17117-x. Epub 2021 Oct 28. PMID: 34713399.

De, A. and N. Roy. 2023. Arsenic in Environment: Sources, Implications and Remedies. DOI: 10.5772/intechopen.1001476.

Du, B., J. Zhou, B. Lu et al. 2020. Environmental and human health risks from cadmium exposure near an active lead-zinc mine and a copper smelter, China. Sci. Total Environ., 720: 137585.

Eriksson, J.E. 1989. The influence of pH, soil type and time on adsorbtion and uptake by plants of Cd added to the soil. Wat. Air Soil Poll., 48: 317–335.

Fatima, G., A.M. Raza, N. Hadi, N. Nigam and A.A. Mahdi. 2019. Cadmium in human diseases: It's more than just a mere metal. Indian J. Clin. Biochem., 34: 371–378.

Fishbein, L. 1981. Sources, transport and alterations of metal compounds: An overview. I. Arsenic, beryllium, cadmium, chromium, and nickel. Environ. Health Perspect., 40: 43–64.

Fu, F. and Q. Wang. 2011. Removal of heavy metal ions from wastewaters: A review. J. Environ. Manage., 92: 407–418.

Gandolff, R. 2023. Lead exposure in childhood and historical land use: A geostatistical analysis of soil lead concentrations in South Philadelphia parks. Environ. Monit. Assess., 195(3): 356. DOI: 10.1007/s10661-022-10871-6. PMID: 36732369; PMCID: PMC9894736.

Gao, Y. and J. Xia. 2011. Chromium contamination accident in China: Viewing environment policy of China. Environ. Sci. Technol., 45(20): 8605–8606. DOI: 10.1021/es203101f. Epub. Sep 19.

Gautam, P.K., R.K. Gautam, M.C. Chattopadhyaya et al. 2016. Heavy metals in the environment: Fate, transport, toxicity and remediation technologies. In: Pathania, D. (ed). Heavy Metals. Nava Science Publishers, Inc ISBN: 978-1-63484-740-7

Godt, J., F. Scheidig, C. Grosse-Siestrup et al. 2006. The toxicity of cadmium and resulting hazards for human health. J. Occup. Med. Toxicol., 1(1): 1–6.

Gorny, J., G. Billon, C. Noiriel, D. Dumoulin, L. Lesven and B. Madé. 2016. Chromium behavior in aquatic environments: A review. Environ. Rev., 24(4): 503–516.

Gustin, K., F. Tofail, M. Vahter and M. Kippler. 2018. Cadmium exposure and cognitive abilities and behavior at 10 years of age: A prospective cohort study. Environ. Int., 113: 259–268.

He, X. and P. Li. 2020. Surface water pollution in the middle Chinese Loess Plateau with special focus on hexavalent chromium (Cr^{6+}): Occurrence, sources and health risks. Exp. Health, 12(3): 385–401.

He, Z.L., X.E. Yang and P.J. Stoffella. 2005. Trace elements in agroecosystems and impacts on the environment. J. Trace Elem. Med. Biol., 19: 125–140.

Herawati, N., S. Suzuki, K. Hayashi et al. 2000. Cadmium, copper, and zinc levels in rice and soil of Japan, Indonesia, and China by soil type, Bull. Environ. Contam. Toxicol., 64: 33–39.

Hull, E.A., M. Barajas, K.A. Burkart et al. 2021. Human health risk from consumption of aquatic species in arsenic-contaminated shallow urban lakes. Sci. Total Environ., 770(20): 145318. doi: 10.1016/j.scitotenv.2021.145318

Hussain, M. M., J. Wang, I. Bibi et al. 2021. Arsenic speciation and biotransformation pathways in the aquatic ecosystem: The significance of algae. J. Hazard. Mater., 403: 124027. DOI: 10.1016/j.jhazmat.2020.124027

Islam, M.S., R. Proshad and S. Ahmed. 2018. Ecological risk of heavy metals in sediment of an urban river in Bangladesh. Hum. Ecol. Risk Assess., 24(3): 699–720.

Jarup, L. 2003. Hazards of heavy metal contamination. Br. Med. Bull., 68: 167–182. DOI: 10.1093/bmb/ldg

Johnson, J., L. Schewel and T.E. Graedel. 2006. The contemporary anthropogenic chromium cycle. Environ. Sci. Technol., 40(22): 7060–7069.

Khan, M.N., M. Mobin, Z.K. Abbas et al. 2018. Fertilizers and their contaminants in soils, surface and groundwater. Encycl. Anthropocene, 5: 225–240.

Kim, H., H.J. Lee, J.Y. Hwang et al. 2010. Blood cadmium concentrations of male cigarette smokers are inversely associated with fruit consumption. J. Nutr., 140(6): 1133–1138.

Kubier, A., R.T. Wilkin, and T. Pichler. 2019. Cadmium in soils and groundwater: A review. Appl. Geochem., 108: 104388.

Kumar, V., I.S. Thakur, A.K. Singh et al. 2020. Application of metagenomics in remediation of contaminated sites and environmental restoration. In: Shah, M.P., Rodriguez-Couto, S., Şengör, S.S. (eds). Emerging Technologies in Environmental Bioremediation. Elsevier, 197–232. https://doi.org/10.1016/B978-0-12-819860-5.00008-0

Lian, G., B. Wang, X. Lee et al. 2019. Enhanced removal of hexavalent chromium by engineered biochar composite fabricated from phosphogypsum and distillers grains. Sci. Total Environ., 697: 134119.

Lilli, M.A., D. Moraetis, N.P. Nikolaidis et al. 2015. Characterization and mobility of geogenic chromium in soils and river bed sediments of Asopos basin. J. Hazard. Mater., 281: 12–19.

Liu, D., J. Zou, M. Wang and W. Jiang. 2008. Hexavalent chromium uptake and its effects on mineral uptake, antioxidant defence system and photosynthesis in *Amaranthus viridis* L. Bioresour. Technol., 99(7): 2628–2636.

Liu, Y., T. Xiao, P.C. Baveye, J. Zhu, Z. Ning and H. Li. 2015. Potential health risk in areas with high naturally-occurring cadmium background in southwestern China. Ecotoxicol. Environ. Saf., 112: 122–131.

López-Luna, J., M.C. González-Chávez, F.J. Esparza-García and R. Rodríguez-Vázquez. 2009. Toxicity assessment of soil amended with tannery sludge, trivalent chromium and hexavalent chromium, using wheat, oat and sorghum plants. J. Hazard. Mater., 163(2–3): 829–834.

Lux, A., M. Martinka, M. Vaculík and P.J. White. 2011. Root responses to cadmium in the rhizosphere: A review. J. Exp. Bot., 62(1): 21–37.

Mandal, B.K. and K.T. Suzuki. 2002. Arsenic round the world: A review. Talanta, 58(1): 201–235. DOI: 10.1016/S0039-9140(02)00268-0

Mann, S.S. and G.S.P. Ritchie. 1995. Forms of cadmium in sandy soils after amendment with soils of higher fixing capacity. Environ. Poll., 87(1): 23–29.

Manzoor, J., M. Sharma and K.A. Wani. 2018. Heavy metals in vegetables and their impact on the nutrient quality of vegetables: A review. J. Plant Nutr. DOI: 10.1080/01904167.2018.1462382

Masindi, V. and K.L. Muedi. 2018. Environmental contamination by heavy metals. In: Heavy Metals, 10, 115–132. InTech. DOI: 10.5772/intechopen.76082

Masuda, H. 2018. Arsenic cycling in the Earth's crust and hydrosphere: Interaction between naturally occurring arsenic and human activities. Prog. Earth Planet. Sci., 5: 68. DOI: 10.1186/s40645-018-0224-3

Nouchi, I. 2002. Plants as bioindicators of air pollutants. In: Omasa, K, Saji, H., Youssedian, S., kondo, N. (eds). Air Pollution and Plant Biotechnology. Springer, Tokyo. DOI: 10.1007/978-4-431-68388-9_2

Nriagu, J.O., P. Bhattacharya, A.B. Mukherjee et al. 2007. Arsenic in soil and groundwater: An overview. In: Trace Metals and Other Contaminants Environment, 9, Elsevier, 3–60. DOI: 10.1016/S1875-1121(06)09001-8

Nriagu, J.O. and E. Nieboer. 1988. Production and uses of chromium. In: Chromium in the Natural and Human Environments, 20, Elsevier, 81–104.

NRS. 1993. National Research Council (US) committee on measuring lead in critical populations. In: Measuring Lead Exposure in Infants, Children, and Other Sensitive Populations. National Academies Press (US), Washington, DC.

Oliveira, H. 2012. Chromium as an environmental pollutant: Insights on induced plant toxicity. J. Bot., 2012, Article ID: 375843. DOI: 10.1155/2012/375843

Pacyna, J.M. and E.G. Pacyna. 2001. An assessment of global and regional emissions of trace metals to the atmosphere from anthropogenic sources worldwide. Environ. Rev., 9(4): 269–298.

Parmar, T.K., R. Deepak and Y.K. Agrawal. 2016. Bioindicators: The natural indicator of environmental pollution. Front. Life Sci., 9(2): 110–118. DOI: 10.1080/21553769.2016.1162753

Prasad, S., K.K. Yadav, S. Kumar et al. 2021. Chromium contamination and effect on environmental health and its remediation: A sustainable approaches. J. Environ. Manage., 285: 112174.

Qaiser, F.U.R., F. Zhang, R.R. Pant et al. 2023. Characterization and health risk assessment of arsenic in natural waters of the Indus River Basin, Pakistan, Sci. Total Environ., 857: 159408. DOI: 10.1016/j.scitotenv.2022.159408

Raju, N.J. 2022. Arsenic in the geo-environment: A review of sources, geochemical processes, toxicity, and removal technologies. Environ. Res., 203: 111782. DOI: 10.1016/j.envres.2021.111782

Rashmi, I., T. Roy, K.S. Kartika et al. 2020. Organic and inorganic fertilizer contaminants in agriculture: Impact on soil and water resources. In: Naeem, M., Ansari, A., Gill, S. (eds). Contaminants in Agriculture: Sources, Impacts and Management. Springer, Cham, 3–41.

Rosaaen, D. 2017. Hexavalent chromium exposure to military aircraft painters. Graduate Theses & Non-Theses. 135. http://digitalcommons.mtech.edu/grad_rsch/135. Material Science. Corpus. I.D., 139386845.
Saha, B., A. Amine and F. Verpoort. 2022. Hexavalent chromium: Sources, toxicity, and remediation. Chem. Africa, 5(6): 1779–1780.
Sanyal, T., P. Bhattacharjee, S. Paul et al. 2020. Recent advances in arsenic research: Significance of differential susceptibility and sustainable strategies for mitigation. Front. Public Health, 8: 464. DOI: 10.3389/fpubh.2020.00464.
Sepe, A., L. Ciaralli, M. Ciprotti et al. 2003. Determination of cadmium, chromium, lead and vanadium in six fish species from the Adriatic Sea. Food Addit. Contam., 20(6): 543–552.
Shahid, M., M. Imran, S. Khalid et al. 2020. Arsenic environmental contamination status in South Asia. In: Srivastava, S. (ed). Arsenic in Drinking Water and Food. Springer, Singapore, 13–39.
Shahid, M., S. Shamshad and M. Rafiq. 2017. Chromium speciation, bioavailability, uptake, toxicity and detoxification in soil-plant system: A review. Chemosphere, 178: 513–533.
Shaji, E., M. Santosh, K.V. Sarath et al. 2021. Arsenic contamination of groundwater: A global synopsis with focus on the Indian Peninsula. Geosci. Front., 12(3): 101079.
Shanker, A.K. and B. Venkateswarlu. 2011. Chromium: Environmental pollution, health effects, and mode of action A2.In: Nriagu, J.O. (ed). Encyclopedia of Environmental Health. Elsevier, Burlington, 650–659. DOI: 10.1016/B978-0-444-52272-6.00390-1
Shankar, S., U. Shanker and Shikha. 2014. Arsenic contamination of groundwater: A review of sources, prevalence, health risks, and strategies for mitigation. Sci. World J., 1–18. DOI: 10.1155/2014/304524
Sharma, V.K. and M. Sohn. 2009. Aquatic arsenic: Toxicity, speciation, transformations, and remediation. Environ. Int., 35: 743–759. DOI: 10.1016/j.envint.2009.01.005
Shrivastava, A., A. Barla, S. Singh et al. 2017. Arsenic contamination in agricultural soils of Bengal deltaic region of West Bengal and its higher assimilation in monsoon rice. J. Hazard. Mater., 324: 526–534. DOI: 10.1016/j.jhazmat.2016.11.022
Singh, H.P., P. Mahajan, S. Kaur et al. 2013. Chromium toxicity and tolerance in plants. Environ. Chem. Lett., 11: 229–254.
Smedley, P. L. and D. G. Kinniburgh. 2002. A review of the source, behaviour and distribution of arsenic in natural waters. Appl. Geochem., 17: 517–568. DOI: 10.1016/S0883-2927(02)00018-5
Sundaramoorthy, P., A. Chidambaram, K.S. Ganesh, P. Unnikannanand and L. Baskaran. 2010. Chromium stress in paddy:(i) nutrient status of paddy under chromium stress;(ii) phytoremediation of chromium by aquatic and terrestrial weeds. Comptes. Rendus Biol., 333(8): 597–607.
Tanaka, K., S. Fujimaki, T. Fujiwara, T. Yoneyama and H. Hayashi. 2007. Quantitative estimation of the contribution of the phloem in cadmium transport to grains in rice plants (*Oryza sativa* L.). Soil Sci. Plant Nutr., 53(1): 72–77.
Tiwari, K.K., S. Dwivedi, N.K. Singh, U.N. Rai and R.D. Tripathi. 2009. Chromium (VI) induced phytotoxicity and oxidative stress in pea (*Pisum sativum* L.): biochemical changes and translocation of essential nutrients. J. Environ. Biol., 30(3): 389–394.
Tumolo, M., V. Ancona, D. De Paola et al. 2020. Chromium pollution in European water, sources, health risk, and remediation strategies: An overview. Int. J. Environ. Res. Public Health, 17(15): 5438.
U.S. EPA. 1986. Air Quality Criteria for Lead (Final Report, 1986). U.S. Environmental Protection Agency, Washington, D.C., EPA/600/8-83/028AF (NTIS PB87142386).
UN News. 2021. End of leaded fuel use a "milestone for multilateralism" press release. https://news.un.org/en/story/2021/08/1098792.
United Nations Environment Programme (UNEP). 2008. Draft final review of scientific information on cadmium. Chemicals Branch DTIE November. 36–37.
United Nations Environment Programme (UNEP). 2022. Lead is a harmful heavy metal. Lead | UNEP – UN Environment Programme.
UNEP and Pure Earth. 2019. UN Environment Annual Report (unep.org).
Uraguchi, S., S. Mori, M. Kuramata, A. Kawasaki, T. Arao and S. Ishikawa. 2009. Root-to-shoot Cd translocation via the xylem is the major process determining shoot and grain cadmium accumulation in rice. J. Exp. Bot., 60(9): 2677–2688.
Vaiopoulou, E. and P. Gikas. 2020. Regulations for chromium emissions to the aquatic environment in Europe and elsewhere. Chemosphere, 254: 126876.
Vogel, C., M. Radtke and U. Reinholz. 2015. Chemical state of chromium, sulfur, and iron in sewage sludge ash based phosphorus fertilizers. ACS Sust. Chem. Eng., 3(10): 2376–2380.

Walker, C.H., R. M. Sibly, S.P. Hopkin et al. 2012. Principles of Ecotoxicology. 4th Edition, CRC Press. DOI: 10.1201/b11767

WHO (World Health Organization). 2001. Arsenic and arsenic compounds, World Health Organization, International Programme on Chemical Safety, Geneva. Environmental Health Criteria: Arsenic (who.int).

WHO (World Health Organization). 2011. Arsenic in Drinking-water. World Health Organization, Geneva, Switzerland. Microsoft Word – Fourth Edition Arsenic final 14 June 2011.doc (who.int).

WHO (World Health Organization). 2022. Arsenic. Arsenic (who.int).

Wong, M.H., S.C. Wu, W.J. Deng et al. 2007. Export of toxic chemicals–a review of the case of uncontrolled electronic-waste recycling. Environ. Poll., 149(2): 131–140.

Xia, S., Z. Song, P. Jeyakumar et al. 2019. A Critical review on bioremediation technologies for Cr(VI)-contaminated soils and wastewater. Critical Reviews in Environmental Science and Technology, 49(12): 1027–1078.

Zeng, X., J. Tang, H. Yin, X. Liu, P. Jiang and H Liu. 2010. Isolation, identification and cadmium adsorption of a high cadmium-resistant *Paecilomyces lilacinus*. Afr. J. Biotechnol., 9(39): 6525–6533.

Zhao, F.J., Y. Ma, Y.G. Zhu, Z. Tang and S.P. McGrath. 2015. Soil contamination in China: Current status and mitigation strategies. Environ. Sci. Technol., 49(2): 750–759.

Zwolak, A., M. Sarzvnska, E. Szpyrka et al. 2019. Sources of soil pollution by heavy metals and their accumulation in vegetables: A review. Water Air and Soil Poll., 230(7). DOI: 10.1007/s11270-019-4221-y

3 Health Impacts Associated with Heavy Metals Exposure

Bhavani R
Department of Biotechnology and Medical Engineering, National Institute of Technology, Rourkela, Odisha, India

*Swati Sharma and Kamal Singh**
Viral Research and Diagnostic Laboratory, Department of Virology, Rajendra Memorial Research Institute of Medical Sciences (RMRIMS), Patna, Bihar, India

3.1 INTRODUCTION

Heavy metals (HMs) are naturally occurring components on the Earth. The lower permitted concentrations are necessary for the proper operation of numerous biochemical and physiological processes in any living organism. However, when taken more than the threshold levels, these heavy metals become hazardous compounds that can have detrimental impacts on the environment and human health (Verma, Bhatt et al. 2021). Industries that use heavy metals regularly include the automotive industry, which encompasses all vehicles, machinery and transportation; agriculture and farming, which uses pesticides and fertilizers; mining; and construction and infrastructure management (Masindi and Muedi 2018, Ren, Li et al. 2021). All these have led to the exposure of heavy metals by humans in larger quantities. Heavy metal pollution is a common issue encountered in recent days which is due to exceeding anthropogenic activities such as industrial applications, domestic and agricultural purposes and many other natural phenomena involving soil erosion, corrosion of metals, volcanic breakout and weathering (Ruggieri, Saavedra et al. 2010, Cimboláková, Uher et al. 2020).

The toxic properties of heavy metals due to their non-biodegradability can lead to severe contamination in freshwater bodies which ultimately impacts the survival rate of aqueous life and is also carried over by the food chain process (Gautam, Gautam et al. 2016, El Ati-Hellal and Hellal 2021). This even affects the quality of drinking water and risks human lives (Chowdhury, Mazumder et al. 2016). With the increasing urbanization process, the health risk assessment ratio is elevated to higher levels (Rai 2008). These traces of Heavy metals (HMs) can enter the human body by various means such as food, water and air which are contaminated by these components. The major reason behind the intoxication of these heavy metals is due to their bond formation with the hydroxyl groups present in the biological systems and producing higher levels of reactive oxygen species (ROS)(Rehman, Fatima et al. 2018, Fu and Xi 2020). This interferes with many physicochemical and biological pathways that alter the body's functions resulting in detrimental effects on the health conditions (Fu and Xi 2020).

The bioaccumulation of increasing concentrations of HMs such as lead, zinc, cadmium, chromium, iron, manganese, arsenic, thallium, antimony and many other compounds can cause defective problems in living systems by creating morphological alterations (Khan, Khan et al. 2015). Genetic mutations, abnormalities in the external features of the body and degradative effects are the substantial outcomes associated with extensive exposure to HMs (Bhargava, Gupta et al. 2017). Human exposure to these components mainly depends on the food chain (Sonone, Jadhav et al. 2020) and breathing of the intoxicated air (Jaishankar, Tseten et al. 2014). Soil pollution is caused

* **Corresponding Author:**kamal.ihbt@gmail.com

 DOI: 10.1201/9781032685793-3

by the usage of mineral fertilizers and many other chemical pesticides and air pollution is due to the emission of toxic compounds as lead into the environment (Mortvedt 1995).

Chronic exposure to certain HMs' such as manganese and cadmium can lead to neurological illness and they are extremely hazardous to the living systems (Martin and Griswold 2009). Many other carcinogenic properties are also deliberately present in these heavy metals which cause severe effects when consumed at higher concentrations (Rehman, Fatima et al. 2018). The various elemental forms of chromium show different side effects and the toxicity is seen higher in the chromium (VI) elemental form (Shrivastava, Upreti et al. 2002). The higher abundance of cobalt in water and rocks implicates the elevated levels of harmful effects on human health conditions (Abdul-Wahab and Marikar 2012). The proportion of industrial utilization of nickel is very high these days which leads to many disease conditions associated with kidney, lungs and heart (Abdul-Wahab and Marikar 2012).

The toxicity levels of heavy metals vary based on the quantity of exposure. Zinc is one such component that acts as a cofactor in various enzymatic reactions but in higher concentrations deteriorates cellular functioning (El Ati-Hellal and Hellal 2021). Even the bare minimum concentration of antimony creates impairing effects on the reproduction mechanism and also causes other side effects (Cooper and Harrison 2009). Thallium toxicity is also more common in the case of industrial emissions, elevating the harmful impacts on the living system (Peter and Viraraghavan 2005).

Both the essential and non-essential HMs have adverse impacts on the human immune system, hormonal regulation, metabolic functioning and cell proliferation (Marth, Jelovcan et al. 2001). The different ways of exposure to these components include the gastrointestinal tract, inhalation and dietary routes. The average absorption of cadmium via the digestive tract ranges from 3% to 7%, as informed by the World Health Organization (WHO) and Institutional Repository for Information Sharing (**IRIS**). Various animal studies have found that these HMs' exposures can cause severe nephrotoxicity, neurotoxicity, dermal allergies, blood intoxication, neuropsychiatric conditions and encephalopathy in humans, including infants and foetus (Dobbs 2011).

Considering the significance of this topic, this book chapter mainly focuses on the characteristic features of different HMs and their mode of exposure to humans and their associated health impacts due to excessive subjection from the environment.

3.2 TOXICITY OF DIFFERENT HMs IN HUMANS

3.2.1 Mercury

This HM is the most common air pollutant due to industrial emissions (Mukherjee, Bhattacharya et al. 2009). This is more toxic in the organic form which is called methylmercury and gets accumulated in the fishes, as a result into the food chain (Al-Sulaiti, Soubra et al. 2022). Different sectors such as the public electricity and heat production, manufacturing and construction industries are involved in the maximum contribution of mercury to the atmosphere. Humans are mostly exposed to this element by means of fish consumption, topical creams and soaps containing little amount of mercury (Pandey and Madhuri 2014). The mercury concentration in air is about 1.5–2.0 ng/m^3 (Slemr, Brunke et al. 2003). The other sources of emission include coal power plants (Klein and Russell 1973). The elemental mercury has the ability to easily penetrate through the blood-brain barrier and also placental barriers (Aschner and Aschner 1990), thus showing higher rates of vulnerability in the pregnant women (Al-Saleh, Shinwari et al. 2011). The half-life of this element is also very high in human body which is nearly 60 days (Yaginuma-Sakurai, Murata et al. 2012). Methylmercury can easily cause bleeding ulcers and the inorganic mercury can be easily transmitted to the infants by breast-feeding (Björnberg, Vahter et al. 2005).

Nearly 10% of the methylmercury is passed into the brain and 5% is absorbed by the red blood cells (Kershaw, Dhahir et al. 1980). Vapours of mercury when inhaled at higher concentrations of more than 1000 μg/m^3 can cause pneumonitis, nasal blockage and damage to the lungs. In severe cases it can also cause brain impairment and lead to coma stages (McFarland and Reigel 1978).

3.2.2 Lead

Many natural calamities and anthropogenic activities lead to the accumulation of this HM (Bradl 2005). Volcanic eruptions, forest fires, combustion of fossil fuels, emission from vehicles, cement manufacturing industrial emissions, steel industries all these are the contributors of mercury emission and dispersion into the environment (MacKenzie and Canil 2008). Some other unfamiliar sources also include metal manufacturing industries. The estimated particle size of atmospheric lead shows about 0.2–1.0 μm in diameter (Milford and Davidson 1985). The other compositional forms of lead include lead sulphate and lead oxide (McKenzie 1980). The solubility of the different elemental forms of lead gives different rates of chemical absorption in the environment (Martınez and Motto 2000). Food and water are the major sources of lead ingestion in adults, whereas toddlers, soil, paints can be the major sources (Chisolm 1980).

Many food sources such as fruits and vegetables are also intoxicated with lead with the range of 6–650 mg/kg (WHO/IPCS, 1995), but this range of uptake is very little as compared to other exposures. Bones are one of the reservoirs of lead in the human body (Zhang, Sun et al. 2020). The most affecting negative impacts due to lead exposure include renal failure, low haemoglobin count, hormonal imbalance and behavioural changes in infants (Lentini, Zanoli et al. 2019). Lead intoxication in breastmilk cannot be excreted from the body, thus leading to transfer of this HM into the infants causing damage to their brains (Karri, Schuhmacher et al. 2016). Neurodevelopment is retarded in case of the infants and the nervous system gets damaged.

3.2.3 Chromium

This is one of the most abundant heavy metals with different oxidation forms. Both the trivalent and hexavalent forms of chromium are toxic to living organisms (Alvarez, Gómez et al. 2021). Fertilizers are the important source for the chromium emission; along with this sewage and burning of coal also play a vital role in this (Tumolo, Ancona et al. 2020). Many other industries such as automobiles, paint manufacturing, paper and pulp sector are important sectors associated with the chromium pollution caused to the environment (Jaishankar, Tseten et al. 2014). All these oxidized forms of chromium are highly water-soluble and thus can be easily intoxicated into humans (Costa 1997). The hexavalent form of chromium is a toxic carcinogen and it leads to severe issues in human health.

Cough, nasal passage irritation, wheezing, redness and severe blemish formation are the most common negative impacts caused to the human body (Bhargava, Gupta et al. 2017). This affects all the parts of the body, including liver, kidney and even nervous system. The permissible amount of chromium content that is acceptable in bottled water is at most 1 mg/L (Martin and Griswold 2009), as per the FDA guidelines.

3.2.4 Arsenic

Arsenic is present in all the forms of the environment, including soil, air and water. The different types of arsenic compounds existing in both organic and inorganic structures are mostly intoxicated in water and fishes, respectively, which in turn leads to human exposure (Chung, Yu et al. 2014, Mahurpawar 2015). The higher usage of pesticides and fertilizers has increased the levels of arsenic concentrations in the soil (Singh, Gupta et al. 2017, Kato, Ferrari et al. 2020, Alengebawy, Abdelkhalek et al. 2021). One of the other biggest sources of arsenic exposure by humans is seafood consumption.

Hyperkeratosis is experienced by people affected by arsenic contamination (Rajiv, George et al. 2023). The elevation of carcinogenic cells in the body occurs with the repeated exposure of this compound from the environment (Mahurpawar 2015). According to WHO, 50–100 μg/L of arsenic exposure can lead to the formation of skin cancer (Järup 2003). Mental breakdown, difficult psychological conditions, kidney failure are the other symptoms associated with chronic exposure of arsenic (Farkhondeh, Naseri et al. 2021).

3.2.5 Copper

Copper exists in different elemental forms and this is also found in various alloy forms based on its composition and mixture with other metals (Karim 2018). The major routes of copper into the body are via skin exposure, respiratory tract, directly through oral contact (Mahurpawar 2015). The major sources of copper contamination into the environment are industries. Acute toxicity leads to severe ill health conditions such as skin irritation, muscle retardation and allergic reactions (Alengebawy, Abdelkhalek et al. 2021), whereas the chronic exposure to copper can cause diseases associated with liver (hepatotoxicity) and neuronal deformities (Vellingiri, Suriyanarayanan et al. 2022). Different concentrations of copper in the water bodies and sources make it vulnerable to accessible people and cause severe health issues. The amount of copper in freshwater bodies comes under the range of 0.5–1000 (Karim 2018). This heavy metal has the ability to bind with vital enzymes such as glutathione and other amino acid residues. The transportation and elimination of copper from the body do not depend on the dosage levels and thus it is highly hazardous to the body.

It is also associated with malfunctioning of the cytochrome P450 enzyme present in the liver and thus results in oxidative stress leading to abnormalities in the cellular body parts functioning (Tang, Xu et al. 2018). Exceeding levels of copper with the range of 50–120 mg in the human body can lead to severe health-related diseases (Karim 2018).

3.2.6 Nickel

Nickel is one of the most important cofactors involved in many enzymatic reactions occurring inside the body. The major route of administration of nickel exposure is by pulmonary absorption (Das, Das et al. 2008). The settlement of the nickel ions into the spinal cord is commonly associated with the oral intoxication (Bar-Sela, Reingold et al. 2001). This heavy metal is used in the production of batteries and metallurgical sectors. Nearly 35% of the externally inhaled nickel is usually absorbed by the blood and reintroduced into the other organs (Das, Das et al. 2008).

This heavy metal causes toxicity because of its competitive binding with other enzymes and proteins. It competes with other important cofactors such as zinc, calcium and magnesium and does not allow the binding of the original cofactors with the receptors. All the composition of nickel is announced to be toxic and carcinogens by the International Agency for Research on Cancer (IARC) (Rehman, Fatima et al. 2018). The toxic form of nickel is said to have the particle size less than 0.1 mm. The general symptoms associated with the acute exposure of nickel are headache, chest pain, congestion, vertigo, fatigue and delayed vision (Bhargava, Gupta et al. 2017).

A little more exposure of nickel can cause diarrhoea, lack of coordination, reduced RBC count, liver dysfunction, fibrosis, lack of immune response against it (Al-fatlawi and Al-Murshedi 2015). Moreover, the chronic exposure of nickel for more than 100 days can cause sinusitis, bronchitis, asthma etc. There are many types of toxicity, including genotoxicity, malfunctioning of the body parts, blood toxicity, allergic reactions and teratogenic toxicity. Some of the inhaled uranium remains stored in the renal tubules itself and thus causing problems associated with kidney failure and urinary infection and irritation (Nemery 1990).

3.2.7 Uranium

Uranium is one of the heaviest metals in the Earth. This is most commonly found in the cigarette smoking, dust particles, aerosols rich with uranium components and drinking water (Brugge and Buchner 2011). The major target system for uranium toxicity is the respiratory system. Acute respiratory distress syndrome (ARDS) (Gwaltney-Brant 2013), skeletal deposition, necrosis, hyaline degeneration are the most common symptoms associated with the heavy exposure of uranium. In some rare cases, embryotoxicity is also seen with higher uranium exposure (Shaki, Zamani et al. 2019). In the cases of the bone decay patients, the deposition of uranium can lead to the development of osteosarcoma cells leading to severe cancer in the body (Priest 1990).

3.2.8 Antimony

The major toxicity associated with the antimony depends on the exposure concentrations, duration of exposure, age, lifestyle and habits (Cooper and Harrison 2009). It has both the positive and negative effects on human health and related disease conditions. The major available components associated with antimony are paints, alloys, semiconductor devices, airborne dust particles, etc. Constipation, dizziness, loss of appetite and glycosuria are the most common symptoms observed with the low concentration exposure of antimony (Karalliedde and Brooke 2012; Bradberry and Vale 2014). There are many types of cancers associated with the exposure of antimony, such as lung cancer, and also some diseases that are non-malignant such as jaundice and cardiac arrhythmias (Wild, Bourgkard et al. 2009).

3.3 HUMAN HEALTH IMPACTS ASSOCIATED WITH EXPOSURE TO DIFFERENT HMs

3.3.1 Neurotoxicity

The prevalence rate of neurotoxicity-related diseases is most commonly found in heavy metal-intoxicated humans. Many disease conditions are associated with heavy metal intoxication. These heavy metals can damage the nervous system and thus lead to mental retardation in severe conditions. Many organs are associated with heavy metal toxicity and affect the cellular functioning of specific organs. Long-term side effects include neurodegenerative disorders such as Parkinson's disease, Alzheimer's disease and tremors. Continuous exposure damages the DNA, and cellular components, causes mutation and creates hormonal imbalance. The prolonged accumulation of HMs can cause ageing of the brain with major health conditions. Most of the symptoms are seen in the peripheral nervous system (PNS), central nervous system (CNS), renal tubules, and astroglial cells in the brain (Table 3.1).

TABLE 3.1
Heavy Metals and Their Toxicity

Heavy Metal Names	Neurotoxicity Symptoms	References
Iron	• Lower birth weight of newborn babies. • Decreased motor activities in rat models. • Lack of habituation in mice model.	• Yip (2000) • Goldenberg, Tamura et al. (1996) • Fredriksson, Schröder et al. (1999)
Lead	• Intellectual disabilities in children. • IQ scores of the children decreased by 2–3 points per 10 µg/dL increment in the lead concentrations in blood. • Affects prefrontal cerebral cortex, cerebellum, and causes neuropsychiatric conditions.	• World Health Organization (2021) • Bellinger (2011) • Sanders, Liu et al. (2009)
Copper	• Chronic exposure causes Wilson's disease. • Causes neurodegenerative diseases also.	• Bandmann, Weiss et al. (2015) • Giampietro, Spinelli et al. (2018)
Zinc	• Triggers mitochondria to produce higher levels of reactive oxygen species. • Leads to plaque formation in the neurones and aggregation of amyloids.	• Belyaeva, Sokolova et al. (2012) • Wright and Baccarelli (2007)
Mercury	• Causes deficiency in the neurodevelopmental processes in infants. • Impaired behavioural conditions were observed in children.	• Crespo-López, Herculano et al. (2005)

3.3.2 Cardiovascular Effects

Cardiovascular effects due to heavy metals are less common with exposure to heavy metals, but it is considered an important health condition since it leads to fatality (Table 3.2). The factors that influence toxicity levels are based on genetics, dose, dietary habits, co-health conditions, age, sex, environmental factors etc. A heavy metal density of more than 5 g/cm^3 is considered toxic to the environment and also health conditions. Joint exposure to these metals can cause differential diseases such as atherogenesis, black foot disease, stroke, increased blood pressure, atherosclerosis and hypercholesterolemia. Certain toxic heavy metals such as arsenic, cadmium, lead and mercury dysfunction enzymatic activities, peroxisome activity, increment of blood pressure and oxidative stress.

3.3.3 Cancer Risks

Heavy metal exposure has the mechanistic effect of carcinogenic properties. Heavy exposure to cadmium leads to deadly symptoms causing pancreas and breast cancer. Table 3.3 depicts the side effects of heavy metals and risk factors leading to cancer. Arsenic as a heavy metal in higher concentrations blocks the activity of tumour-suppressing genes in the human body, thus leading to elevated levels of cancer-causing genes. The DNA binding domain in the zinc finger motif will be

TABLE 3.2
Cardiovascular Effects of Heavy Metal Ions

Heavy Metal Names	Cardiovascular Effects	References
Arsenic	• Affects the central peripheral arterial disease (PAD).	Alissa and Ferns (2011)
Lead	• Disrupts the regulators of calcium ions and causes heart diseases and stroke.	Yang, Lo et al. (2020)
Cadmium	• Increased blood pressure.	Houston (2007); Tellez-Plaza, Guallar et al. (2013)
Mercury	• Effect on cardiac parasympathetic activities.	Houston (2007); Houston (2011)

TABLE 3.3
Heavy Metal-Mediated Cancer Risk

Heavy Metal Names	Cancer Risks	References
Arsenic	• Disruption of DNA repair mechanism. • This binds with methyl-transferase enzyme and downregulates the production of tumour-suppressing genes.	Chiarelli and Roccheri (2012) Palma-Lara, Martínez-Castillo et al. (2020)
Cadmium	• It replaces the binding of zinc in the DNA zinc-binding domain. • This gets accumulated over the endometrial layer of tissues in females. • This mimics the endogenous oestrogen and leads to breast cancer.	Malgieri, Palmicri ct al. (2014) Sapmaz-Metin, Topcu-Tarladacalisir et al. (2017) McElroy, Shafer et al. (2006)
Nickel	• Other compositional forms of nickel can cause lung cancer and nasal cancer.	Khlifi, Olmedo et al. (2013)

replaced by cadmium due to higher concentrations and thus blocks the functions of apoptosis, gene expression, transcriptional inactivation, protein misfolding and many other dysfunctions. Nickel causes a potential negative impact on DNA by damaging the chromosomal integrity. The process of detoxification and antioxidant-based DNA protection is completely lost in case of heavy metal intoxication.

3.3.4 Reproductive Toxicity

The careless usage of these heavy metals can cause damage to different organs of humans. This leads to the disruption of various enzymes and hormones (Table 3.4). Most of the reproductive toxicity associated in men due to heavy metal intoxication is due to the damage it creates in the DNA and mimicking activity of these heavy metals against the pathways associated with cellular growth and development. Several symptoms such as oocyte maturation, sperm motility, and viable sperm count are reduced significantly with continuous exposure to heavy metals. Stillbirths and premature delivery show the most defective health conditions associated with this.

3.3.5 Teratogenicity

Most of the heavy metals cause destructive effects on human embryos and also other animal embryos. Active or passive smoking can cause defects in pregnant women along with the prenatal embryos. The cadmium levels get increased in such cases and pass to the newborn baby via breast milk. This causes malformations in embryos due to the accumulation of cadmium in the anterior visceral endoderm (Mobarak 2008).

Similarly, over-exposure of the foetus to lead created damage in the tibia bone and patella bone deformities, thus resulting in the weight loss of the babies.

3.4 CASE STUDIES OF HM TOXICITY IN HUMAN HEALTH

Different case studies were carried out all over the world to exhibit HM toxicity in humans. Different sources of heavy metals and different populations and the risk factors linked to it are depicted clearly in Table 3.5 with proper examples of each case. Most of the case studies taken into account show the carcinogenic and non-carcinogenic damages caused to the human body after the high exposure to HM.

TABLE 3.4
Heavy Metal-Mediated Reproductive Toxicity

Heavy Metal Names	Reproductive Toxicity	References
Lead	• Lead gets deposited in the vas deferens and testes and thus infertility in males. • The activity of sperm is also retarded in several cases.	Chowdhury (2009)
Chromium	• Chronic exposure leads to infertility due to the degradation of semen quality. • Spermatogenic cell count will be reduced.	Pizent, Tariba et al. (2012)
Arsenic	• Damages the androgen-binding protein in males.	Gerhard, Monga et al. (1998)

TABLE 3.5
Cases Associated with Heavy Metal Toxicity in Parts of the World

Case Study Number	Type of Sample and Associated Pollutants	Place of Assessment	The Heavy Metals Detected from the Samples	Associated Health Issues	References
1	Drinking water sources intoxicated with pesticides (atrazine and acetochlor).	Dalian, China.	Copper, zinc, cadmium, nickel, arsenic, mercury.	Both the carcinogenic risk and non-carcinogenic risk were identified in the drinking water.	Dong, Zhang et al. (2020)
2	Urban soils.	Urban areas of South India.	Lead, chromium, copper, zinc, cobalt, nickel.	Soil was highly polluted by copper with mean concentration of 40.64mg/kg of soil samples.	Adimalla, Chen et al. (2020)
3	Leather industry.	Dhaka industrial area, Bangladesh.	Chromium.	Industrial workers were analysed with nearly 21.85–483 mg/kg of chromium which was observed in the hair and nail samples.	Hasan, Hosain et al. (2019)
4	Fresh fruits.	Bangalore city, India.	Cadmium, cobalt, chromium, copper, iron, manganese, nickel, lead, zinc.	The detected heavy metals ranged higher than the acceptable range said by the Indian Prevention of Food Adulteration Act, 1954.	Mahdavian and Somashekar (2008)
5	Cultured fish species, *Labeo rohita.*	Urban pond near a coal mining region, Dhanbad, India.	Mercury, arsenic, lead, cadmium.	The estimated concentration of As, Cd, Hg and Pb was 0.33, 0.26, 0.19 and 3.9 mg/kg, respectively, which is higher than the proposed limits given by the Food and Agriculture Organisation (FAO) and the World Health Organization (WHO).	Pal and Maiti (2019)

3.5 CONCLUSION

This book chapter majorly focuses on the heavy metal intoxication in the human body due to the exposure of it in day-to-day life. The prevalence of heavy metals is very high and it can be a vulnerable threat to the human population in case of acquiring various diseases. All age groups of people are vulnerable to this exposure, including the embryo which is present in the mother's womb. These heavy metals can easily pass through the placenta and blood-brain barrier and thus cause malfunction in various organs of the body. Drinking water, vegetables, fruits and other consumables are intoxicated with heavy metals and thus have to be treated before consumption. Future perspectives include the development of various techniques and methods to detect heavy metal contamination in the products that are consumed.

ACKNOWLEDGEMENTS

Authors acknowledge the support provided by the Department of Health Research under the Ministry of Health & Family Welfare (Government of India), New Delhi, and ICMR, Rajendra Memorial Research Institute of Medical Sciences (RMRIMS), Patna, Bihar. The authors would also like to thank Dr. Veer Singh for giving the opportunity and helping us to write this book chapter.

REFERENCES

Abdul-Wahab, S., & Marikar, F. (2012). "The environmental impact of gold mines: Pollution by heavy metals." Open Engineering, **2**(2): 304–313.

Adimalla, N., Chen, J., & Qian, H. (2020). "Spatial characteristics of heavy metal contamination and potential human health risk assessment of urban soils: A case study from an urban region of South India." Ecotoxicology and Environmental Safety, **194**: 110406.

Al-fatlawi, A. C., & Al-Murshedi, M. H. (2015). "The effects of Heavy metal (Nickel) on Hematological parameters of laboratory male mice." International Journal of Advanced Research, **3**(9): 598–601.

Al-Saleh, I., Shinwari, N., Mashhour, A., Mohamed, G. E. D., & Rabah, A. (2011). "Heavy metals (lead, cadmium and mercury) in maternal, cord blood and placenta of healthy women." International Journal of Hygiene and Environmental Health, **214**(2): 79–101.

Al-Sulaiti, M. M., Soubra, L., & Al-Ghouti, M. A., (2022). "The causes and effects of mercury and methylmercury contamination in the marine environment: A review." Current Pollution Reports, **8**(3): 249–272.

Alengebawy, A., Abdelkhalek, S. T., Qureshi, S. R., & Wang, M. Q. (2021). "Heavy metals and pesticides toxicity in agricultural soil and plants: Ecological risks and human health implications." Toxics, **9**(3): 42.

Alissa, E. M., & Ferns, G. A. (2011). "Heavy metal poisoning and cardiovascular disease." Journal of Toxicology, **2011**.

Alvarez, C. C., Gómez, M. E. B., & Zavala, A. H. (2021). "Hexavalent chromium: Regulation and health effects." Journal of Trace Elements in Medicine and Biology, **65**: 126729.

Aschner, M., & Aschner, J. L. (1990). "Mercury neurotoxicity: Mechanisms of blood-brain barrier transport." Neuroscience & Biobehavioral Reviews, **14**(2): 169–176.

Bandmann, O., Weiss, K. H., & Kaler, S. G. (2015). "Wilson's disease and other neurological copper disorders." The Lancet Neurology, **14**(1): 103–113.

Bar-Sela, S., Reingold, S., & Richter, E. D. (2001). "Amyotrophic lateral sclerosis in a battery-factory worker exposed to cadmium." International Journal of Occupational and Environmental Health, **7**(2): 109–112.

Bellinger, D. C. (2011). "The protean toxicities of lead: New chapters in a familiar story." International Journal of Environmental Research and Public Health, **8**(7): 2593–2628.

Belyaeva, E. A., Sokolova, T. V., Emelyanova, L. V., & Zakharova, I. O. (2012). "Mitochondrial electron transport chain in heavy metal-induced neurotoxicity: Effects of cadmium, mercury, and copper." The Scientific World Journal, **2012**.

Bhargava, P., Gupta, N., Vats, S., & Goel, R. (2017). "Health issues and heavy metals." Austin Journal of Environmental Toxicology, **3**(1): 3018.

Björnberg, K. A., Vahter, M., Berglund, B., Niklasson, B., Blennow, M., & Sandborgh-Englund, G. (2005). "Transport of methylmercury and inorganic mercury to the fetus and breast-fed infant." Environmental Health Perspectives, **113**(10): 1381–1385.

Bradberry, S. M., & Vale, J. A. (2014). Poisoning due to metals and their salts. *Oxford Desk Reference: Toxicology*, Oxford Desk Reference Series, Oxford (Online edn, Oxford Academic). https://doi.org/10.1093/med/9780199594740.003.0010, accessed 28 Sept. 2024.

Bradl, H. B. (2005). Sources and origins of heavy metals. *Interface Science and Technology*, Elsevier: **6**: 1–27.

Brugge, D., & Buchner, V. (2011). *Health Effects of Uranium: New Research Findings.*

Chiarelli, R., & Roccheri, M. C. (2012). "Heavy metals and metalloids as autophagy inducing agents: Focus on cadmium and arsenic." Cells, **1**(3): 597–616.

Chisolm, J. J. (1980). "Poisoning from heavy metals (mercury, lead, and cadmium)." Pediatric Annals, **9**(12): 28–42.

Chowdhury, A. R. (2009). "Recent advances in heavy metals induced effect on male reproductive function—A retrospective." Al Ameen Journal of Medical Sciences, **2**(2): 37–42.

Chowdhury, S., Mazumder, M. J., Al-Attas, O., & Husain, T. (2016). "Heavy metals in drinking water: Occurrences, implications, and future needs in developing countries." Science of the Total Environment, **569**: 476–488.

Chung, J. Y., Yu, S. D., & Hong, Y. S. (2014). "Environmental source of arsenic exposure." Journal of Preventive Medicine and Public Health, **47**(5): 253.

Cimboláková, I., Uher, I., Laktičová, K. V., Vargová, M., Kimáková, T., & Papajová, I. (2019). Heavy metals and the environment. Environmental factors affecting human health. *Environmental Factors Affecting Human Health*, **10**. https://doi.org/10.5772/intechopen.86876

Cooper, R. G., & Harrison, A. P. (2009). "The exposure to and health effects of antimony." Indian Journal of Occupational and Environmental Medicine, **13**(1): 3–10.

Costa, M. (1997). "Toxicity and carcinogenicity of Cr (VI) in animal models and humans." Critical Reviews in Toxicology, **27**(5): 431–442.

Crespo-López, M. E., Herculano, A. M., Corvelo, T. C., & Do Nascimento, J. L. (2005). "Mercury and neurotoxicity." Revista de Neurologia, **40**(7): 441–447.

Das, K. K., Das, S. N., & Dhundasi, S. A. (2008). "Nickel, its adverse health effects & oxidative stress." Indian Journal of Medical Research, **128**(4): 412–425.

Dobbs, M. R. (2011). Toxic encephalopathy. *Seminars in Neurology*, Thieme Medical Publishers.

Dong, W., Zhang, Y., & Quan, X. (2020). "Health risk assessment of heavy metals and pesticides: A case study in the main drinking water source in Dalian, China." Chemosphere, **242**: 125113.

El Ati-Hellal, M., & Hellal, F. (2021). "Heavy metals in the environment and health impact." Environmental Health, **51**.

Farkhondeh, T., Naseri, K., Esform, A., Aramjoo, H., & Naghizadeh, A. (2021). "Drinking water heavy metal toxicity and chronic kidney diseases: A systematic review." Reviews on Environmental Health, **36**(3): 359–366.

Fredriksson, A., Schröder, N., Eriksson, P., Izquierdo, I., & Archer, T. (1999). "Neonatal iron exposure induces neurobehavioural dysfunctions in adult mice." Toxicology and Applied Pharmacology, **159**(1): 25–30.

Fu, Z., & Xi, S. (2020). "The effects of heavy metals on human metabolism." Toxicology Mechanisms and Methods, **30**(3): 167–176.

Gautam, P. K., Gautam, R. K., Banerjee, S., Chattopadhyaya, M. C., & Pandey, J. D. (2016). Heavy metals in the environment: Fate, transport, toxicity and remediation technologies. *Heavy Metals: Sources Toxicity and Remediation Techniques*, Nova Science Publishers, **60**: 101–130.

Gerhard, I., Monga, B., Waldbrenner, A., & Runnebaum, B. (1998). "Heavy metals and fertility." Journal of Toxicology and Environmental Health Part A, **54**: 593–612.

Giampietro, R., Spinelli, F., Contino, M., & Colabufo, N. A. (2018). "The pivotal role of copper in neurodegeneration: A new strategy for the therapy of neurodegenerative disorders." Molecular Pharmaceutics, **15**(3): 808–820.

Goldenberg, R. L., Tamura, T., DuBard, M., Johnston, K. E., Copper, R. L., & Neggers, Y. (1996). "Plasma ferritin and pregnancy outcome." American Journal of Obstetrics and Gynecology, **175**(5): 1356–1359.

Gwaltney-Brant, S. M. (2013). Heavy metals. *Haschek and Rousseaux's Handbook of Toxicologic Pathology*, Elsevier: 1315–1347. Academic Press.

Hasan, M. M., Hosain, S., Poddar, P., Chowdhury, A. A., Katengeza, E. W., & Roy, U. K. (2019). "Heavy metal toxicity from the leather industry in Bangladesh: A case study of human exposure in Dhaka industrial area." Environmental Monitoring and Assessment, **191**: 1–9.

Houston, M. C. (2007). "The role of mercury and cadmium heavy metals in vascular disease, hypertension, coronary heart disease, and myocardial infarction." Alternative Therapies in Health and Medicine, **13**(2): S128–S133.

Houston, M. C. (2011). "Role of mercury toxicity in hypertension, cardiovascular disease, and stroke." The Journal of Clinical Hypertension, **13**(8): 621–627.

Jaishankar, M., Tseten, T., Anbalagan, N., Mathew, B. B., & Beeregowda, K. N. (2014). "Toxicity, mechanism and health effects of some heavy metals." Interdisciplinary Toxicology, **7**(2): 60.

Järup, L. (2003). "Hazards of heavy metal contamination." British Medical Bulletin, **68**(1): 167–182.

Karalliedde, L., & Brooke, N. (2012). Toxicity of heavy metals and trace elements. *Essentials of Toxicology for Health Protection: A Handbook for Field Professionals.* Oxford Academic: 168–186.

Karim, N. (2018). "Copper and human health-a review." Journal of Bahria University Medical and Dental College, **8**(2): 117–122.

Karri, V., Schuhmacher, M., & Kumar, V. (2016). "Heavy metals (Pb, Cd, As and MeHg) as risk factors for cognitive dysfunction: A general review of metal mixture mechanism in brain." Environmental Toxicology and Pharmacology, **48**: 203–213.

Kato, L. S., Ferrari, R. G., Leite, J. V. M., & Conte-Junior, C. A. (2020). "Arsenic in shellfish: A systematic review of its dynamics and potential health risks." Marine Pollution Bulletin, **161**: 111693.

Kershaw, T. G., Dhahir, P. H., & Clarkson, T. W. (1980). "The relationship between blood levels and dose of methylmercury in man." An International Journal, **35**(1): 28–36.
Khan, A., Khan, S., Khan, M. A., Qamar, Z., & Waqas, M. (2015). "The uptake and bioaccumulation of heavy metals by food plants, their effects on plants nutrients, and associated health risk: A review." Environmental Science and Pollution Research, **22**: 13772–13799.
Khlifi, R., Olmedo, P., Gil, F., Hammami, B., Chakroun, A., Rebai, A., & Hamza-Chaffai, A. (2013). "Arsenic, cadmium, chromium and nickel in cancerous and healthy tissues from patients with head and neck cancer." Science of the Total Environment, **452**: 58–67.
Klein, D. H., & Russell, P. (1973). "Heavy metals. Fallout around a power plant." Environmental Science & Technology, **7**(4): 357–358.
Lentini, P., Zanoli, L., de Cal, M., Granata, A., & Dell'Aquila, R. (2019). Lead and heavy metals and the kidney. *Critical Care Nephrology*, Elsevier: 1324–1330.e1321.
MacKenzie, J. M., & Canil, D. (2008). "Volatile heavy metal mobility in silicate liquids: Implications for volcanic degassing and eruption prediction." Earth and Planetary Science Letters, **269**(3–4): 488–496.
Mahdavian, S. E., & Somashekar, R. K. (2008). "Heavy metals and safety of fresh fruits in Bangalore city, India—A case study." Kathmandu University Journal of Science, Engineering and Technology, **4**(1): 17–27.
Mahurpawar, M. (2015). "Effects of heavy metals on human health." International Journal of Research-GRANTHAALAYAH, **530**(516): 1–7.
Malgieri, G., Palmieri, M., Esposito, S., Maione, V., Russo, L., Baglivo, I., … & Isernia, C. (2014). "Zinc to cadmium replacement in the prokaryotic zinc-finger domain." Metallomics, **6**(1): 96–104.
Marth, E., Jelovcan, S., Kleinhappl, B., Gutschi, A., & Barth, S. (2001). "The effect of heavy metals on the immune system at low concentrations." International Journal of Occupational Medicine and Environmental Health, **14**(4): 375–386.
Martin, S., & Griswold, W. (2009). "Human health effects of heavy metals." Environmental Science and Technology Briefs for Citizens, **15**(5).
Martınez, C. E., & Motto, H. L. (2000). "Solubility of lead, zinc and copper added to mineral soils." Environmental Pollution, **107**(1): 153–158.
Masindi, V., & Muedi, K. L. (2018). "Environmental contamination by heavy metals." Heavy Metals, **10**: 115–132.
McElroy, J. A., Shafer, M. M., Trentham-Dietz, A., Hampton, J. M., & Newcomb, P. A. (2006). "Cadmium exposure and breast cancer risk." Journal of the National Cancer Institute, **98**(12): 869–873.
McFarland, R. B., & Reigel, H. (1978). "Chronic mercury poisoning from a single brief exposure." Journal of Occupational and Environmental Medicine, **20**(8): 532–534.
McKenzie, R. M. (1980). "The adsorption of lead and other heavy metals on oxides of manganese and iron." Soil Research, **18**(1): 61–73.
Milford, J. B., & Davidson, C. I. (1985). "The size of particulate trace elements in the atmosphere—a review." Journal of the Air Pollution Control Association, **35**(12): 1249–1260.
Mobarak, Y. (2008). "Review of the developmental toxicity and teratogenicity of three environmental contaminants (cadmium, lead and mercury)." The International Journal of Environmental Sciences, **3**(1): 31–43.
Mortvedt, J. J. (1995). "Heavy metal contaminants in inorganic and organic fertilizers." Fertilizer Research, **43**: 55–61.
Mukherjee, A. B., Bhattacharya, P., Sarkar, A., & Zevenhoven, R. (2009). Mercury emissions from industrial sources in India and its effects in the environment. *Mercury Fate and Transport in the Global Atmosphere: Emissions, Measurements and Models*, Springer: 81–112.
Nemery, B. (1990). "Metal toxicity and the respiratory tract." European Respiratory Journal, **3**(2): 202–219.
Pal, D., & Maiti, S. K. (2019). "Evaluation of potential human health risks from toxic metals via consumption of cultured fish species *Labeo rohita*: A case study from an urban aquaculture pond." Exposure and Health, **11**: 33–46.
Palma-Lara, I., Martínez-Castillo, M., Quintana-Pérez, J. C., Arellano-Mendoza, M. G., Tamay-Cach, F., Valenzuela-Limón, O. L., … & Hernández-Zavala, A. (2020). "Arsenic exposure: A public health problem leading to several cancers." Regulatory Toxicology and Pharmacology, **110**: 104539.
Pandey, G., & Madhuri, S. (2014). "Heavy metals causing toxicity in animals and fishes." Research Journal of Animal, Veterinary and Fishery Sciences, **2**(2): 17–23.
Peter, A. J., & Viraraghavan, T. (2005). "Thallium: A review of public health and environmental concerns." Environment International, **31**(4): 493–501.
Pizent, A., Tariba, B., & Živković, T. (2012). "Reproductive toxicity of metals in men." Arhiv za higijenu rada i toksikologiju, **63**(Supplement 1): 35–45.
Priest, N. D. (1990). The distribution and behaviour of heavy metals in the skeleton and body: Studies with bone-seeking radionuclides. *Trace Metals and Fluoride in Bones and Teeth*, Routledge: 83–139.

Rai, P. K. (2008). "Heavy metal pollution in aquatic ecosystems and its phytoremediation using wetland plants: An eco-sustainable approach." International Journal of Phytoremediation, **10**(2): 133–160.

Rajiv, S. V., George, M., & Nandakumar, G. (2023). "Dermatological manifestations of arsenic exposure." Journal of Skin and Sexually Transmitted Diseases, **5**(1): 14–21.

Rehman, K., Fatima, F., Waheed, I., & Akash, M. S. H. (2018). "Prevalence of exposure of heavy metals and their impact on health consequences." Journal of Cellular Biochemistry, **119**(1): 157–184.

Ren, X., Li, P., He, X., Su, F., & Elumalai, V. (2021). "Hydrogeochemical processes affecting groundwater chemistry in the central part of the Guanzhong Basin, China." Archives of Environmental Contamination and Toxicology, **80**: 74–91.

Ruggieri, F., Saavedra, J., Fernandez-Turiel, J. L., Gimeno, D., & Garcia-Valles, M. (2010). "Environmental geochemistry of ancient volcanic ashes." Journal of Hazardous Materials, **183**(1–3): 353–365.

Sanders, T., Liu, Y., Buchner, V., & Tchounwou, P. B. (2009). "Neurotoxic effects and biomarkers of lead exposure: A review." Reviews on Environmental Health, **24**(1): 15–46.

Sapmaz-Metin, M., Topcu-Tarladacalisir, Y., Kurt-Omurlu, I., Karaoz Weller, B., & Unsal-Atan, S. (2017). "A morphological study of uterine alterations in mice due to exposure to cadmium." Biotechnic & Histochemistry, **92**(4): 264–273.

Shaki, F., Zamani, E., Arjmand, A., & Pourahmad, J. (2019). "A review on toxicodynamics of depleted uranium." Iranian Journal of Pharmaceutical Research: IJPR, **18**(Suppl 1): 90.

Shrivastava, R., Upreti, R. K., Seth, P. K., & Chaturvedi, U. C. (2002). "Effects of chromium on the immune system." FEMS Immunology and Medical Microbiology, **34**(1): 1–7.

Singh, N., Gupta, V. K., Kumar, A., & Sharma, B. (2017). "Synergistic effects of heavy metals and pesticides in living systems." Frontiers in Chemistry, **5**: 70.

Slemr, F., Brunke, E. G., Ebinghaus, R., Temme, C., Munthe, J., Wängberg, I., … Berg, T. (2003). "Worldwide trend of atmospheric mercury since 1977." Geophysical Research Letters, **30**(10).

Sonone, S. S., Jadhav, S., Sankhla, M. S., & Kumar, R. (2020). "Water contamination by heavy metals and their toxic effect on aquaculture and human health through food Chain." Letters in Applied NanoBioScience, **10**(2): 2148–2166.

Tang, H., Xu, M., Shi, F., Ye, G., Lv, C., Luo, J., … & Li, Y. (2018). "Effects and mechanism of nano-copper exposure on hepatic cytochrome P450 enzymes in rats." International Journal of Molecular Sciences, **19**(7): 2140.

Tellez-Plaza, M., Guallar, E., Howard, B. V., Umans, J. G., Francesconi, K. A., Goessler, W., … & Navas-Acien, A. (2013). "Cadmium exposure and incident cardiovascular disease." Epidemiology, **24**(3): 421.

Tumolo, M., Ancona, V., De Paola, D., Losacco, D., Campanale, C., Massarelli, C., & Uricchio, V. F. (2020). "Chromium pollution in European water, sources, health risk, and remediation strategies: An overview." International Journal of Environmental Research and Public Health, **17**(15): 5438.

Vellingiri, B., Suriyanarayanan, A., Selvaraj, P., Abraham, K. S., Pasha, M. Y., Winster, H., …& Venkatesan, D. (2022). "Role of heavy metals (copper (Cu), arsenic (As), cadmium (Cd), iron (Fe) and lithium (Li)) induced neurotoxicity." Chemosphere, **301**: 134625.

Verma, S., Bhatt, P., Verma, A., Mudila, H., Prasher, P., & Rene, E. R. (2021). "Microbial technologies for heavy metal remediation: Effect of process conditions and current practices." Clean Technologies and Environmental Policy, 1–23.

Wild, P., Bourgkard, E., & Paris, C. (2009). Lung cancer and exposure to metals: The epidemiological evidence. *Cancer Epidemiology: Modifiable Factors*, Humana Press: 139–167.

World Health Organization (2021). *Report on Activities during the Eighth International Lead Poisoning Prevention Week, 25–31 October 2020.* https://pesquisa.bvsalud.org/portal/resource/pt/who-345169?lang=en

Wright, R. O., & Baccarelli, A. (2007). "Metals and neurotoxicology." The Journal of Nutrition, **137**(12): 2809–2813.

Yaginuma-Sakurai, K., Murata, K., Iwai-Shimada, M., Nakai, K., Kurokawa, N., Tatsuta, N., &Satoh, H. (2012). "Hair-to-blood ratio and biological half-life of mercury: Experimental study of methylmercury exposure through fish consumption in humans." The Journal of Toxicological Sciences, **37**(1): 123–130.

Yang, A. M., Lo, K., Zheng, T. Z., Yang, J. L., Bai, Y. N., Feng, Y. Q., … & Liu, S. M. (2020). "Environmental heavy metals and cardiovascular diseases: Status and future direction." Chronic Diseases and Translational Medicine, **6**(04): 251–259.

Yip, R. (2000). "Significance of an abnormally low or high hemoglobin concentration during pregnancy: Special consideration of iron nutrition." The American Journal of Clinical Nutrition, **72**(1): 272S–279S.

Zhang, S., Sun, L., Zhang, J., Liu, S., Han, J., & Liu, Y. (2020). "Adverse impact of heavy metals on bone cells and bone metabolism dependently and independently through anemia." Advanced Science, **7**(19): 2000383.

4 Heavy Metal Pollutants and Sources of Their Contamination

Pankaj Kumar Chaurasia, Nagraj, and Nagendra Sharma*
L.S. College, Muzaffarpur, India

*Shashi Lata Bharati**
North Eastern Regional Institute of Science and Technology, Nirjuli, Arunachal Pradesh, India

4.1 INTRODUCTION

Entry of the pollutants or any contaminating materials into the environment generating adverse impacts directly or indirectly on either humans or other living beings is termed pollution and such pollution-causing materials may be in any form like solids, liquids or gases. The composition of the pollution-causing materials may be either foreign things or naturally generated contaminants. Although pollution may also be due to natural processes and humans have no control on that, but the term pollution is mostly used for the pollution created due to the different human activities (anthropogenic activities) like pollution generated from various types of industries, improper management of wastes, agricultural activities, transportations, mining, factories, synthetic and dye industries and many other sectors. Pollution may be designated as point sources or nonpoint sources. Point sources are related to the highly dense specific sites like mines or factories, while nonpoint sources belong to widely distributed sources, i.e. agricultural runoff or microplastics (https://en.wikipedia.org/wiki/Pollution).

Based on the availability of different pollutants in various conditions/forms, pollutions may be categorized with different names. Air pollution which is defined as pollution of air caused by the gaseous chemicals and different types of particulates like fine dusts, smog, carbon dioxide, oxides of nitrogen and sulphur, chlorofluorocarbons, smokes and ashes. Water pollution is caused by different industrial, commercial or home-based pollutants released in aquatic environments like ponds and rivers. Noise pollution is due to the various types of noises from industrial sectors, roadways, aircraft, etc. Soil pollution is due to the either accidental release of chemicals/oils/industrial materials into the soil or commonly used chemicals like pesticides, insecticides, herbicides, heavy metals and hydrocarbons. Thermal pollution is variation in the temperature of natural water bodies due to human industrial or other power plant activities. Electromagnetic pollution is caused due to overabundance of the non-ionizing form of the electromagnetic radiations. Light pollution comprises over-illumination, light trespass and astronomical interference; plastic pollutions (one of the highly serious global issues) is due to the plastics, plastic products or microplastics; radioactive pollution is due to the release of radioactive rays in environment by human nuclear activities. There may be other types pollutant-specific pollutions (https://en.wikipedia.org/wiki/Pollution). Some selected types of pollution have well-described literature and studies like air pollution (Hospido et al., 2023),

* Corresponding authors: pankaj.chaurasia31@gmail.com and shashilatachem@gmail.com

 DOI: 10.1201/9781032685793-4

soil and agricultural pollutions (Chaurasia and Bharati, 2021; Tindwa and Singh, 2023), water pollution (Lin et al., 2022), plastic pollutions (Kibria et al., 2023), pesticides and insecticides (Chaurasia et al., 2022) and many more (Agarwal et al., 2018; Fayomi et al., 2019; Chaurasia and Bharati, 2022; Goya-Heredia et al., 2023; Chaurasia, Kumari et al., 2023; Tripathi et al., 2023).

Pollutants are directly related to the pollution and out of different types of pollutants like polycyclic aromatic hydrocarbons, pesticides, insecticides, pharmaceutical components, dust particles and particles present in air, metals, especially heavy metals, are one of the noteworthy toxic pollutants (Chaurasia, Nagraj et al., 2023). Aziz et al. (2023) described about the aquatic pollution by heavy metals and discussed the low-cost elimination approaches. Spatial study of heavy metal pollution linked with road-deposited sediments was done by Goya-Heredia et al. (2023). Similarly, there are various articles which have described the heavy metal pollutions, associated concerns and their management approaches adopted to eliminate or control such pollutions (Ahmed et al., 2022; Aziz et al., 2023). Like non-essential heavy metals, essential heavy metals are also toxic in nature and have adverse effects on humans as well as other living beings if exceeds their required extents. This chapter describes about various pollutants in brief by giving more emphasis on sources of heavy metal pollutants and their impacts on human health.

4.2 POLLUTANTS

A substance or energy entered into the environment having negative influences or negatively affect the resource's usefulness is known as pollutant or novel entity (Persson et al., 2022) and their nature of origin may be natural (e.g. extracted materials or minerals) or anthropogenic (e.g. factory-made substances or by-products). Pollutants develop human concern when they are available in a concentration which is able to cause adverse impacts on human or environmental lives. Long- or short-term harmful effects may be imposed by the pollutants by changing the plants' and animals' growth rates, or by meddling with resources utilized by humans. Many pollutants are biodegradable which have not long life in environment and biologically degradable; some are non-degradable, while many pollutants themselves produce another pollutant after degradation, like degradation of dichlorodiphenyltrichloroethane produces dichlorodiphenyldichloroethylene and dichlorodiphenyldichloroethane (https://en.wikipedia.org/wiki/Pollutant). Pollution develops adverse impacts on the environment (Persson et al., 2022), and after analysis from perspective of planetary boundaries, novel entities have been released by human society that well exceed the safe level (Persson et al., 2022).

4.2.1 Types of Pollutants and Their Effects

4.2.1.1 Persistent Organic Pollutants

Those organic compounds which cannot be degraded by different processes like chemical, biological and photolytic techniques are known as persistent organic pollutants (Ritter et al., 2007). These molecules are very toxic and worldwide negatively affect the health of humans and they have adverse effects on humans and wildlife far from where they are utilized and released because wind and water help in the transportation of these chemical compounds. Many of the persistent organic pollutants are pesticides and insecticides, while some are pharmaceuticals, solvents and industrial chemicals (Ritter et al., 2007). Some persistent organic pollutants have natural origin but mostly are human-generated (El-Shahawi et al., 2010). The Stockholm Convention recognizes the main persistent organic pollutants as 'dirty dozen' which comprises aldrin, dieldrin, heptachlor, chlordane, endrin, toxaphene, HCB (hexachlorobenzene), mirex, dioxins, PCBs (polychlorinated biphenyls), DDT (dichlorodiphenyltrichloroethane) and polychlorinated dibenzofurans. With time, many new persistent organic pollutants have been added to this list like per- and polyfluoroalkyl substances (PFASs) (https://en.wikipedia.org/wiki/Persistent_organic_pollutant).

Developmental defects, chronic disease and even death may be caused by the exposure of persistent organic pollutants (https://en.wikipedia.org/wiki/Persistent_organic_pollutant). A few of them are carcinogens in nature as per IARC (International Agency for Research on Cancer) and possibly may be responsible for breast cancer (Ritter et al., 2007), while several persistent organic pollutants are able to cause endocrine disruption within reproductive system, immune system or CNS (central nervous system). In most of the cases, diet is responsible for the human or animals' exposure to persistent organic pollutants or during the time of growth in the womb (Ritter et al., 2007). In the case of humans having no exposure to such pollutants *via* the means of accidental or occupational ways, above 90% of the contact arises from the food products of animals because of the bioaccumulation processes in fat tissues. Generally, serum levels of persistent organic pollutants grow with age, and in females, it tends to be greater than in males (Vallack et al., 1998). For the assessment of disease risk because of the persistent organic pollutants in specific location, human health risk assessment may be produced by the government considering the bioavailability of pollutants and dose-response relationship (Szabo and Loccisano, 2012).

4.2.1.2 Polycyclic Aromatic Hydrocarbon

Those organic compounds which are made of more than one aromatic ring are called polycyclic aromatic hydrocarbons and naphthalene (two aromatic rings) is its simplest example (https://en.wikipedia.org/wiki/Polycyclic_aromatic_hydrocarbon). Other few simple examples are phenanthrene (three aromatic rings), anthracene (three aromatic rings), biphenyl, pyrene, etc. (Figure 4.1). These compounds may contain rings of different sizes and some rings of them may not be aromatic in nature. Those consist of only six-membered rings, which are called 'alternant' (Harvey, 1998). Coal and oil deposits have many of these compounds and organic matter's incomplete combustion also produces such types of compounds.

Polycyclic aromatic hydrocarbons typically scatter from urban and suburban sources *via* sewage, runoff, atmospheric flow and subsequent particulate air pollution's deposition (Hylland, 2006; Davis et al., 2019). High amount of PAH contamination may be found in the soil and river dregs near sites of industries like creosote manufacturing services (Ravindra et al., 2008). PAHs may also be widely distributed by creosote, coal mining dust, oil spills and other fossil fuel-related sources (Ravindra et al., 2008; Achten and Hofmann, 2009). Slow transformation of PAHs generates degradation products' wide range. In the environment, microbial degradation of PAHs is a leading way of PAHs transformation (Johnsen et al., 2005; Haritash and Kaushik, 2009). Degradation of PAHs can be performed with speed by earthworms (soil-consuming invertebrates) either *via* direct metabolism or by refining the environment for microbial transformations (Johnsen et al., 2005). In the surface waters' top layer and atmosphere, hydroxylated, nitrogenated, oxygenated and halogenated PAHs may be produced by the abiotic degradation and many of them may be greater water soluble, mobile and toxic in nature in comparison with their parent polycyclic aromatic hydrocarbons (Hylland, 2006; Lundstedt et al., 2007; Fu et al., 2012). PAH exposure causes cancer as primary health risk (Boström et al., 2002). Exposure of adults to these pollutants is associated with cardiovascular illness (Korashy and El-Kadi, 2006). Polyaromatic hydrocarbons are among the contaminants' complex suite in the smoke of tobacco and particulate air pollution and can pay to cardiovascular-related illness due to such exposures (Lewtas, 2007) (https://en.wikipedia.org/wiki/Polycyclic_aromatic_hydrocarbon).

4.2.1.3 Persistent Pharmaceutical Pollutants

Persistent pharmaceutical pollutants are those drugs/medicines which cannot be degraded easily or completely by chemical or biological means. Examples of some persistent pharmaceutical pollutants are oestradiol (oestrogen, synthetic hormone), neuro-psychiatric and cardiovascular medicines (e.g. propranolol, cholesterol and triglycerides lowering drug gemfibrozil, etc.), citalopram/fluoxetine, antibiotics, etc. (https://en.wikipedia.org/wiki/Environmental_persistent_pharmaceutical_pollutant). Due to the technological developments in the field of the measurement and detection of pharmaceuticals,

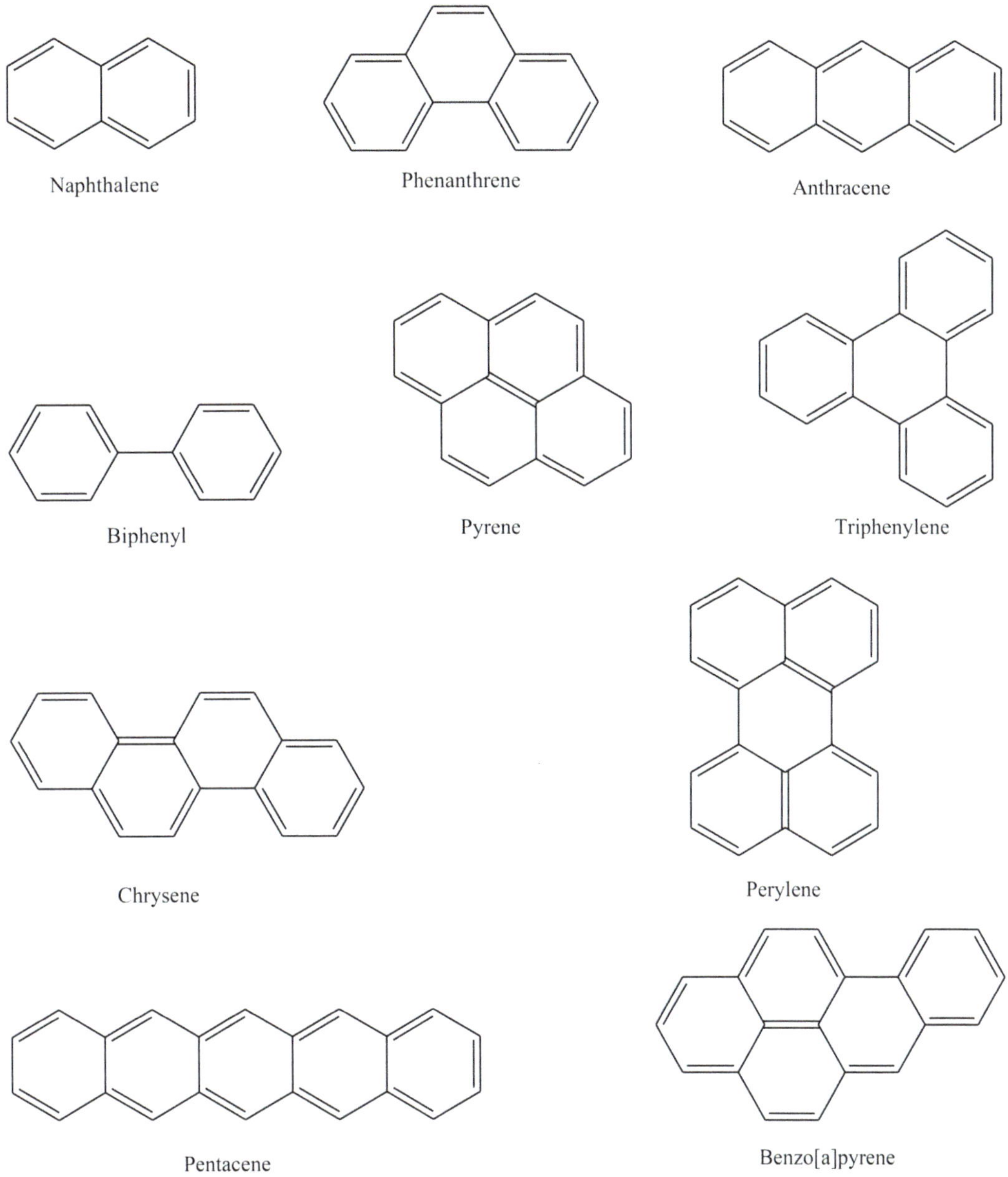

FIGURE 4.1 Chemical structures of some polyaromatic hydrocarbons.

today it has been easier to detect such compounds present in even very less concentrations which was not possible many years ago. Antibiotics and other pharmaceuticals' big concentrations were detected downstream from sewage treatment plants in sub-catchments (where the main role is played by the hospital wastewater discharge) (noPILLS in waters, 2015) or in catchments with industries of pharmaceuticals (https://en.wikipedia.org/wiki/Environmental_persistent_pharmaceutical_pollutant).

4.2.1.4 Volatile Organic Compounds

Organic molecules having high pressure at room temperature are specified as volatile organic compounds (VOCs) (Carroll and Kirschman, 2022). Scents, perfumes and/or pollutants odour are due to these VOCs (https://en.wikipedia.org/wiki/Volatile_organic_compound). An important role in communication between animal and plant communications is played by these VOCs, e.g. protection from predation (Kessler and Baldwin, 2001), attractants for pollinators (Pichersky and Gershenzon, 2002) and even inter-plant interactions (Baldwin et al., 2006).

A few of the VOCs may be harmful for the environment or dangerous for the health of humans (https://en.wikipedia.org/wiki/Volatile_organic_compound). Law regulates the anthropogenic VOCs, especially indoors. Most of these compounds are not acutely toxic but long-term chronic health effects may be shown. They are medicinally valuable, so many of them have been pharmaceutically utilized, while because of their recreational application others are administrative controls' target. In infants or children, allergic effects, respiratory or immune effects are linked with VOCs made by humans or other outdoor or indoor air pollutants (Mendell, 2007). New oxidation products as well as secondary aerosols can be generated due to the reaction of a few VOCs like styrene and limonene with oxides of nitrogen or with O_3 (ozone) which may produce symptoms of sensory irritations (Wolkoff et al., 2006). A few of them may be responsible for cancer in animals, while a few may be responsible for cancer in humans. Conjunctival irritation, headache, discomfort in nose and throats, skin allergy, vomiting, fatigue, dyspnoea, bleeding of nose, declines in serum cholinesterase levels and dizziness are the main signs related to the VOC exposures (US EPA, OAR [18 August 2014]).

4.2.1.5 Particulates or Particle in Air

Microscopic solid particles or liquid matter are generally found to be suspended in air. These particles are generally termed particulates. Mixture of particulate/air is termed aerosol, as opposed to particulate matter (PM) alone (Seinfeld and Pandis, 1998). These particulates may be generated naturally or anthropogenic means (Plainiotis et al., 2010). They have direct or indirect impacts on human health. Atmospheric particles' types comprise suspended PM; thoracic and respirable particles (Brown et al., 2013); inhalable coarse particles, titled PM_{10}, which are coarse particles with 10 μm in diameter or less; fine particles, titled $PM_{2.5}$, with 2.5 μm in diameter or less (US EPA, OAR [19 April 2016], 'Particulate Matter (PM) Basics', US EPA); ultrafine particles, with a 100 nm in diameter or less; and soot. Airborne particulates are designated by IARC and WHO (World Health Organization) as Group 1 carcinogen (Hamra et al., 2014). The most harmful form of air pollution is particulates (except ultra-fines) (Wasley et al., 2019) because of the capability of deep penetration into the lungs and brain *via* the flow of the blood, generating various health issues like disease of lung, illness of heart and premature death (US EPA, OAR [26 April 2016]). The 4.1 million death from illness of heart, lung cancer, stroke, chronic lung issues and breathing infections was contributed by exposure to $PM_{2.5}$ in 2016 (State of Global Air/2018 a Special Report on Global Exposure to Air Pollution and Its Disease Burden. Health Effects Institute. 2018) (https://en.wikipedia.org/wiki/Particulates).

4.3 TOXIC HEAVY METAL, THEIR SOURCES AND ADVERSE EFFECTS

4.3.1 Toxic Heavy Metals Sources

Any comparatively dense metals or metalloids are considered as toxic heavy metals which are defined for their potential toxicity, especially in the context of environment (Srivastava and Goyal, 2010; Zhang et al., 2019). The word has specific use to cadmium (Cd), mercury (Hg) and lead (Pb) (Brathwaite and Rabone, 1985) which are also present in the ten chemicals' list of WHO of main public concern. Chromium (Cr), manganese (Mn), cobalt (Co), nickel (Ni), copper (Cu), zinc (Zn), silver (Ag), antimony (Sb) and thallium (Tl) ('10 chemicals of public health concern'. www.who.int, 2020). Arsenic (As), Hg and Pb toxic influences were recognized to the olden, systematic studies of a few heavy metals seem to period from only 1868. Chelating agents' administration is the treatment technique for the poisoning of heavy metals in humans. There are some metals which are essentially required in small proportion for the health of humans but toxic in nature above the limit (https://en.wikipedia.org/wiki/Toxic_heavy_metal).

Anthropogenic actions are responsible for the heavy metals' existence in the ecosystem, which are human exposure's chief sources to these materials (Verma et al., 2023). In the Earth, ores are the source of heavy metals that are extracted in the course of mining activities as minerals. Heavy metals are found in the majority of the ores, comprising but not restricted to arsenic, lead, iron (Fe), zinc, gold (Au), nickel, cobalt and silver in the sulphide form, while metals like manganese, aluminium (Al) and selenium (Se) are present in the oxide-based ores. Release of heavy metals takes place from the ores during the extraction process of mining activities, which are consequently disseminated into the surroundings. These metals are found to be present in soil, and through the flow of air or water, these metals may shift towards different locations. Furthermore, some elements may be released in environment *via* combustion or in soil or aquatic systems as effluents as a result of the heavy metals used in industrial activities. Moreover, heavy metal origin may also be generated from the industrial commodities like cosmetics, herbicides, pesticides and paints (Abd Elnabi et al., 2023).

Both natural and anthropogenic acts contribute to the heavy metal pollution of the environment (Aziz et al., 2023). Volcanic eruptions, wildfires, biogenic processes, rocks' weathering and many other natural processes contribute to the heavy metal entry in the environment (Zaynab et al., 2022) and pollute it, whereas heavy metal pollution is considerably contributed by mining, industrial activities, industrial water-based irrigation of agricultural lands and many other industries as well as agricultural-based practices (Li et al., 2022). Toxic heavy metals bearing wastewater's considerable quantity come into environment primarily through electronic device production, electroplating, smelting, mining, metallurgy, nuclear fuel, manufacturing of paper, power plants, production of fertilizer and chemical etching (Saini and Dhania, 2020). Industrial release of wastewater contaminated with heavy metals into the environment occurs either *via* the direct exoneration in water system or *via* the seepage or overflow from industrial sites, leading to serious contamination of water (Zaynab et al., 2022). Cd, Hg, Pb, Ni, Zn and As in environment are contributed in the majority by the agricultural and industrial activities, which do not have particular recognizable source also called nonpoint source of pollution. Natural sources like atmospheric deposits are also responsible for the origin of heavy metals' significant quantity, which may be conveyed to the Earth's surface *via* the process of precipitation. Sources discussed above are the chief reasons behind the environmental contamination by heavy metals (Niu et al., 2020; Fan et al., 2022; Miranda et al., 2022).

In the absence of proper management to control, heavy metal release into aquatic environment from anthropogenic activities is dangerous for health of humans as well as for aquatic life (Otunola and Ololade, 2020). Due to the precipitation changes and human activity change, fluctuation has been found in level of the aquatic systems' contamination with heavy metals (Elzwayie et al., 2017; Yuan et al., 2020). For instance, in drinking water, 10 $\mu g\ L^{-1}$ is the highest permissible limit for the As established by the World Health Organization. But arsenic limits have been reported above this permissible limit by several countries like Bangladesh, Saudi Arabia, Pakistan, Mexico, Iran, Latin America, the USA, China (in the Yangtze River and Han River basins) and Ethiopia. Asia has maximum risk of exposure to drinking water contaminated with arsenic (Dilpazeer et al., 2023). So, in-depth study is very essential for the heavy metal pollution resulting from the aquatic products' expansion, particularly in the areas where there is compromised quality of drinking water. The principal givers of the pollution of heavy metals can be identified by this kind of research and better understanding of the environmental effects of expansion of aquatic products can also be provided. Furthermore, in this way, effective strategy development and policy creation can be facilitated to manage and lessen the heavy metal pollutions' harmful effects on public health as well as aquatic environment (Aziz et al., 2023).

4.3.2 Adverse Effects of Some Toxic Heavy Metals

Various aspects of the heavy metal toxicity and elimination have been reviewed by Abd Elnabi et al. (2023). Humans' exposure to the heavy metals is industrial actions' consequences and can manifest *via* inhalation, ingestion and dermal absorption (Ali et al., 2019). Heavy metals may be categorized

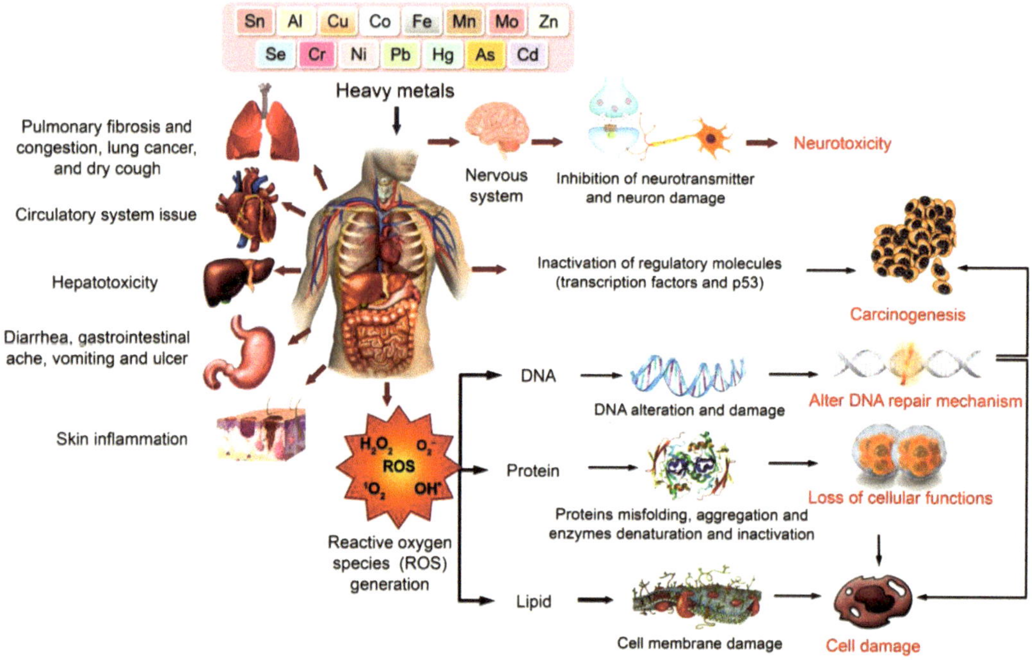

FIGURE 4.2 Presentation of oxidative stress and toxicity of human organ due to the heavy metals' exposure. (Reprinted from Abd Elnabi et al., 2023. An open access article under https://creativecommons.org/licenses/by/4.0/.)

as essential or non-essential heavy metals which is based on their biological activities. But, excess quantity of any types of heavy metals either essential or non-essential may generate disorder of different types (like morphological or physiological issue) or genetic mutations (Mitra et al., 2022).

Erosion, acid rain or runoff are responsible factors which transport the heavy metals to the different locations of soil and water system. Heavy metal origins' trajectory and subsequent effects on humans are shown in Figure 4.2 (Abd Elnabi et al., 2023). Heavy metals' toxicity is generated by their tendency to make bond with sulfhydryl groups and produce ROS (reactive oxygen species). Crucial macromolecules' deactivation is the result of the above-discussed phenomenon, causing to the oxidative stress' manifestation and glutathione depletion. A crowd of processes happen upon the exposure to noxious metals and their subsequent introduction in body, including certain metabolic pathways' potential interaction or inhibition (Rehman et al., 2021). So, deleterious impacts' plethora is observed on human as well as animal populations. The above-described conditions include several medical concerns like organ dysfunction, congenital anomalies, hormonal imbalances, compromised immune system function, metabolic irregularities and malignancy (Li et al., 2017; Deng et al., 2019). Therefore, regulations were established by various international organizations pertaining to the metals' existence in foods, drinking water and environment (Abd Elnabi et al., 2023).

Historical lead applications, industrial doings and leaded gasoline are the sources of lead and children are most commonly affected by the lead poisoning. Inhalation *via* nose and ingestion *via* drinking water and soil are its ways of exposure (Jyothi, 2020). Humans' major exposure to the Cd takes place *via* the polluted water and food material. Cigarette smoke is the exposure's way *via* route of inhalation (Jyothi, 2020). The cadmium toxicity is because of the accumulation in animals as well as plants for approximately 25–30 years. One of the nice methods to eliminate Cd from food is microbial fermentation (Genchi et al., 2020). Phosphate-based fertilizers and the waste incineration practice are the major Cd sources in environment. Levels of Cd in blood are having a great dissimilarity between cigarette smokers and non-smokers (Jarup, 2003). Pb and Cd in the body of human can travel to the brain and may produce disease of Alzheimer (Bakulski et al., 2020). In humans, it may gather at kidney after the exposure to the cadmium; because of this, levels of urinary cadmium have been taken as biomarkers for the levels of cadmium in humans (Suzzi et al., 2016).

Hg is known for its very toxic nature and has tendency to enter into the food chain. Regarding exposure, sources and toxicity of Hg, a detailed review has been done by Jyothi and Mohamed Farook (2020). It is known as global pollutant because of its easy transportation (Driscoll et al., 2013). After entering into the aquatic system, life of animals is adversely affected by Hg, and through them, Hg enters into the human beings *via* food chain (Jyothi, 2020).

Due to the carcinogenic nature of arsenic, even after being a metalloid, it has been discussed under the heading of heavy metals (Jyothi, 2020). Arsenic's health impacts were reviewed by Abdul et al. (2015). Major population is exposed to the arsenic *via* environmental air, some types of foods and groundwater. Its health-related issues are not restricted to harm to renal, cardiovascular, reproductive and endocrine system. Carcinogenic effects of arsenic on prostate gland generate prostate cancer, causing leukaemia and lesions in the hepatic area producing the liver cancer (Jyothi, 2020).

Several essential heavy metals may also cause many health issues above their necessary limits, i.e. gastrointestinal disorders, kidney and liver abnormal functioning may be caused by zinc, haematological, cardiovascular, hepatic, endocrine-related health effects with cobalt, abdominal disorders and metabolic activity abnormalities due to copper and vomiting, diarrhoea, abdominal pain, dehydration and lethargy due to iron's high uptake as supplements and oral consumptions (Kim et al., 2019; Jyothi, 2020). On the other hand, health effects of non-essential amino acids may also be of different types, i.e. lung disorders, renal and reproductive system may be affected from chromium, effects on the dermal region (cancer), brain and cardiac problems due to arsenic, severe result on mental health (Alzheimer's illness) and nervous system due to lead, osteo-associated issues, prostate cancer, lung illnesses and renal problems due to cadmium and sclerosis, gastric issues, Minamata disease, blindness, deafness, renal disorders due to mercury (Kim et al., 2019; Jyothi, 2020).

4.4 CONCLUDING REMARKS

This chapter has compact discussions on the various types of pollutants along with detailed discussion on the various sources of heavy metals responsible for the toxicity and pollution of environment and aquatic systems and their adverse effects. Some specific heavy metal toxicity has also been discussed specifically like lead, arsenic, cadmium, mercury, along with many essential and non-essential metals, and their adverse effects have been explored. Adverse effects of these heavy metals can be minimized by minimizing the contact of living system with metal-contaminated site or by proper management of their release into open system. Proper management at government level may be helpful in reducing the toxicity of such heavy metals, but through the involvement of private sectors and by social responsibility and awareness, many of the heavy metal toxicity may also be reduced at great level.

ACKNOWLEDGEMENTS

Dr. Pankaj Kumar Chaurasia, Mr. Nagraj and Mr. Nagendra Sharma are thankful to the P.G. Department of Chemistry, L.S. College, B.R. Ambedkar Bihar University, Muzaffarpur (India), and Dr. Shashi Lata Bharati is thankful to the Department of Chemistry, NERIST, Nirjuli (Itanagar), India, for any support during the work.

REFERENCES

10 chemicals of public health concern. 2020. www.who.int. https://www.who.int/news-room/photo-story/photo-story-detail/10-chemicals-of-public-health-concern

Abd Elnabi, M.K., Elkaliny, N.E., Elyazied, M.M., Azab, S.H., Elkhalifa, S.A., Elmasry, S., Mouhamed, M.S., Shalamesh, E.M., Alhorieny, N.A., Abd Elaty, A.E. et al. 2023. Toxicity of heavy metals and recent advances in their removal: A review. *Toxics*, 11: 580. https://doi.org/10.3390/toxics11070580

Abdul, K.S.M., Jayasinghe, S.S., Chandana, E.P.S., Jayasumana, C., and De Silva, P.M.C.S. 2015. Arsenic and health effects: A review. *Environ. Toxicol. Pharmacol.*, 40(3): 828–846. https://doi.org/10.1016/j.etap.2015.09.016

Achten, C., and Hofmann, T. 2009. Native polycyclic aromatic hydrocarbons (PAH) in coals – A hardly recognized source of environmental contamination. *Sci. Total Environ.*, 407(8): 2461–2473. https://doi.org/10.1016%2Fj.scitotenv.2008.12.008

Agarwal, S., Gupta, K.K., Chaturvedi, V.K., Kushwaha, A., Chaurasia, P.K., and Singh, M.P. 2018. The Potential Application of Peroxidase Enzyme for the Treatment of Industry Wastes. In: Research Advancements in Pharmaceutical, Nutritional and Industrial Enzymology, Bharati S.L. and Chaurasia P.K. (eds), 278–293, IGI Global. https://doi.org/10.4018/978-1-5225-5237-6.ch012

Ahmed, S.F., Kumar, P.S., Rozbu, M.R., Chowdhury, A.T., Nuzhat, S., Rafa, N., Mahlia, T.M.I., Ong, H.C., and Mofijur, M. 2022. Heavy metal toxicity, sources, and remediation techniques for contaminated water and soil. *Environ. Technol. Innov.*, 25: 102114. https://doi.org/10.1016/j.eti.2021.102114

Ali, H., Khan, E., and Ilahi, I. 2019. Environmental chemistry and ecotoxicology of hazardous heavy metals: Environmental persistence, toxicity, and bioaccumulation. *J. Chem.*, 1–14: Article ID 6730305, 14 pages. https://doi.org/10.1155/2019/6730305

Aziz, K.H.H., Mustafa, F.S., Omer, K.M., Hama, S., Hamarawf, R.F., and Rahman, K.O. 2023. Heavy metal pollution in the aquatic environment: Efficient and low-cost removal approaches to eliminate their toxicity: A review. *RSC Adv.*, 13: 17595–17610. https://doi.org/10.1039/D3RA00723E

Bakulski, K.M., Hu, H., and Park, S.K. 2020. Lead, cadmium and Alzheimer's disease. In: Genetics, Neurology, Behavior, and Diet in Dementia. The Neuroscience of Dementia, Volume 2, Martin, C.R. and Preedy, V.R. (eds), 813–830, Academic Press. https://doi.org/10.1016/B978-0-12-815868-5.00051-7

Baldwin, I.T., Halitschke, R., Paschold, A., von Dahl, C.C., and Preston, C.A. 2006. Volatile signaling in plant-plant interactions: "Talking Trees" in the genomics era. *Science*, 311(5762): 812–815. https://doi.org/10.1126/science.1118446

Boström, C.-E., Gerde, P., Hanberg, A. et al. 2002. Cancer risk assessment, indicators, and guidelines for polycyclic aromatic hydrocarbons in the ambient air. *Environ. Health Perspect.*, 110(Suppl 3): 451–488. https://doi.org/10.1289/ehp.110-1241197

Brathwaite, R.L., and Rabone, S.D.C. 1985. Heavy metal sulphide deposits and geochemical surveys for heavy metals in New Zealand. *J. Roy. Soc. New Zealand*, 15: 363–370. https://doi.org/10.1080/03036758.1985.10421713

Brown, J.S., Gordon, T., Price, O., and Asgharian, B. 2013. Thoracic and respirable particle definitions for human health risk assessment. *Part. Fibre Toxicol.*, 10: 12. https://doi.org/10.1186/1743-8977-10-12

Carroll, G.T., and Kirschman, D.L. 2022. A peripherally located air recirculation device containing an activated carbon filter reduces VOC levels in a simulated operating room. *ACS Omega*, 7(50): 46640–46645. https://doi.org/10.1021/acsomega.2c05570

Chaurasia, P.K., and Bharati, S.L. 2021. Applicability of Fungi in Agriculture and Environmental Sustainability. In: Microbes in Land Use Change Management, J.S. Singh, S. Tiwari, C. Singh and A.K. Singh (eds), 155–172, Elsevier. https://doi.org/10.1016/B978-0-12-824448-7.00010-3

Chaurasia, P.K., and Bharati, S.L. 2022. Applicability of Plants in Detoxification of Dyes. In: Bioremediation: Green Approaches for a Clean and Sustainable Environment, Bharagava, R.N. et al. (eds), 249–262, CRC Press. https://doi.org/10.1201/9781003181224-16

Chaurasia, P.K., Kumari, S., Singh, S., and Bharati, S.L. 2023. The fate of fungi in the bioremediation of pharmaceutical drug wastes. *Curr. Green Chem.*, 10(2): 100–104. http://dx.doi.org/10.2174/2213346110666230416152330

Chaurasia, P.K., Nagraj, Sharma, N., Rudakia, D.M., Singh, S., and Bharati, S.L. 2022. Fungal-assisted bioremediation of agricultural organic pollutants (Pesticides and Herbicides). *Curr. Green Chem.*, 9(1): 14–25. https://doi.org/10.2174/2213346109666220927121948

Chaurasia, P.K., Nagraj, Sharma, N., Kumari, S., Yadav, M., Singh, S., Mani, A., Yadava, S., and Bharati, S.L. 2023. Fungal assisted bio-treatment of environmental pollutants with comprehensive emphasis on noxious heavy metals: Recent updates. *Biotechnol. Bioeng.*, 120(1): 57–81. https://doi.org/10.1002/bit.28268

Davis, E., Walker, T.R., Adams, M. et al. 2019. Source apportionment of polycyclic aromatic hydrocarbons (PAHs) in small craft harbor (SCH) surficial sediments in Nova Scotia, Canada. *Sci. Total Environ.*, 691: 528–537. https://doi.org/10.1016%2Fj.scitotenv.2019.07.114

Deng, Y., Wang, M., Tian, T., Lin, S., Xu, P., Zhou, L., Dai, C., Hao, Q., Wu, Y., Zhai, Z. et al. 2019. The effect of hexavalent chromium on the incidence and mortality of human cancers: A meta-analysis based on published epidemiological cohort studies. *Front. Oncol.*, 9: 24. https://doi.org/10.3389/fonc.2019.00024

Dilpazeer, F., Munir, M., Baloch, M.Y.J., Shafiq, I., Iqbal, J., Saeed, M., Abbas, M.M., Shafique, S., Aziz, K.H.H., Mustafa, A. et al. 2023. A comprehensive review of the latest advancements in controlling arsenic contaminants in groundwater. *Water*, 15: 478. https://doi.org/10.3390/w15030478

Driscoll, C.T., Mason, R.P., Chan, H.M., Jacob, D.J., and Pirrone, N. 2013. Mercury as global pollutant: Sources, pathways and effects. *Environ. Sci. Technol.*, 47: 4967–4983. https://doi.org/10.1021/es305071v

El-Shahawi, M.S., Hamza, A., Bashammakh, A.S., and Al-Saggaf, W.T. 2010. An overview on the accumulation, distribution, transformations, toxicity and analytical methods for the monitoring of persistent organic pollutants. *Talanta*, 80(5): 1587–1597. https://doi.org/10.1016/j.talanta.2009.09.055

Elzwayie, A., Afan, H.A., Allawi, M.F., and El-Shafie, A. 2017. Heavy metal monitoring, analysis and prediction in lakes and rivers: State of the art. *Environ. Sci. Pollut. Res.*, 24: 12104–12117. https://doi.org/10.1007/s11356-017-8715-0

Fan, Y., Chen, X., Chen, Z., Zhou, X., Lu, X., and Liu, J. 2022. Pollution characteristics and source analysis of heavy metals in surface sediments of Luoyuan Bay, Fujian. *Environ Res.*, 203: 111911. https://doi.org/10.1016/j.envres.2021.111911

Fayomi, G.U., Mini, S.E., Fayomi, O.S.I., Oyeleke, O., Omole, D.O., and Akinwumi, I.I. 2019. Overview of industrial pollution activities and its curbing mechanisms. IOP Conf. Ser.: Mater. Sci. Eng., 640: 012097. https://doi.org/10.1088/1757-899X/640/1/012097

Fu, P.P., Xia, Q., Sun, X., and Yu, H. 2012. Phototoxicity and environmental transformation of polycyclic aromatic hydrocarbons (PAHs)—Light-induced reactive oxygen species, lipid peroxidation, and DNA damage. *J. Environ. Sci. Health, Part C*, 30(1), 1–41. https://doi.org/10.1080/10590501.2012.653887

Genchi, G., Sinicropi, M.S., Lauria, G., Carocci, A., and Catalano, A. 2020. The effects of cadmium toxicity. *Int. J. Environ. Res. Public Health*, 17(11), 3782. https://doi.org/10.3390/ijerph17113782

Goya-Heredia, A.V., Zafra-Mejía, C.A., and Rondón-Quintana, H.A. 2023. Spatial analysis of heavy metal pollution in road-deposited sediments based on the traffic intensity of a megacity. *Atmosphere*, 14: 1033. https://doi.org/10.3390/atmos14061033

Hamra, G.B., Guha, N., Cohen, A. et al. 2014. Outdoor particulate matter exposure and lung cancer: A systematic review and meta-analysis. *Environ. Health Perspect.*, 122(9): 906–911 ehp.niehs.nih.gov. https://doi.org/10.1289/ehp/1408092

Haritash, A.K., and Kaushik, C.P. 2009. Biodegradation aspects of polycyclic aromatic hydrocarbons (PAHs): A review. *J. Hazard. Mater.*, 169(1–3), 1–15. https://doi.org/10.1016%2Fj.jhazmat.2009.03.137

Harvey, R.G. 1998. Environmental Chemistry of PAHs. In: PAHs and Related Compounds. The Handbook of Environmental Chemistry, Neilson, A.H. (ed), 1–54, Springer. https://doi.org/10.1007/978-3-540-49697-7_1

Hospido, L., Sanz, C., and Villanueva, E. 2023. Air pollution: A review of its economic effects and policies to mitigate them. Banco de Espana Occasionally Paper No. 2301, 34 pages. Available at SSRN: https://ssrn.com/abstract=4331604 or https://doi.org/10.2139/ssrn.4331604

Hylland, K. 2006. Polycyclic aromatic hydrocarbon (PAH) ecotoxicology in marine ecosystems. *J. Toxicol. Environ. Health, Part A*, 69(1–2): 109–123. https://doi.org/10.1080/15287390500259327

Jarup, L. 2003. Hazards of heavy metal contamination. *Br. Med. Bull.*, 68(1): 167–182. https://doi.org/10.1093/bmb/ldg032

Johnsen, A.R., Wick, L.Y., and Harms, H. 2005. Principles of microbial PAH-degradation in soil. *Environ. Pollut.*, 133(1): 71–84. https://doi.org/10.1016%2Fj.envpol.2004.04.015

Jyothi, N.R. 2020. Heavy Metal Sources and Their Effects on Human Health. In: Heavy Metals, Nazal, M.K. and Zhao, H. (eds), IntechOpen. https://doi.org/10.5772/intechopen.95370

Jyothi, N.R., and Mohamed Farook, N.A. 2020. Heavy Metal Toxicity in Public Health. IntechOpen. https://doi.org/10.5772/intechopen.90333

Kessler, A., and Baldwin, I.T. 2001. Defensive function of herbivore-induced plant volatile emissions in nature. *Science*, 291(5511), 2141–2144. https://doi.org/10.1126/science.291.5511.2141

Kibria, M.G., Masuk, N.I., Safayet, R. et al. 2023. Plastic waste: Challenges and opportunities to mitigate pollution and effective management. *Int. J. Environ. Res.*, 17: 20. https://doi.org/10.1007/s41742-023-00507-z

Kim, J.J., Kim, Y.S., and Kumar, V. 2019. Heavy metal toxicity: An update of chelating therapeutic strategies. *J. Trace Elem. Med. Biol.*, 54: 226–231. https://doi.org/10.1016/j.jtemb.2019.05.003

Korashy, H.M., and El-Kadi, A.O.S. 2006. The role of aryl hydrocarbon receptor in the pathogenesis of cardiovascular diseases. *Drug Metab. Rev.*, 38(3): 411–450. https://doi.org/10.1080/03602530600632063

Lewtas, J. 2007. Air pollution combustion emissions: Characterization of causative agents and mechanisms associated with cancer, reproductive, and cardiovascular effects. *Mutat. Res. Rev. Mutat. Res.*, 636(1–3): 95–133. https://doi.org/10.1016/j.mrrev.2007.08.003

Li, C., Wang, H., Liao, X., Xiao, R., Liu, K., Bai, J., Li, B., and He, Q. 2022. Heavy metal pollution in coastal wetlands: A systematic review of studies globally over the past three decades. *J. Hazard. Mater.*, 424: 127312. https://doi.org/10.1016/j.jhazmat.2021.127312

Li, R., Wu, H., Ding, J., Fu, W., Gan, L., and Li, Y. 2017. Mercury pollution in vegetables, grains and soils from areas surrounding coal-fired power plants. *Sci. Rep.*, 7: 46545. https://doi.org/10.1038/srep46545

Lin, L., Yang, H., and Xu, X. 2022. Effects of water pollution on human health and disease heterogeneity: A review. *Front. Environ. Sci.*, 10: 880246. https://doi.org/10.3389/fenvs.2022.880246

Lundstedt, S., White, P.A., Lemieux, C.L. et al. 2007. Sources, fate, and toxic hazards of oxygenated polycyclic aromatic hydrocarbons (PAHs) at PAH- contaminated sites. *Ambio: A J. Hum. Environ.*, 36(6): 475–485. https://doi.org/10.1579/0044-7447(2007)36[475:SFATHO]2.0.CO;2

Mendell, M.J. 2007. Indoor residential chemical emissions as risk factors for respiratory and allergic effects in children: A review. *Indoor Air*, 17(4), 259–277. https://doi.org/10.1111/j.1600-0668.2007.00478.x

Miranda, L.S., Deilami, K., Ayoko, G.A., Egodawatta, P., and Goonetilleke, A. 2022. Influence of land use class and configuration on water-sediment partitioning of heavy metals. *Sci. Total Environ.*, 804: 150116. https://doi.org/10.1016/j.scitotenv.2021.150116

Mitra, S., Chakraborty, A.J., Tareq, A.M., Emran, T.B., Nainu, F., Khusro, A., Idris, A.M., Khandaker, M.U., Osman, H., Alhumaydhi, F.A. et al. 2022. Impact of heavy metals on the environment and human health: Novel therapeutic insights to counter the toxicity. *J. King Saud Univ.-Sci.*, 34: 101865. https://doi.org/10.1016/j.jksus.2022.101865

Niu, Y., Jiang, X., Wang, K., Xia, J., Jiao, W., Niu, Y., and Yu, H. 2020. Meta analysis of heavy metal pollution and sources in surface sediments of Lake Taihu, China. *Sci. Total Environ.*, 700: 134509. https://doi.org/10.1016/j.scitotenv.2019.134509

noPILLS in waters. 2015. http://www.no-pills.eu/conference/BS_NoPills_Final%20Report_long_EN.pdf

Otunola, B.O., and Ololade, O.O. 2020. A review on the application of clay minerals as heavy metal adsorbents for remediation purposes. *Environ. Technol. Innov.*, 18: 100692. https://doi.org/10.1016/j.eti.2020.100692

Persson, L., Almroth, B.M.C., Collins, C.D., Cornell, S., de Wit, C.A., Diamond, M.L. et al. 2022. Outside the safe operating space of the planetary boundary for novel entities. *Environ. Sci. Technol.*, 56(3): 1510–1521. https://doi.org/10.1021/acs.est.1c04158

Pichersky, E., and Gershenzon, J. 2002. The formation and function of plant volatiles: Perfumes for pollinator attraction and defense. *Curr. Opin. Plant Biol.*, 5(3), 237–243. https://doi.org/10.1016/S1369-5266(02)00251-0

Plainiotis, S., Pericleous, K.A., Fisher, B.E., and Shier, L. 2010. Application of Lagrangian particle dispersion models to air quality assessment in the Trans-Manche region of Nord-Pas-de-Calais (France) and Kent (Great Britain). *Int. J. Environ. Pollut.*, 40(1/2/3): 160–174. https://doi.org/10.1504/IJEP.2010.030891

Ravindra, K., Sokhi, R., and Grieken, R.V. 2008. Atmospheric polycyclic aromatic hydrocarbons: Source attribution, emission factors and regulation. *Atmos. Environ.*, 42(13): 2895–2921. https://doi.org/10.1016/j.atmosenv.2007.12.010

Rehman, A.U., Nazir, S., Irshad, R., Tahir, K., ur Rehman, K., Islam, R.U., and Wahab, Z. 2021. Toxicity of heavy metals in plants and animals and their uptake by magnetic iron oxide nanoparticles. *J. Mol. Liq.*, 321: 114455. https://doi.org/10.1016/j.molliq.2020.114455

Ritter, L., Solomon, K.R., Forget, J., Stemeroff, M., and O'Leary, C. 2007. *Persistent organic pollutants*, for the international programme on chemical safety (IPCS) within the framework of the inter-organization programme for the sound management of chemicals (IOMC), United Nations Environment Programme.

Saini, S., and Dhania, G. 2020. Cadmium as an Environmental Pollutant: Ecotoxicological Effects, Health Hazards, and Bioremediation Approaches for Its Detoxification from Contaminated Sites. In: Bioremediation of Industrial Waste for Environmental Safety, Bharagava, R. and Saxena, G. (eds), 357–387, Springer. https://doi.org/10.1007/978-981-13-3426-9_15

Seinfeld, J., and Pandis, S. 1998. Atmospheric Chemistry and Physics: From Air Pollution to Climate Change (2nd ed.), 97. John Wiley & Sons. ISBN 978-0-471-17816-3.

Srivastava, S., and Goyal, P. 2010. Novel Biomaterials: Decontamination of Toxic Metals from Wastewater. Springer. https://doi.org/10.1007/978-3-642-11329-1

State of Global Air/2018 a Special Report on Global Exposure to Air Pollution and Its Disease Burden. Health Effects Institute. 2018. https://www.stateofglobalair.org/sites/default/files/soga-2018-report.pdf

Suzzi, C.V., Kruse, D., Harrington, J., Levin, K., and Meliker, J.R. 2016. Is urinary cadmium a biomarker of long-term exposure in humans? A review. *Curr. Environ. Health Rep.*, 3: 450–458. https://doi.org/10.1007/s40572-016-0107-y

Szabo, D.T., and Loccisano, A.E. 2012. POPs and Human Health Risk Assessment, Schecter, A. (ed), 579–618, Dioxins and Health. https://doi.org/10.1002/9781118184141.ch19

Tindwa, H.J., and Singh, B.R. 2023. Soil pollution and agriculture in sub-Saharan Africa: State of the knowledge and remediation technologies. *Front. Soil Sci.*, 2: 1101944. https://doi.org/10.3389/fsoil.2022.1101944

Tripathi, M., Singh, P., Singh, R., Bala, S., Pathak, N., Singh, S., Chauhan, R.S., and Singh, P.K. 2023. Microbial biosorbent for remediation of dyes and heavy metals pollution: A green strategy for sustainable environment. *Front. Microbiol.*, 14: 1168954. https://doi.org/10.3389%2Ffmicb.2023.1168954

US EPA, OAR (18 August 2014). Volatile Organic Compounds' Impact on Indoor Air Quality. *US EPA*. Retrieved September 2023. https://www.epa.gov/indoor-air-quality-iaq/volatile-organic-compounds-impact-indoor-air-quality

US EPA, OAR (19 April 2016). Particulate Matter (PM) Basics. *US EPA*. Retrieved September 2023. https://www.epa.gov/pm-pollution/particulate-matter-pm-basics#PM

US EPA, OAR (26 April 2016). Health and Environmental Effects of Particulate Matter (PM). *US EPA*. Retrieved October 2023. https://www.epa.gov/pm-pollution/health-and-environmental-effects-particulate-matter-pm

Vallack, H.W., Bakker, D.J., Brandt, I., Broström-Lundén, E., Brouwer, A., Bull, K.R. et al. 1998. Controlling persistent organic pollutants–what next? *Environ. Toxicol. Pharmacol.*, 6(3): 143–175. https://doi.org/10.1016/S1382-6689(98)00036-2

Verma, S., Bhatt, P., Verma, A. et al. 2023. Microbial technologies for heavy metal remediation: Effect of process conditions and current practices. *Clean Technol. Environ. Policy*, 25: 1485–1507. https://doi.org/10.1007/s10098-021-02029-8

Wasley, A., Heal, A., Harvey, F., and Lainio, M. 2019. Revealed: UK government failing to tackle rise of serious air pollutant. *The Guardian*. https://www.theguardian.com/environment/2019/jun/13/revealed-uk-government-failing-to-tackle-rise-of-ammonia-serious-air-pollutant

Wolkoff, P., Wilkins, C.K., Clausen, P.A., and Nielsen, G.D. 2006. Organic compounds in office environments – Sensory irritation, odor, measurements and the role of reactive chemistry. *Indoor Air*, 16(1): 7–19. https://doi.org/10.1111/j.1600-0668.2005.00393.x

Yuan, P., Wu, X., Xia, Y., Peng, C., Tong, H., Liu, J., Jiang, L., and Wang, X. 2020. Spatial and seasonal variations and risk assessment for heavy metals in surface sediments of the largest river-embedded reservoir in China. *Environ. Sci. Pollut. Res. Int.*, 27(28): 35556–35566. https://doi.org/10.1007/s11356-020-09868-w

Zaynab, M., Al-Yahyai, R., Ameen, A., Sharif, Y., Ali, L., Fatima, M., Khan, K.A., and Li, S. 2022. Health and environmental effects of heavy metals. *J. King Saud Univ., Sci.*, 34: 101653. https://doi.org/10.1016/j.jksus.2021.101653

Zhang, H., Walker, T.R., Davis, E., and Ma, G. 2019. Ecological risk assessment of metals in small craft harbour sediments in Nova Scotia, Canada. *Mar. Pollut. Bull.*, 146: 466–475. https://doi.org/10.1016/j.marpolbul.2019.06.068

5 Types of Heavy Metals Found in the Environment and Their Toxicological Properties

Divya J Tandel *
Department of Pharmacology, PIP, Parul University, Vadodara, Gujarat, India

Himanshu N Panchal
Department of Pharmaceutical Chemistry, PIP, Parul University, Vadodara, Gujarat, India

Sapna Desai
Department of Pharmacology at Pioneer Pharmacy College, Vadodara, Gujarat, India

5.1 INTRODUCTION

The environment is filled with metals, some of which are extremely poisonous while others are necessary to all living things. Assuming that toxicity and heaviness are correlated, metalloids that may cause toxicity at low exposure levels, like arsenic, are also considered heavy metals.[1] Because the body is unable to produce essential metals, which are necessary for enzyme reactions and other physiological activities, eating a balanced diet is essential to maintaining good health. Because heavy metals are found in a variety of environmental matrices as shown in Figure 5.1, at trace quantities (ppb range to less than 10 ppm), they are also regarded as trace elements.[2]

Pollutants can also pollute the environment by sediment re-suspension, air deposition, erosion of soil, heavy metal leaching, an evaporation of metals from water bodies into soil and groundwater, and metal corrosion as shown in Figure 5.2. The effects of heavy metals on organelles and components in the biological systems have been documented as shown in Figure 5.3. These include the originals like cell membrane, mitochondria, lysosome, reticulum, nucleus, and enzymes those are involved in metabolism, removal of toxins, and resolving the cellular damage. It has been discovered that metal ions interact with genetic material and related components of cell.[3]

Crucial at low quantities, heavy metals aren't as harmful and function as coenzymes in biological processes. For instance, iron is found in haemoglobin and myoglobin, whereas cobalt is found in vitamin B12. Non-vital heavy metals are extremely hazardous to living things, even at very low quantity, and they are not biodegradable.[4]

* **Corresponding Author:** divya.tandel16217@paruluniversity.ac.in

DOI: 10.1201/9781032685793-5

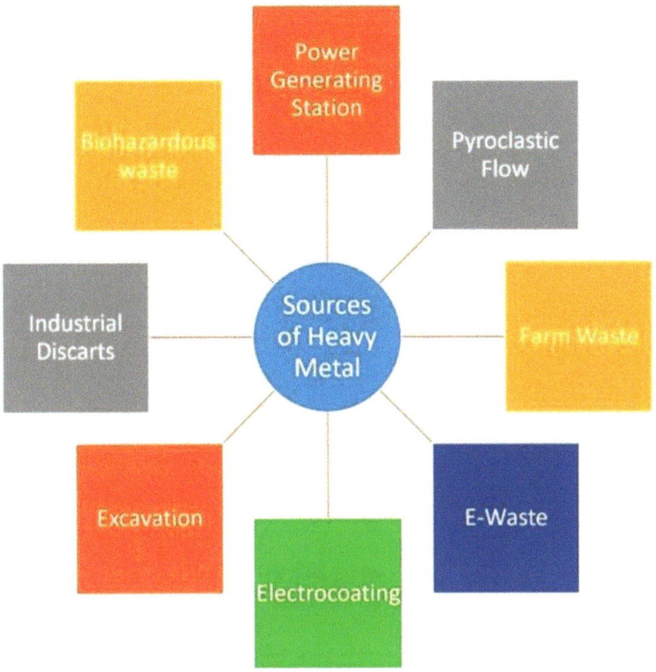

FIGURE 5.1 Sources of heavy metal in environment.

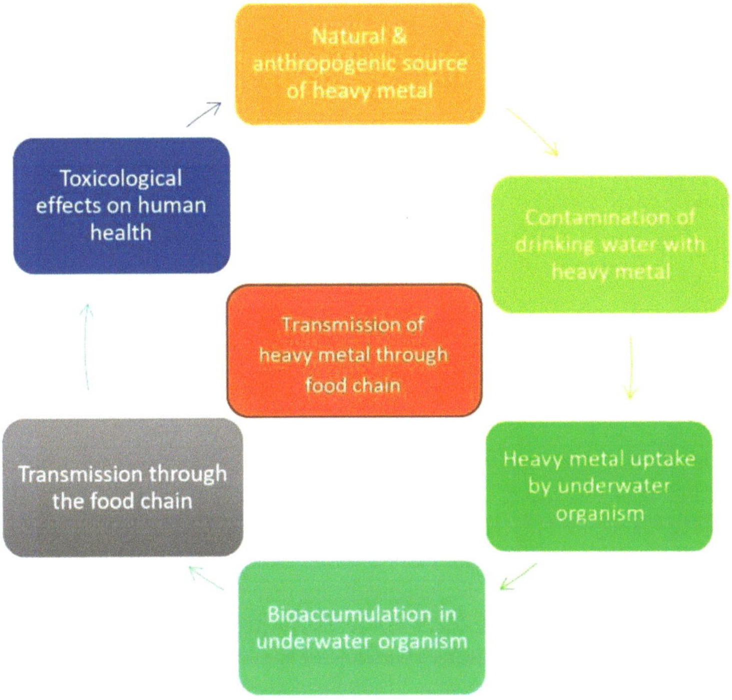

FIGURE 5.2 Transmission of heavy metal in food chain.

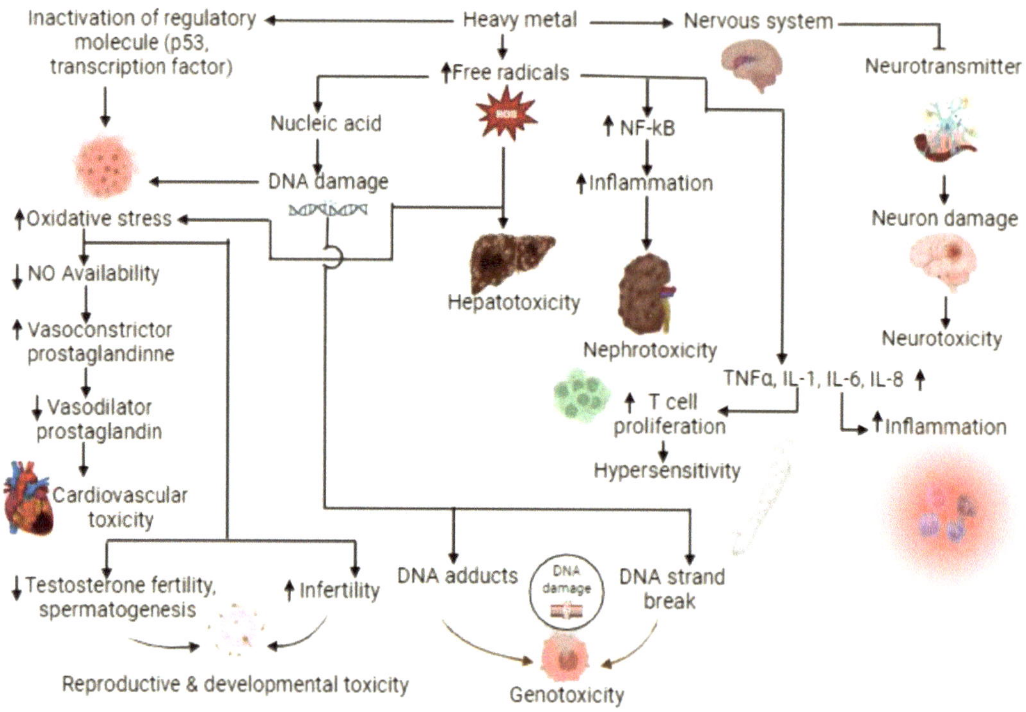

FIGURE 5.3 Toxicological effect of heavy metal in human body.

Following is an array of heavy metals that are more prevalent in daily life and have densities of more than 5 g/cm^3.

- Mercury
- Lead
- Copper
- Zinc
- Chromium
- Cobalt
- Nickel
- Cadmium
- Arsenic

5.2 MERCURY (Hg)

Mercury is the heavy metal having an atomic weight of 200.6 g/mol and a density of 13.6. It can be found in several kinds of ore and is uncommon in the crust of the planet. The main mercurial mineral is cinnabar HgS. It is the only metal which exist in liquid state at ambient temperature, which makes it a special metal. Furthermore, it can be found in three different chemical forms in nature: organic, inorganic, and elemental. At the last era of the 19th century, allergic contact dermatitis was first identified, and mercury is a well-known inductor of the condition. Elevated proteinuria, typically a nephrotic syndrome, was identified in exposure to mercury since the 1800s.[5–9]

Mechanism: The two major ways by which humans are exposed to mercury are by eating fish and dental amalgam. An additional significant route of exposure is through inhaling mercury vapour. Elements such as methylmercury and elemental mercury are highly absorbed in tissues because of their high lipophilicity. Because inorganic mercury salts are insoluble in lipid, which makes them unable to pass through the blood-brain and barriers in placenta. Humans absorb almost 90% of the methylmercury from the gastrointestinal tract, but not greater than 10% of the inorganic mercury salts are absorbed.[10]

Physicochemical Properties: Mercury finds widespread application in a number of industries, including the pulp and paper, wood processing, caustic soda production, and battery manufacturing. Mercury is also used as a catalyst in the chemical sector, as amalgams in the dentistry business, and as a preservative in pharmaceutical products. Mercury is utilized as a metal in thermometers since its coefficient of expansion is linear. Furthermore, mercury is utilized in the production of jewellery, insecticides, barometers, and incandescent lights.[1–13]

5.3 LEAD (Pb)

The Earth's crust naturally contains trace amounts of lead, mining has taken place for ages, causing it to be spread throughout many ecosystems and eventually integrated into the structural tissue of humans, plants, and animals. Pb^{+2} ions are present in lead during the chemical process. Although lead can also be absorbed through the skin, the digestive and respiratory systems are the primary routes for lead absorption. In addition to respiratory, urinary, and cardiovascular issues, lead exposure can result in oxidative, inflammatory, and immune-modulating diseases. Pb may also disrupt the balance of the oxidant-antioxidant system and produce inflammatory responses in several parts of body.[14,15]

Mechanism: Lead's ability to mimic or inhibit the actions of calcium and to interact with proteins is one of the main biochemical processes by means of that it exerts its toxicological effect. Lead replaces calcium as a component of the mineral in the skeleton. Lead binds to biological molecules and through a variety of mechanisms interferes with their ability to function. Enzyme sulfhydryl and amide groups are bound by lead, which changes the structure of the enzymes and reduces their activity. Lead can also interfere with the transport of necessary cations like calcium or by competitive antagonism with essential cations for active site, thereby blocking enzyme function. Numerous researchers have shown that lead poisoning causes damage to the cell that is due to the formation of reactive oxygen species (ROS).[14–16]

Physicochemical Properties: Lead (Pb), one of the first heavy metals, was historically used in hair colours, pottery glazes, and insecticides.[17] When newly cut, this naturally present bluish-grey-coloured metal is very malleable, ductile, and lustrous. Lead is a dense metal with an anatomical mass of 207.2 g/mol and a specific gravity of 11.35. It is mostly found in ores in the Earth's crust, where it may be found in a range of inorganic forms such as acetates, nitrates, carbonates, sulphates, and chlorides. Lead can also be detected in the environment as a result of radioactive decay as it is a by-product of the disintegration of uranium. Currently, native lead is recovered from ore which also includes silver, copper, and zinc (blende), as native lead is rare. The main natural source of lead is galena (PbS).[10]

5.4 COPPER (Cu)

Copperiedus is used to refer the consequences of an excess of copper in the body. Copperiedus condition occurs due to eating acid foods cooked in uncoated copper vessel, or if person drink a water having excess of copper or other surrounding sources. The activity of numerous metalloproteins and enzymes depends on copper. It is essential for controlling the expression of genes. It is also essential for several vital function of body like growth, defence, energy production, blood production, transport of iron, and metabolism. An estimated 80 mg of copper, within a range of 50–120 mg, is present in the body overall. Foods high in copper in the diet include shellfish, meat, beans, and nuts. Dairy products, carbohydrates, and refined cereals are low in nutrients.[18,19]

Mechanism: ROS are formed in part by copper ions. Cu^{+1} (cuprous) and Cu^{+2} (cupric) ROS may participate in the redox processes. Ascorbic acid and glutathione are examples of biological reductants that may convert cupric ROS into cuprous ROS. Cuprous ROS is capable of catalysing the decomposition of hydrogen peroxide by producing OH^- by the Fenton reaction. Hydrogen peroxide can then react with a range of biomolecules. Research has shown that copper may break DNA strands and oxidize bases due to free radicals containing oxygen and hydroxyl. Through the genotoxic benzene metabolite (1,2,4-benzenetriol), the cupric and cuprous forms of copper more strongly than iron cause DNA breaking.[20]

Physicochemical Properties: The 26th most common metal in the environment is copper, an essential element. It is a reddish-gold colour that can be found in minerals like bornite, sometimes referred to as peacock ore, which also contains copper, iron, and sulphur, and chalcopyrite, which is a mineral that contains these three elements. Additionally, copper is a good heat and electrical conductor.

Copper is alloyed with many of the metal, including brass and bronze, copper wires, barrier cream, coins, pipelines, plating, fertilizer, and fabric and wood preservation chemical examinations to determine Fehling's solution's sugar content copper sulphate is used as an algaecide in water treatment and as a treatment for mildew in agriculture.[20,23]

5.5 ZINC (Zn)

Zinc is a component of several metalloenzymes and transcription factors that are involved in many biological activities, including the transcription process, transmission of signals, gene expression, and replication. Based on a systemic effect, zinc supplementation has been shown to inhibit carcinogenesis. Zinc has the strongest antagonistic potential, followed by calcium and magnesium. The degree to which these metals can be antagonistic elevated fidelity Cd-DNA binding likely resembles their ability to enrage in vivo.[10,23–25]

Mechanism: Zinc has also been observed to interact with the metabolism of cyclic nucleotides. It was discovered that the metal mediate the production of second messengers like cyclic guanosine monophosphate (cGMP) and cyclic adenosine monophosphate (cAMP), through the actions of phosphodiesterase (PDE) on cyclic nucleotides.

Protein kinase C (PKC) is also regulated by zinc, which leads to PKC translocation to the cytoskeleton. High quantities of external zinc enhance the activation of protein tyrosine phosphorylation and mitogen-activated protein kinases (MAPKs). Zinc is known to increase mitogenic signalling because it works in concert with calcium to affect the MAPK. Zinc was also discovered to promote DNA synthesis in NIH3T3 cells. Cytological map position 67B5–6 also known as metal response element-binding transcription factor-1 (MTF-1) is a transcription factor that is activated by zinc. MTF-1 stimulates the zinc transporter ZnT-1 as well as the metallothionein genes in the response to cellular zinc. In addition to its involvement in cell proliferation, zinc is a part of structural protein that is involved in growth control. When cells lack zinc, several cell types will undergo apoptosis.

When extracellular concentrations of zinc surpass the capacity of zinc homeostasis, cytotoxicity-induced increased intracellular zinc levels lead to apoptosis. Cells begin to necrotize at extremely high zinc concentrations.[2]

Zinc finger proteins are a type of DNA-binding proteins that preferentially bind to short DNA sequences to regulate the transcription of many genes, so acting as transcription factors. Zinc finger proteins can contain several metals in place of zinc, such as cobalt, cadmium, copper, nickel, and iron. It is predicted that when a redox metal takes the role of zinc in the zinc finger protein, free radicals will be produced that would harm DNA.

Physical Properties: Zinc has a density of 7.134 g/cm^3, making it a necessary element. This is the 4th most common metal. It's a silvery-white metal with a hint of blue. Zinc blende, made of ZnS, and $ZnCO_3$, made of zinc silicate, are the two most often used ores. Zinc tarnishes in the air.

Zinc has been utilized in galvanization and die-casting processes to prevent metal from rusting. Other goods that include zinc oxide include pigments, beauty products, soaps, deodorants, anti-dandruff shampoos, weaponry, electrical devices, rechargeable batteries, plastic, ink, medicines, textiles, and rubber. Zinc sulphide is found in fluorescent lights, X-ray screens, and luminaire paint. Anaemia, fatigue, vomiting, fever, diarrhoea, nausea, and copper insufficiency are all possible side effects of zinc poisoning.[20,21,22]

5.6 CHROMIUM

Fuels produced from iron chromate refractory material, fertilizer, catalysts, chromium steel, drilling oil wells, and metal plating tanneries are among the natural sources of chromium. By using sewage and fertilizers, people discharge chromium into the environment. While Cr(III) is stationary in its reduced form and does not dissolve in water, Cr(VI) is extremely soluble in it when it is in its oxidized state and consequently mobile.[27]

Mechanism: Because of its low membrane permeability, trivalent chromium Cr(III) is typically safe to be around. Hexavalent chromium Cr(VI), on the other hand, is more successful in penetrating the cell membrane through isoelectric and isotypic anion channels, such as $(SO_4)^{-2}$ and $(HPO_4)^{-2}$ channels. These chromates are taken in via phagocytosis. Temporary species of chromium with four and five valency that vary from Cr(III) can be produced by reducing the strong oxidizing agent Cr(VI). When intracellular reduction of Cr(VI) occurs outside of the target area, it is considered a detoxifying process since glutathione stabilizes the pentavalent form. However, if intracellular decrease of Cr(VI) occurs at the target region, Cr may be triggered. When Cr(VI) reacts with biological reductants such thiols and ascorbate, it produces ROS including oxides of super, H_2O_2, and hydroxyl radical. In the end, this leads to oxidative stress within the cell, causing damage to proteins and DNA. Studies of published research have shown that because Cr(VI) penetrates cells more readily and is eventually converted to Cr(III), it is significantly more dangerous than Cr(III).[27,28,29]

Physical Properties: Density of chromium is 7.15 g/cm^3. Chromium is the 21st most prevalent element on Earth's surface. It is exploited as a chromite ore and is also referred to as Siberian red lead. Chromium is a hard, glossy, grey, slightly reactive metal that reacts with most acids. Moreover, it forms a coating of Cr(III) oxide that decreases the corrosiveness of the metal.

Chromium is used in the synthesis of artificial rubies, metal porcelain, alloys, electroplating, tanning leather, and chromium salt-tinted green glass.[30–33]

5.7 NICKLE (Ni)

A genetic component to nickel allergy has been proposed, based in part on previous twin and familial studies as well as the discovery of a link between human leukocyte antigen (HLA) class II genes polymorphism and allergy due to nickel. Nevertheless, there is no proof of associations with this. According to recent research, nickel primarily interacts with HLA-DR before moving on to a beta chain conserved amino acid. However, there is still a chance that other genetic variables unrelated to the HLA system will play a role in nickel allergy development. However, a more recent study using female twins suggests that environmental factors are more responsible than genetic factors in the development of nickel allergy.[33,41–45]

Mechanism: Nickel (Ni) has the ability to induce both worldwide and gene-specific hypermethylation, which have been linked to gene silence.

Histone acetylation reductions that are both global and gene-specific can also result from Ni exposure. Ni has been demonstrated to inhibit DNA methyltransferase (DNMT) activity both in vitro as well as in vivo. Ni lowered the amounts of genomic DNA methylation and DNMT activity in cells, although this suppression was only temporary. Following Ni treatment, following a period of recuperation, DNMT activity somewhat recovered and levels of genomic DNA methylation rose above basal levels. Ni specifically attached to heterochromatin and damaged chromosomes in heterochromatic areas of Chinese hamster ovary cells.

Heterochromatin binding may have a major impact on Ni-induced cellular change and ageing. During Ni-induced cell transformation, the senescence gene(s) in Ni-transformed cells were silenced, suggesting that DNA hypermethylation had a major role in the restoration of senescence after therapy with the DNA methylation inhibitor 5-Azacytidine.

Further research using transgenic Chinese hamster ovary (CHO) cell lines revealed more evidence linking Ni with DNA hypermethylation and deactivation of genes near heterochromatin.

It was shown that Ni could mute a transgene adjacent to a heterochromatic area in CHO cells while preventing transgene deletions or mutations. Instead, greater amounts of chromatin condensation and DNA methylation were seen at the Ni-silenced transgenic location. The demethylating agent 5-AzaC corrected this silence, indicating that DNA hypermethylation was the source of the Ni-induced transgene inactivation.[34–40]

Physicochemical Properties: With a density of 8.9 g/cm^3, nickel is the 28th number element in periodic table. Among the minerals found were pimelite, a green hydrous nickel silicate complex, and pentlandite, an $(NiFe)_3S_2$, high quantity of nickel from meteorites; metallic silver; resistant to rusting at high temperatures.

Nickel is used to make coins, jewellery, and other items. Other metals are plated to prevent corrosion: alloy made of stainless steel, welding, armour plating, OAT engines and propeller shafts, and heat-using appliances use chrome alloy, which is non-corrosive.[18,46–49]

5.8 ARSENIC (As)

Low concentrations of arsenic are a prevalent element that may be found on practically all environmental surfaces. The two primary chemical forms of arsenic are pentavalent arsenate and trivalent arsenate. The forms of organic matter are the methylated metabolites monomethylarsonic acid (MMA), dimethylarsinic acid (DMA), and trimethylarsine oxide (TMAO). Both natural phenomena like volcanic eruptions and degraded soil and man-made activities can lead to environmental arsenic poisoning.[50] Many compounds containing arsenic are produced by industrial processes and are used to manufacture agricultural products such as herbicides, insecticides, fungicidal agents, algaecides, sheep dips, timber preservation agents, and dyes. They have also been used in animal health care to treat sheep and cattle for tapeworm infestations.[51] Additionally, for at least ten decades, trypanosomiasis, syphilis, yaws, and amoeba dysentery and several veterinary disorders have all been treated using compounds containing arsenic.[52] Arsenic trioxide has just been licenced by the US FDA to treat acute promyelocytic leukaemia as an anticancer medication. Its therapeutic efficacy has been associated with its capacity to cause leukaemia cells to go through apoptosis.[53,54]

5.8.1 Possibility of Heavy Metals to Human Contact

Millions of people are thought to be chronically affected by arsenic worldwide, particularly in nations where substantial levels of arsenic contamination are found in the ground water, such as Bangladesh, India, Chile, Uruguay, Mexico, and Taiwan.[50,51,55] The clinical health limit in rural area is 0.000001–0.000003 mg/m^3 and in city 0.00002–0.0001 mg/m^3. The level of arsenic is higher in soil near mines and agricultural land due to the pesticide sprinkling.[56,57] Arsenic can be ingested orally, inhaled, via skin, and sometimes enter the bloodstream.[50,51,55] The main way that most people are exposed is through their diet, which typically contains 50 μg daily. Although exposure via the air, water, and soil is typically considerably lower, it can become considerable in regions where arsenic poisoning is present. Workers may be exposed to far higher levels of arsenic in sectors such as vineyards, ceramics, glassmaking, smelting, metallic ore refining, pesticide manufacturing and application, wood conservation, and the production of semiconductors.[58] Given that arsenic may have a number of detrimental effects on human health, significant levels of arsenic pollution are cause for worry. Numerous epidemiological studies have linked arsenic exposure to a higher risk of carcinogenic and systemic health consequences.[59]

Interest in arsenic toxicity has increased due to recent reports of large populations exposed to high concentrations of arsenic in drinking water in the states of West Bengal, Bangladesh, Thailand, Inner Mongolia, Taiwan, China, Mexico, Argentina, Chile, Finland, and Hungary. These populations are exhibiting a variety of clinicopathological conditions, including developmental anomalies, cancer, Type 2 diabetes, hearing fall, portal fibrosis, blood disorders (anaemia, leukopenia, and eosinophilia), cardiovascular and peripheral vascular diseases, and blood disorders.[50,52,57,58]

Almost every organ system is impacted by arsenic exposure, including the neurological, hepatobiliary, respiratory, cardiovascular, dermatological, and gastrointestinal systems.[59] Mortality ratio is higher in patient suffering from malignancies of the skin, liver, kidney, bladder, and in many regions where arsenic contamination is present.[60]

5.8.2 Mechanism

It is difficult to analyse the toxic effects of arsenic since a variety of intrinsic and extrinsic variables, like the metal's oxidative state and solubility, can have a significant impact on toxicity.[61] A number of factors, including age, gender, dose, frequency of exposure, genetic composition, individual susceptibilities, and dietary variables, can affect how toxic arsenic is.[62] The majority of arsenic poisoning incidents in humans have been linked to inorganic arsenic exposure. Ten to twenty times as poisonous as pentavalent arsenate As(V) is inorganic trivalent arsenite As(III).[63] As(III) can bind to sulfhydryl groups (R-SH) on proteins, rendering over 200 enzymes inert. This is the probable mechanism by which arsenic has a wide range of impacts on many organ systems. As(V) can replace phosphate since it is an element of several metabolic pathways.[63,64] Arsenic causes harm by inhibiting many enzymes in the mitochondria and by decoupling oxidative phosphorylation, which both affect cellular respiration. The majority of arsenic's toxicity stems from its capacity to bind with protein and enzyme sulfhydryl groups and to function as a phosphorus replacement in a range of metabolic activities.[65] In vitro, arsenic combines with sulfhydryl groups in proteins to inactivate enzymes like thiolase and dihydrolipoyl dehydrogenase, which reduce pyruvate oxidation and fatty acid beta-oxidation; the primary metabolic pathway via which inorganic arsenic reaches the body is methylation. The first of two major metabolites that arsenic trioxide undergoes methylation is MMA, which is further methylated by enzymes to DMA before elimination in the urine. This process occurs non-enzymatically.[59,64,66] Genotoxicity experiments reveal that arsenic compounds induce chromosomal aberrations, sister-chromatid exchanges, the formation of micronuclei, and the suppression of DNA repair in both cultured human and rodent cells as well as in human cells that are exposed to them.[67–70] Based on the comet test, it has been discovered that arsenic trioxide oxidizes DNA in human cells and mice white blood cells. This results in gene amplification, stops cell division, hinders the process of gene repair, and expresses the c-fos gene. Arsenic trioxide is toxic to cell and can transcriptionally activate several stress genes and related proteins in human hepatocellular carcinoma.[68,71–73] Numerous studies suggest that arsenic can trigger cancer by promoting aberrant gene expression through DNA hypomethylation. Furthermore, arsenic was demonstrated to be a potent activator of extracellular signal-regulated protein kinase Erk1 and AP-1 transactivation activities, as well as an efficient inducer of the expression of the c-fos and c-jun genes. The stimulation of c-jun and c-fos by arsenic is associated with JNK activation.[74–76] Extended exposure to elevated arsenic concentrations may increase cellular susceptibility to mitogenic activation, and modifications in mitogenic signalling proteins may augment arsenic's carcinogenic properties. All in all, a number of recent investigations have demonstrated that arsenic may interfere with cell signalling pathways, including the p53 signalling pathway, which is frequently connected to the development and growth of a variety of tumour types in lab animals as well as some cancers in humans.[77–79] In tumours taken from individuals who had previously consumed arsenic, the p53 gene's expression and mutation were also studied. p53 is involved in many biological processes, such as the control of the cell cycle, repair of DNA, distinction, genomic plasticity, and programmed cell death. Numerous research have offered more evidence in favour of the theory that arsenic can alter gene expression.[80,81]

5.9 CADMIUM (Cd)

Atomic weight of 112.4 u and density of 8.65 g/cm^3 characterize cadmium, a soft, bluish-white metal. At 594°K, it melts. Friedrich Stromeyer made the initial discovery of cadmium in Germany in 1817. It is an uncommon element that isn't found in nature in its pure form. As 2.5–3 kg of cadmium may be obtained from 1 tonne of zinc, it is commonly associated with zinc. Cadmium may also be found in lead and copper metals, as well as in naturally occurring phosphates (26 ppm in Morocco and 40 ppm in Tunisia). In this latter case, other methods of examination might be applied.[1] Cadmium has a biological $t_{1/2}$ of 10–25 years after it enters the environment.[82]

5.9.1 Possibility of Human Contact

Consuming tainted food will result in significant Cd exposure. Smoking is another activity that can expose one to cadmium and raise blood and urine cadmium contents. The body's essential processes may be disrupted by the presence of Cd in tainted water, potentially leading to either acute or chronic illnesses.[83,84] Seafood, cereals, and rice have all been reported to be contaminated with Cd. Nevertheless, a small amount of Cd is absorbed upon oral consumption. Regrettably, widespread cadmium poisoning of sources of water and food was the reason of the ouch-ouch disease outbreak in Japan. The patients experienced excruciating lung and gastrointestinal disorders, renal failure, and degenerative bone disease. Industrial dust is a more effective way for Cd to be removed from the lungs than low GI absorption. Renal tubular dysfunction and pulmonary damage may result from either acute or chronic lead exposure in industrial regions. Smokers' blood content of cadmium is about double that of non-smokers.[85–87]

5.9.2 Mechanism of Toxicity

If consumed or breathed in, cadmium may be extremely irritating to the lungs and digestive system and even lethal. Within 15–30 minutes following acute intake, symptoms such as nausea, vomiting, salivation, cramping in the muscles, vertigo, shock, loss of mindfulness, and scissors typically manifest. Acute cadmium poisoning can also cause coma, lung, hepatic, or renal damage, as well as gastrointestinal tract erosion, depending on the mode of poisoning.[88,89] An extended period of cadmium exposure lowers norepinephrine, serotonin, and acetylcholine levels.[90] Cadmium exposure over time results in lung adenocarcinomas.[91,92] Additionally, it may cause prostatic proliferative lesions, including adenocarcinomas, following systemic or direct exposure.[93] Although the mechanisms of cadmium toxicity are poorly understood, it has been hypothesized that cadmium primarily affects cells by creating ROS, which damage ssDNA and interfere with the proteins production and nucleic acids.[94,95] At 1 mg/kg body weight, cadmium influences male reproduction in a mouse model. However, in comparison to other cancer causing metals, it is a poor mutagen.[96,97] Cadmium has an impact on signal transduction pathways through calcium channel blockage, cytoplasmic free calcium accumulation in a diverse cell types, and inositol polyphosphate production.[97,98]

ACKNOWLEDGEMENT

Authors are thankful to the Parul University, Vadodara, for necessary support.

REFERENCES

1. Yu M-H, Tsunoda H, Tsunoda M. Environmental Toxicology: Biological and Health Effects of Pollutants. Third Edition. Boca Raton, London, New York: Taylor & Francis Group, CRC Press; 2011.
2. Duffus JH. Heavy metals-a meaningless term? Pure Appl Chem. 2002; 74(5):793–807.
3. Clancy HA, Sun H, Passantino L, Kluz T, Muñoz A, Zavadil J et al. Gene expression changes in human lung cells exposed to arsenic, chromium, nickel or vanadium indicate the first steps in cancer. Metallomics. 2012; 4(8):784–793.
4. Kim JJ, Kim YS, Kumar V. Heavy metal toxicity: An update of chelating therapeutic strategies. J Trace Elem Med Biol. 2019; 54:226–231.

5. Boyd AS, Seger D, Vannucci S et al. Mercury exposure and cutaneous disease. J Am Acad Dermatol. 2000; 43:81–90.
6. Tchounwou PB, Yedjou CG, Patlolla AK, Sutton DJ. Heavy metal toxicity and the environment. Exp Suppl. 2012; 101:133–164.
7. Tiller KG. Heavy Metals in Soils and Their Environmental Significance. In: Stewart, BA, editor. Advances in Soil Science. New York: Springer; 1989; pp. 113–142.
8. Forstner U, Wittmann GTW. Metal Pollution in the Aquatic Environment. Berlin, Heidelberg, New York: Springer-Verlag; 1981.
9. Enghag P. Encyclopedia of the Elements: Technical Data - History - Processing - Applications. Weinheim: Wiley -VCH Verlag GmbH & Co. KGaA; 2004.
10. Valko M, Morris H, Cronin MT. Metals, toxicity and oxidative stress. Curr Med Chem. 2005; 12(10):1161–1208.
11. Jaishankar M, Tseten T, Anbalagan N, Mathew BB, Beeregowda KN. Toxicity, mechanism and health effects of some heavy metals. Interdiscip Toxicol. 2014; 7(2):60–72.
12. Genchi G, Sinicropi MS, Carocci A, Lauria G, Catalano A. Mercury exposure and heart diseases. Int J Environ Res Public Health. 2017; 14(1):74.
13. Agency for Toxic Substances and Disease Registry (ATSDR). Toxicological Profile for Lead. Public Health Service. Atlanta: U.S. Department of Health and Human Services; 1999.
14. Flora SJS, Saxena G, Gautam P, Kaur P, Gill KD. Lead induced oxidative stress and alterations in biogenic amines in different rat brain regions and their response to combined administration of DMSA and MiADMSA. Chem Biol Interac. 2007; 170:209–220.
15. Hermes-Lima M, Pereira B, Bechara EJ. Are free radicals involved in lead poisoning? Xenobiotica. 1991; 8:1085–1090 [PubMed: 1776279].
16. Ye X, Wong O. Lead exposure, lead poisoning, and lead regulatory standards in China, 1990–2005. Regul Toxicol Pharmacol. 2006; 46(2):157–162.
17. WHO. Copper. In: WHO, editor. Trace Elements in Human Nutrition and Health. Geneva, Switzerland: World Health Organization; 1996; pp. 123–143.
18. Milne DB. Copper intake and assessment of copper status. Am J Clin Nutr. 1998; 67(5 Suppl):1041S–1045S.
19. Toxicological Profile for Copper, ATSDR, 2004. https://www.atsdr.cdc.gov/toxprofiles/tp132.pdf
20. Kaye GWC, Laby TH. Tables of Physical and Chemical Constants, Fifteenth Edition. London, UK: John Wiley and Sons Ltd; 1986.
21. Roels HA, Ghyselen P, Buchet JP, Ceulemans E, Lauwerys RR. Assessment of the permissible exposure level to manganese in workers exposed to manganese dioxide dust. Br J Ind Med. 1992; 49:25–34.
22. Tsafrir J. Copper toxicity: A common cause of psychiatric symptoms. 2017. Available online: https://www.psychologytoday.com/intl/blog/holistic-psychiatry/201709/copper-toxicity-common-cause-psychiatric-symptoms. (Accessed 7 July 2018).
23. Berg JM, Shi Y. The galvanization of biology: A growing appreciation for the roles of zinc. Science. 1996; 271:1081–1085.
24. Kasprzak KS, Waalkes MP. The role of calcium, magnesium, and zinc in carcinogenesis. Adv Exp Med Biol. 1986; 206:497–515.
25. Waalkes MP, Poirier LA. In vitro cadmium-DNA interactions: Cooperativity of cadmium binding and competitive antagonism by calcium, magnesium, and zinc. Toxicol Appl Pharmacol. 1984; 75:539–546.
26. Ghani A. Effect of chromium toxicity on growth, chlorophyll and some mineral nutrients of *Brassica juncea* L. Egyptian Acad J Biol Sci. 2011; 2(1):9–15.
27. Stohs SJ, Bagchi D. Oxidative mechanisms in the toxicity of metal ions. Free Radic Biol Med. 1995; 18(2):321–336.
28. Wolińska A, Stępniewska Z, Włosek R. The influence of old leather tannery district on chromium contamination of soils, water and plants. Nat Sci. 2013; 5(2A):253–258.
29. World Health Organization (WHO). Air Quality Guidelines for Europe. 2000 (91). https://iris.who.int/bitstream/handle/10665/107335/9789289013581-eng.pdf?sequence=1
30. Occurrences Barnhart J. Occurrences, uses, and properties of chromium. Regul Toxicol Pharmacol. 1997; 26:S3–S7.
31. Abadin H, Fay M, Ingerman L, Tencza B, Yu D, Wilbur SB. Toxicological Profile for Chromium. CDC stacks, 2012. https://stacks.cdc.gov/view/cdc/12385/cdc_12385_DS1.pdf
32. Martin S, Wendy G. Human health effects of heavy metals. Environ Sci Technol Brief Citizens. 2009; 15:1–6.
33. Dayan AD, Paine AJ. Mechanisms of chromium toxicity, carcinogenicity and allergenicity: Review of the literature from 1985 to 2000. Hum Exp Toxicol. 2001; 20:439–451.

34. Lee YW, Broday L, Costa M. Effects of nickel on DNA methyltransferase activity and genomic DNA methylation levels. Mutat Res. 1998; 415:213–218.
35. Lee YW, Klein CB, Kargacin B et al. Carcinogenic nickel silences gene expression by chromatin condensation and DNA methylation: A new model for epigenetic carcinogens. Mol Cell Biol. 1995; 15:2547–2557.
36. Lee YW, Pons C, Tummolo DM et al. Mutagenicity of soluble and insoluble nickel compounds at the gpt locus in G12 Chinese hamster cells. Environ Mol Mutagen. 1993; 21(4):365–371.
37. Klein CB, Costa M. DNA methylation, heterochromatin and epigenetic carcinogens. Mutat Res. 1997; 386:163–180.
38. Klein CB, Kargacin B, Su L et al. Metal mutagenesis in transgenic Chinese hamster cell lines. Environ Health Perspect. 1994a; 102(Suppl 3):63–67.
39. Klein CB, Su L, Rossman TG et al. Transgenic gpt+ V79 cell lines differ in their mutagenic response to clastogens. Mutat. Res. 1994b; 304:217–228.
40. Klein CB, Conway K, Wang XW et al. Senescence of nickel-transformed cells by a mammalian X chromosome: Possible epigenetic control. Science. 1991; 251:796–799.
41. Fleming CJ, Burden AD, Forsyth A. The genetics of allergic contact hypersensitivity to nickel. Contact Dermatitis. 1999; 41:251–253.
42. Menne T, Holm NV. Nickel allergy in a female twin population. Int J Dermatol. 1983; 22:22–28.
43. Olerup O, Emtestam L. Allergic contact dermatitis to nickel is associated with a Taq I HLA-DQA allelic restriction fragment. Immunogenetics. 1988; 28:310–313.
44. Gamerdinger K, Moulon C, Karp DR et al. A new type of metal recognition by human T cells: Contact residues for peptide-independent bridging of T cell receptor and major histocompatibility complex by nickel. J Exp Med. 2003; 197:1345–1353.
45. Bryld LE, Hindsberger C, Kyvik KO et al. Genetic factors in nickel allergy evaluated in a population-based female twin sample. J Invest Dermatol. 2004; 123:1025–1029.
46. Haynes WM, editors. CRC Handbook of Chemistry and Physics, Ninety Second Edition. Boca Raton, FL: CRC Press; 2011.
47. Denkhaus E, Salnikow K. Nickel essentiality, toxicity, and carcinogenicity. Crit Rev Oncol Hematol. 2002; 42:35–56.
48. Sharma AD. Low nickel diet in dermatology. Indian J Dermatol. 2013; 58:240.
49. Agency for Toxic Substances and Disease Registry (ATSDR). Toxicological Profile for Arsenic TP92/09. Atlanta: Center for Disease Control; 2000.
50. Tchounwou PB, Wilson B, Ishaque A. Important considerations in the development of public health advisories for arsenic and arsenic-containing compounds in drinking water. Rev Environ Health. 1999; Oct–Dec; 14(4):211–229. doi: 10.1515/reveh.1999.14.4.211. PMID: 10746734.
51. Centeno JA, Tchounwou PB, Patlolla AK, Mullick FG, Murakata L, Meza E, TodorTodorov DL, Yedjou CG. Environmental Pathology and Health Effects of Arsenic Poisoning. In: Naidu R, Smith E, Owens G, Bhattacharya P, editors. Managing Arsenic in the Environment: From Soil to Human Health. Clayton, Australia: CSIRO Publishing; 2006; Feb 7; pp. 311–327.
52. Yedjou CG, Tchounwou PB. In-vitro cytotoxic and genotoxic effects of arsenic trioxide on human leukemia (HL-60) cells using the MTT and alkaline single cell gel electrophoresis (Comet) assays. Mol Cell Biochem. 2007; Jul; 301(1–2):123–130. doi: 10.1007/s11010-006-9403-4. Epub 2007 Jan 10. PMID: 17216127; PMCID: PMC3760965.
53. Rousselot P, Labaume S, Marolleau JP, Larghero J, Noguera MH, Brouet JC, Fermand JP. Arsenic trioxide and melarsoprol induce apoptosis in plasma cell lines and in plasma cells from myeloma patients. Cancer Res. 1999; Mar 1; 59(5):1041–1048. PMID: 10070961.
54. National Research Council of Canada (NRCC). Effects of Arsenic in the Environment. Ottawa: National Research Council of Canada Publication; 1978; pp. 1–349.
55. Morton WE, Dunnette DA. Health effects of environmental arsenic. Adv Environ Sci Technol. New York. 1994; 27:17–34.
56. Tchounwou PB, Centeno JA, Patlolla AK. Arsenic toxicity, mutagenesis, and carcinogenesis–A health risk assessment and management approach. Mol Cell Biochem. ; 255:47–55.
57. Khan MA, Ho YS. Arsenic in drinking water: A review on toxicological effects, mechanism of accumulation and remediation. Asian Journal of Chemistry. 2011 May 1;23(5):1889.
58. Tchounwou PB, Patlolla AK, Centeno JA. Invited reviews: Carcinogenic and systemic health effects associated with arsenic exposure—A critical review. Toxicol Pathol. 2003; 31(6):575–588.
59. Tchounwou PB, Wilson BA, Abdelghani AA, Ishaque AB, Patlolla AK. Differential cytotoxicity and gene expression in human liver carcinoma ($HepG_2$) cells exposed to arsenic trioxide, and monosodium acid methanearsonate (MSMA). Int J Mol Sci. 2002; Nov 30; 3(11):1117–1132.

60. Centeno JA, Gray MA, Mullick JG, Tchounwou PB, Tseng C. Arsenic in drinking water and health issues. Treatments and effects in ecology and human health. 2014 Sep;47(5):245–52
61. Abernathy CO, Liu YP, Longfellow D, Aposhian HV, Beck B, Fowler B, Goyer R, Menzer R, Rossman T, Thompson C, Waalkes M. Arsenic: Health effects, mechanisms of actions, and research issues. Environ Health Perspect. 1999; 107(7):593–597.
62. Goyer RA, Clarkson TW. Toxic effects of metals. In: Klaassen CD editor. Casarett and Doull's Toxicology: The Basic Science of Poisons. New York: McGraw-Hill; 1996; pp. 691–736.
63. Hughes MF. Arsenic toxicity and potential mechanisms of action. Toxicol Lett. 2002; 133(1):1–6.
64. Wang Z, Rossman TG. The Toxicology of Metals. Boca Raton: LW Chang; 1996.
65. Belton JC, Benson NC, Hanna ML, Taylor RT. Growth inhibitory and cytotoxic effects of three arsenic compounds on cultured Chinese hamster ovary cells. J Environ Sci Health Part A. 1985; 20(1):37–72.
66. Li JH, Rossman TG. Inhibition of DNA ligase activity by arsenite: A possible mechanism of its comutagenesis. Mol Toxicol. 1989; 2(1):1–9.
67. Jha AN, Noditi M, Nilsson R, Natarajan AT. Genotoxic effects of sodium arsenite on human cells. Mutat Res-Fundam Mol Mech Mutagen. 1992; Dec 16; 284(2):215–221.
68. Hartmann A, Speit G. Comparative investigations of the genotoxic effects of metals in the single cell gel (SCG) assay and the sister chromatid exchange (SCE) test. Environ Mol Mutagen. 1994; 23(4): 299–305.
69. Patlolla AK, Tchounwou PB. Cytogenetic evaluation of arsenic trioxide toxicity in Sprague–Dawley rats. Mutat Res/Genet Toxicol Environ Mutagen. 2005; Nov 10; 587(1–2):126–133.
70. Anderson DY, Yu TW, Phillips BJ, Schmezer P. The effect of various antioxidants and other modifying agents on oxygen-radical-generated DNA damage in human lymphocytes in the COMET assay. Mutat Res-Fundam Mol Mech Mutagen. 1994; 307(1):261–271.
71. Banu BS, Danadevi K, Jamil K, Ahuja YR, Rao KV, Ishaq M. In vivo genotoxic effect of arsenic trioxide in mice using comet assay. Toxicology. 2001; May 21; 162(3):171–177.
72. Tchounwou PB, Yedjou CG, Dorsey WC. Arsenic trioxide induced transcriptional activation and expression of stress genes in human liver carcinoma cells (HepG2). Cell Mol Biol. 2003; 49(7):1071–1079.
73. Zhao CQ, Young MR, Diwan BA, Coogan TP, Waalkes MP. Association of arsenic-induced malignant transformation with DNA hypomethylation and aberrant gene expression. Proc Nat Acad Sci. 1997; Sep 30; 94(20):10907–10912.
74. Liu Y, Guyton KZ, Gorospe M, Xu Q, Lee JC, Holbrook NJ. Differential activation of ERK, JNK/SAPK and P3/CSBP/RK map kinase family members during the cellular response to arsenite. Free Radic Biol Med. 1996; 21(6):771–781.
75. Ludwig S, Hoffmeyer A, Goebeler M, Kilian K, Häfner H, Neufeld B, Han J, Rapp UR. The stress inducer arsenite activates mitogen-activated protein kinases extracellular signal-regulated kinases 1 and 2 via a MAPK kinase 6/p38-dependent pathway. J Biol Chem. 1998; Jan 23; 273(4):1917–1922.
76. Trouba KJ, Wauson EM, Vorce RL. Sodium arsenite-induced dysregulation of proteins involved in proliferative signaling. Toxicol Appl Pharmacol. 2000; Apr 15; 164(2):161–170.
77. Vogt BL, Rossman TG. Effects of arsenite on p53, p21 and cyclin D expression in normal human fibroblasts—A possible mechanism for arsenite's comutagenicity. Mutat Res-Fundam Mol Mech Mutagen. 2001; 478(1–2):159–168.
78. Porter AC, Fanger GR, Vaillancourt RR. Signal transduction pathways regulated by arsenate and arsenite. Oncogene. 1999; 18(54):7794–7802.
79. Tully DB, Collins BJ, Overstreet JD, Smith CS, Dinse GE, Mumtaz MM, Chapin RE. Effects of arsenic, cadmium, chromium, and lead on gene expression regulated by a battery of 13 different promoters in recombinant HepG2 cells. Toxicol Appl Pharmacol. 2000; Oct 15; 168(2):79–90.
80. Lu T, Liu J, LeCluyse EL, Zhou YS, Cheng ML, Waalkes MP. Application of cDNA microarray to the study of arsenic-induced liver diseases in the population of Guizhou, China. Toxicol Sci. 2001; 59(1):185–192.
81. Roberts TL. Cadmium and phosphorous fertilizers: The issues and the science. Procedia Eng. 2014; 83:52–59.
82. Jiang JH, Ge G, Gao K, Pang Y, Chai RC, Jia XH, Kong JG, Yu AC. Calcium signaling involvement in cadmium-induced astrocyte cytotoxicity and cell death through activation of MAPK and PI3K/Akt signaling pathways. Neurochem Res. 2015; 40:1929–1944.
83. Richter P, Faroon O, Pappas RS. Cadmium and cadmium/zinc ratios and tobacco-related morbidities. Int J Environ Res Public Health. 2017; 14(10):1154.
84. Chunhabundit R. Cadmium exposure and potential health risk from foods in contaminated area, Thailand. Toxicol Res. 2016; 32:65–72.

85. Nishijo M, Nakagawa H, Suwazono Y, Nogawa K, Kido T. Causes of death in patients with Itai-Itai disease suffering from severe chronic cadmium poisoning: A nested case–control analysis of a follow-up study in Japan. BMJ Open. 2017; 7(7):e015694.
86. Batáriová A, Spěváčková V, Beneš B, Čejchanová M, Šmíd J, Černá M. Blood and urine levels of Pb, Cd and Hg in the general population of the Czech Republic and proposed reference values. Int J Hyg Environ Health. 2006; Jul 19; 209(4):359–366.
87. Baselt RC, Cravey RH. Disposition of Toxic Drugs and Chemicals in Man. Davis: Biomedical publications; 1982.
88. Baselt RC. Disposition of Toxic Drugs and Chemicals in Man. Fifth Edition. Foster City: Chemical Toxicology Institute; 2000.
89. Singhal RL, Merali Z, Hrdina PD. Aspects of the biochemical toxicology of cadmium. Fed Proc. 1976; 35(1):75–80.
90. Waalkes MP, Berthan G, editors. Handbook on Metal-Ligand Interactions of Biological Fluids. Vol. 2. New York: Marcel Dekker; 1995; pp. 471–482.
91. Waalkes MP, Misra RR, Chang LW, editors. Toxicology of Metals. Boca Raton: CRC Press; 1996; pp. 231–244.
92. Waalkes MP, Rehm S. Carcinogenicity of oral cadmium in the male Wistar (WFNCr) rat: Effect of chronic dietary zinc deficiency. Fundam Appl Toxicol. 1992; 19(4):512–520.
93. Stohs SJ, Bagchi D. Oxidative mechanisms in the toxicity of metal ions. Free Radic Biol Med. 1995; 18(2):321–336.
94. Mitra RS. Protein synthesis in *Escherichia coli* during recovery from exposure to low levels of Cd^{2+}. Appl Environ Microbiol. 1984; 47(5):1012–1016.
95. Rossman TG, Roy NK, Lin WC. Is cadmium genotoxic? IARC Sci Pub. 1992; Jan 1(118):367–375.
96. Smith JB, Dwyer SD, Smith L. Lowering extracellular pH evokes inositol polyphosphate formation and calcium mobilization. J Biol Chem. 1989; May 25; 264(15):8723–8728.
97. Thevenod F, Jones SW. Cadmium block of calcium current in frog sympathetic neurons. Biophys J. 1992; 63(1):162–168.
98. Dally H, Hartwig A. Induction and repair inhibition of oxidative DNA damage by nickel (II) and cadmium (II) in mammalian cells. Carcinogenesis. 1997; 18(5):1021–1026.

6 Various Sources of Heavy Metals Contamination, Including Industrial Activities, Mining, Agriculture, and Urbanization

Sumit Mishra and Asmita De*
Birla Institute of Technology, Mesra, Ranchi, India

6.1 INTRODUCTION

Heavy metals are primarily defined as elements that are characterized by their high atomic weight (>20) and high specific gravity (>5) (i.e., density greater than 5 g/cm^3) [1]. Metalloids, which exhibit the properties of both metals and non-metals, are sometimes included in the category of heavy metals [2]. Heavy metals can be categorized into three groups: (i) precious metals such as palladium (Pd), platinum (Pt), silver (Ag) and gold (Au); (ii) toxic metals such as mercury (Hg), zinc (Zn), chromium (Cr), nickel (Ni), cadmium (Cd) and lead (Pb); and (iii) radionuclides such as uranium (U), thorium (Th), radium (Ra) and americium (Am) [3]. Heavy metal contamination is defined as the existence of high concentration of heavy metals in the environment which includes air, water, soil and living beings, exceeding established permissible limits.

Although several heavy metals serve as essential nutrients for plants, animals and humans at low concentrations, higher levels of certain metals can be toxic, posing a threat to life [4, 5]. For example, Cu and Zn play a critical role in normal metabolism but are toxic at higher concentrations. Higher concentrations of Cu and Zn in the body cause damage and dysfunction in human beings [6].

The list of various essential and non-essential heavy metals is shown in Table 6.1.

TABLE 6.1
Classification of Essential Metals and Non-Essential Heavy Metal Ions

Category	Classification	Metals
Essential metals	Micronutrient metals	Cr, Cu, Co, Mn, Zn, Ni, Fe, Mo
	Macronutrient metals	K, Ca, Mg
Non-essential metals	Precious heavy metals	Pt, Pd, Au, Ag, Ru
	Extremely toxic heavy metals	Cd, As, Hg, Sn, Pb, Se, Tl
	Radionuclides metals	Th, Pr, U, Am, Ra

* Corresponding Author: sumitmishra1@gmail.com, smishra@bitmesra.ac.in

DOI: 10.1201/9781032685793-6

The widespread emergence of heavy metal contamination has posed a significant threat to both environment and human health. Human body can be exposed to heavy metals through four ways: by inhalation from atmosphere, by ingestion of contaminated food, by drinking contaminated water and by contact with skin in areas related to industries, manufacturing, mining, agriculture and pharmaceuticals. While heavy metals naturally exist in the environment through weathering, several anthropogenic activities like rapid advancements in agriculture and industries also significantly contribute to environmental degradation by releasing heavy metals. Wastewater discharged from various commercial industries such as textile, leather, tanning, electroplating, printing, nuclear power plants, mining and metal refineries contains traces of various heavy metals [7].

Heavy metals do not easily degrade within the environment and can persist for longer periods accumulating in soil and water [8]. Metals may undergo changes in their chemical forms (speciation) and bioavailability. The speciation of metals has significant effects on ecosystem, human health and the environment. Heavy metals tend to accumulate within living organisms, a phenomenon referred to as bioaccumulation [9]. This can result in the degradation of the food web or food chain. Heavy metals impact agricultural productivity and food safety. Consuming food contaminated with heavy metals can significantly reduce the essential nutrients in the body, leading to weak immune system, stunted growth and malnutrition [10].

Exposure to heavy metals in humans can cause harm to vital organs such as eyes, liver, lungs and kidneys as well as compromise biological systems such as nervous system, immune system, respiratory system and reproductive system. This can ultimately lead to the development of various health conditions such as nausea, anaemia, organ failure, cancer, reduction in bone mineral density and the induction of tumours [11]. Heavy metal ions can also accumulate within the body cells and interact with proteins and nucleic acids causing damage to DNA and biological functions [12]. Thus, understanding and preventing heavy metal contamination is imperative due to its far-reaching consequences.

In recent years, various remediation methods including physical, chemical and biological approaches have been employed to address heavy metal-contaminated water, soil and sediments. These methods include membrane separation, adsorption, distillation, ion-exchange, electrokinetic, electrodialysis and chemical precipitation [13, 14]. However, these methods have some disadvantages like less efficiency, high maintenance costs, large energy consumption and chemical sludge accumulation [15].

This chapter focuses on heavy metal pollution and various sources by which heavy metals are introduced into our environment, their impact and control measures.

6.2 SOURCES OF HEAVY METALS CONTAMINATION

Environmental chemistry suggests the occurrence of heavy metals in the form of oxides, hydroxides, silicates, carbonates, phosphates, sulphides, sulphates and other organic matrices [16]. Heavy metal contamination can arise from both natural and anthropogenic sources (Figure 6.1).

6.2.1 Natural Sources

Heavy metals naturally occur in the Earth's crust, with approximately 95% of the crust being composed of igneous rocks and the remaining 5% comprising sedimentary rocks [17]. Basaltic igneous rocks are abundant in heavy metals like Co, Ni, Cu and Cd, whereas sedimentary rocks contain significant quantities of Mn, Cu, Pb, Zn and Cd [1]. The process of soil formation from lithogenic sources also leads to the accumulation of heavy metals in the environment. Volcanic eruptions, weathering of rocks, geothermal activity, natural fires, biological activities, erosion and sedimentation also contribute to heavy metal contamination [18].

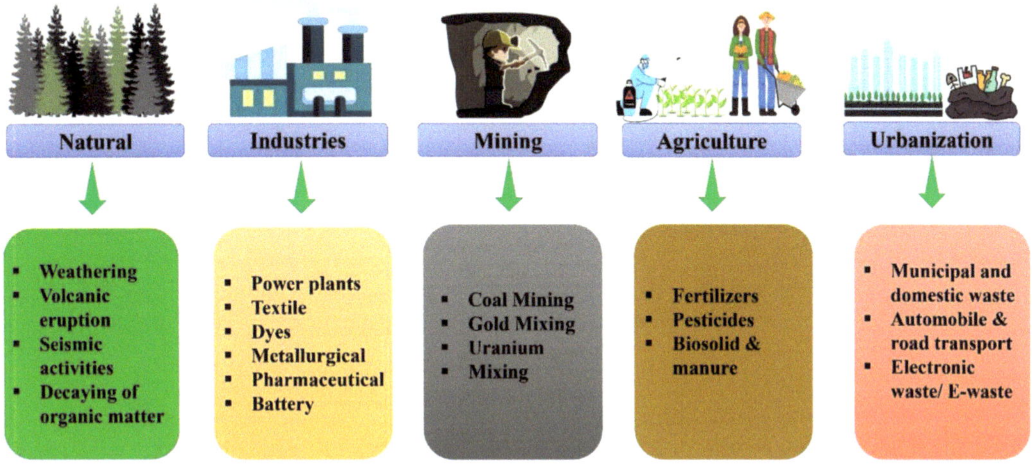

FIGURE 6.1 Sources of heavy metal contamination.

6.2.1.1 Weathering

Weathering is the gradual decomposition of rocks, soil and minerals through exposure to the atmosphere, water and living organisms. Weathering can occur through physical, chemical and biological processes. Physical weathering involves the disintegration of minerals containing heavy metals through direct exposure to climatic conditions like heat, water, ice and pressure [19]. In contrast, chemical weathering involves the interaction of rock chemicals with microclimatic conditions, resulting in heavy metal release from rock into the surroundings. Physical and chemical weathering can happen simultaneously or sequentially.

Biological weathering involves the decomposition of rocks and soil molecules by flora and fauna. These organisms seek nutrients for growth or for their survival. The disintegration and weathering of rocks are also caused by the fracturing of rocks by plant roots. Minerals derived from weathered rocks consist of various heavy metals that are harmful to the surroundings [20].

6.2.1.2 Volcanic Eruptions

A volcanic eruption is a natural phenomenon that results in the release of particles to the Earth's surface. Volcanic ash is the most widely distributed outcome of volcanic eruptions, which can spread to about a hundred kilometres distance from the erupting volcano [21]. The ash contains heavy metals like Cu, Pb, Cr, Fe, Al, Zn and Cd that pose toxicity risks to the environment and living beings [22].

6.2.1.3 Seismic Activities

Earthquakes and other seismic activities contribute to the release of heavy metals into the environment. Due to earthquake, the underground geology gets fractured, leading to the exposure of materials that are susceptible to oxidation under oxidizing conditions [23]. Incidences of rainfall or underground leakages percolate water through these fractures present in the soil, resulting in the release of heavy metals into the surroundings [24, 25].

6.2.1.4 Degradation of Organic Matter

Living organisms take in metals as essential nutrients, which are vital for their growth. Upon death and decomposition of these plants, the contaminants are reintroduced to the soil, contributing to the release of metals [26]. The commonly encountered heavy metals include Cr, As, Pb, Zn, Cu, Hg and Cd [27]. These metals have the potential for bioaccumulation and biomagnification in the food chain due to their presence in the body matrices of decomposing organic matter [28, 29].

6.2.2 Anthropogenic Sources

The rise in socio-economic activities resulting from population increase has contributed to a swift surge in heavy metal levels in the ecosystem. During the Industrial Revolution, metals were extracted from natural sources and subjected to industrial processing resulting in the emission of heavy metals into the environment. Heavy metals are also found in areas such as mining, smelters and foundries. Other sources of heavy metals in the environment include domestic activities, agriculture and pharmaceuticals.

6.2.2.1 Industrial Activities

Industrial activities are one of the main contributors of heavy metals in the environment.

6.2.2.1.1 Power Plants

Fossil fuel and nuclear power plants facilitate release of considerable thermal pollution into the water body, leading to alterations in the water temperature [30]. The Environment Protection Agency (EPA) reports that the thermoelectric power plants are alone responsible for substantial portion accounting for 50–60% of all toxic pollutants discharged into water [31]. Of all the different types of thermoelectric generating units, coal-fired plants stand out as the largest contributor of toxic pollutants. Coal-based power generation represents a significant point source of heavy metal pollution. Thermal power plants produce a significant quantity of fly ash as solid coal combustion residue. Fly ash contains various toxic heavy metals like Mn, Ni, Pb, Hg, Cd, Cr and Zn. The heavy metal concentration found in fly ash depends upon the type of coal (anthracite, bituminous, lignite) utilized and the combustion conditions [32]. Fly ash is reused and utilized in the construction sector. However, the production rate exceeds its consumption rate. Thus, a considerable amount of ash is disposed of in settling ponds with effluent outlets that enter nearby water bodies [33]. If there is no lining in the ash pond, a significant quantity of leachate can percolate down the water aquifers. It can lead to the contamination of water table through the accumulation of metals. Power plant industries produce substantial quantities of ash which may contain high concentration of many heavy metals.

6.2.2.1.2 Textile and Dye Industry

The effluents released by various industries like textile, dye, paper and pulp, paint and tannery consist of a diverse range of organic pollutants. Among these, textile industries contribute the largest share of 54%, generating half of the global dye effluents, followed by dyeing industries at 21%, paper and pulp industries at 10%, tannery and paint industries at 8% and dye manufacturing industries at 7% [34]. The textile manufacturing sector is a vast global industry that produces a substantial volume of wastewater. Textile industry produces approximately 10,000 distinct textile dyes according to colour index, leading to the global production of around 7,00,000 tons of dyes [35]. India ranks among the top textile producers, making a 5% contribution to its gross domestic product (GDP). Other textile-producing countries are the United States, China, Brazil, Taiwan, Pakistan, Turkey, South Korea and Bangladesh. Textile effluents comprise dyes and toxic heavy metals such as Cu, Cr, Ni, Zn, Cd, Hg, Pb and As. The heavy metals can originate from four primary dyes: chrome acid dyes, reactive metal containing dyes, metal complex acid dyes and vat dyes [34]. Additionally, heavy metals can be introduced through chemical treatments applied to fibres to enhance their resistance to light and moisture, as well as during bleaching of fabric where metals serve as catalysts for oxidizing agents [36].

China is the world's largest producer of pulp and paper currently. India is positioned as the 20th country in the list of paper production. According to the Ministry of Environment and Forests, India (MOEF), the paper and pulp industry ranks sixth most polluting industry among 17 polluting industries [37]. Annually, the worldwide production and consumption of paper is approximately 400 million metric tons [38]. The primary challenge faced by the pulp and paper industry is the extensive use of water and the release of a large quantity of wastewater which contains colour, chlorine, heavy metals and sulphur. In the process of paper manufacturing, black liquor, a liquid comprising 70% solid, is introduced into the digester [39]. Black liquor consists of several heavy metals like As, Cd, Hg, Cr and Pb.

6.2.2.1.3 *Metallurgical Industry*

Metallurgy refers to the extraction of metals from ores using various processing methods like mining, cleaning, smelting and rolling. After mining operations, the minerals are transported to metallurgical house where minerals are extracted and purified for subsequent processing. During this procedure, various byproducts known as residues are generated which include heavy metals.

Aluminium (Al) is one of the most extensively used non-ferrous metals globally. Al is mixed with other metals to form alloys that are strong and light in weight. The Bayer process is commonly used for extraction of aluminium from bauxite to generate aluminium or alumina [40]. Bauxite comprises only 30–40% Al_2O_3 (alumina), with remaining content consisting of a combination of silica, different iron oxide and TiO_2 (titanium dioxide) [41]. Purification of Al_2O_3 is necessary before it undergoes the process of refining into aluminium metal. The complete procedure of aluminium extraction from bauxite generates numerous hazards that can adversely impact living beings. Red mud is a hazardous solid waste byproduct produced during the industrial process of producing alumina through bauxite ore digestion. Red mud usually comprises Fe_2O_3 (30–60% w/w), Al_2O_3 (10–20% w/w), SiO_2 (3–50% w/w), Na_2O (2–8% w/w), CaO (2–8% w/w) and TiO_2 (trace-25% w/w) [42]. The red mud contains various minor heavy metals like Mn, Pb, P, As, Ba, Cr, K and V. [43].

Copper (Cu) is the second most utilized non-ferrous metal after aluminium (Al) [44]. As per the data from the United States Geological Survey (USGS), the worldwide copper production in 2019 accounted to approximately 20 million tons. According to USGS, the annual mining capacity of Chile was recorded 5.6 million tons, followed by Peru at 2.4 million tons and China at 1.6 million tons in 2020 [45]. Copper production is a multifaceted and multi-step process that comprises mining, crushing, milling, ore concentration, smelting, refining and waste management. Copper ore mining sector produces 10 billion tons of byproducts, including dusts, slags and aerosols containing diverse metals and metalloids such as Pb, Cd, Hg and As [46]. Copper smelters serve as a source of As, Cd and Pb which are frequently dispersed by wind and can penetrate deeply into the human respiratory system [47].

Steel manufacturing industries emit a diverse range of harmful chemicals, which cause severe hazard to human beings and the environment [48]. The blast furnace operation releases suspended solids, cyanides, fluorides and some heavy metals like Zn and Pb [49]. The combustion of coal for energy generation results in the release of toxic effluents like V, As, Hg, Cd and Pb. Steel slag is a residual waste or byproduct originating from steel manufacturing predominantly composed of lime (CaO), silica (SiO_2), alumina (Al_2O_3), iron oxide (FeO, Fe_2O_3), manganese oxide (MnO_2), magnesia (MgO) and phosphorus oxide (P_2O_5) and several toxic heavy metals like Zn, Ni, Cd, As and Pb [50]. Similarly, the process of steel refining leads to effluents with elevated levels of suspended solids, oils and heavy metals like Cu, Cr, Ni and Pb [51].

6.2.2.1.4 *Pharmaceutical Industry*

Pharmaceutical effluents refer to the waste produced by pharmaceutical industries during drug manufacturing. The increasing demand for pharmaceuticals has led to the generation of numerous manufacturing units worldwide. During the synthesis of drugs, active pharmaceutical ingredients (APIs) are released into the environment which shows negative effects on living organisms [52]. It is estimated that approximately half of the pharmaceutical industry's global waste is directly discharged without the necessary treatment. Pharmaceutical effluents contain a variety of chemicals, including inorganic and organic substances, catalysts, spent solvents and heavy metals like Cr, Fe, Co, Ni and Cd [53]. These effluents have the potential to harm surface water, groundwater and soil with adverse consequences for both human health and the environment [54].

6.2.2.1.5 Battery Industry

Lithium-ion batteries have gained prominence due to their high energy and long product life [55]. Inadequate recycling or improper disposal of batteries and accumulators commonly found in portable devices, automobiles and industrial applications may result in heavy metal contamination. Battery waste usually contains Pb, Hg, Cd and Ni and improper disposal and handling can result in harmful environmental and health effects.

6.2.2.2 Mining

The mining industry is one of the most significant contributors to the development and economic growth of various countries. The mining sector extracts precious minerals and various geological materials from the Earth's crust [56]. Mining is one of the significant contributors of heavy metal contamination. At present, the mining process consists of open-pit and underground mining operations, along with on-site mineral processing. The ores undergo a sequence of operations, including crushing, grinding and leaching with sulphuric acid. Subsequently, ion-exchange process is applied, followed by precipitation. Mining activities along with grinding, ore crushing and disposal of tailings in open areas are dispersed by wind and floodwater leading to environmental contamination [57]. In mining regions, the acceleration of heavy metal release is attributed to the increased oxidation rates. This acceleration occurs because of providing greater air accessibility through mining operations, waste rock and tailings facilitated by mineral processing [57].

Water pollution is a significant issue in mining operations. Large volume of water released from the mine drainage, aqueous extraction and other mining processes elevate the risk of chemical contamination in both ground and surface water [58]. Surface water quality primarily depends upon accidental chemical spills, waste material erosion and the release of contaminated water from mines [59]. The acid mine drainage can vary acidity, basicity or neutrality based on the weathered minerals. Acid mine drainage phenomenon, also known as acid metalliferous drainage (AMD), is identified by a pH level ranging from 1 to 4 [60].

AMD is considered a significant challenge in mining regions. The oxidation of iron pyrite (FeS_2) and other disulphide results in the generation of acidic water [61]. The generation of AMD from pyrite can be described through three fundamental chemical reactions: iron hydrolysis, pyrite oxidation and ferrous oxidation [62]. Iron pyrite (FeS_2) represents the predominant sulphur mineral in coal and gold lithology.

In the course of pyrite (FeS_2) oxidation, sulphur undergoes oxidation to sulphate (SO_4^{2-}), while ferrous (Fe(II)) ion is liberated.

$$2FeS_2 + 7O_2 + 2H_2O \rightarrow 2Fe^{2+} + 4SO_4^{2-} + 4H^+$$

The Fe(II) reacts with O_2 and H^+ to produce ferric (Fe(III)) ions and water (H_2O).

$$4Fe^{2+} + O_2 + 4H^+ \rightarrow 4Fe^{3+} + 2H_2O$$

The hydrolysis of Fe(III) results in the production of $Fe(OH)_3$ consequently forming additional moles of acidity.

$$4Fe^{3+} + 12H_2O \rightarrow 4Fe(OH)_3 + 12H^+$$

Similarly, other metals like Cu, Zn and Al may also experience hydrolysis depending on pH. The acidity of AMD results in metal leaching from nearby rocks and lithologies. This process contaminates the surrounding environment with Mn, Fe, Al and several harmful heavy metals such as Cu, Hg, Cd, Zn and Pb.

6.2.2.2.1 Coal Mining

Coal is a non-renewable fossil fuel with approximately 100–500 million years old remains [63]. About 36% of the world's energy requirements rely on coal. Opencast coal mines lead to more pollution of soil, water and the environment. Underground coal mining on the other hand causes risk of land subsidence and fire accidents in coal storage [64]. Coal mining leads to the generation of large amounts of metal dust from coal extraction, soil dumping and transportation in the surrounding mine areas. Coal dust consists of various heavy metals like Pb, Zn, Mn, Fe, Cr, Zr, As, Cu and Sr which mixes with the topsoil of the mining areas [65]. Surface runoff from the coal mine spoil dumps mobilizes heavy metals in the topsoil, causing them to leach into the subsurface soil [66].

6.2.2.2.2 Gold Mining

Gold mining industries generate waste through three primary activities: mining, mineral processing and metallurgical extraction. The primary byproducts generated by extraction of gold are tailings which contain substantial concentration of heavy metals. Gold mining involves either open-pit operations or deep shaft mining often accompanied by the extraction of other heavy metals such as Pb, Cu and Ag. Gold mine tailings are characterized by high levels of toxic heavy metals, including Ni, Cu, Zn, As, Cd, Pb, Co and Hg [67]. Water permeating through sulphide containing tailings results in the leaching of substantial quantities of heavy metals and sulphate ions into the streams and rivers [68]. Surface water and groundwater are significantly impacted by the presence of tailings and rock dumps. Chemicals employed in the gold extraction process represent an additional contributor to environmental pollution from gold mines. In 2011, artisanal and small-scale gold mining (ASGM) utilized approximately 1400 metric tons of mercury, with an annual average discharge of 1000 metric tons of inorganic mercury [69]. About one-third of this calculated quantity is released into the air, while the remainder becomes distributed among tailings heaps, soil and waterways. The reprocessing of certain old gold tailings dumps may result in the emission of Hg in the surrounding.

6.2.2.2.3 Uranium Mining

Uranium (U) is extensively used as a primary source of fuel for nuclear energy [70]. While uranium mining contributes to the economic development of society, it simultaneously leads to pollution. In the last 60 years, a large quantity of uranium tailings has been generated through uranium mining and processing [71]. This uranium tailings slag is accumulated in an open pit within the uranium tailings dam for disposal. The leachable and the mobile toxic metals present in the slag can readily infiltrate the surrounding ecological environment by natural processes or human activities leading to pollution of the environment [72]. Uranium tailings retain approximately 85% of the original ore radioactivity because uranium is primarily extracted leaving other heavy metals as waste [73]. The waste includes long-lived radionuclides such as radium (Ra) and thorium (Th) isotopes along with non-radioactive heavy metals such as Cr, Pb, Cd, As and Hg leading to heavy metal pollution.

6.2.2.3 Agriculture

Agroecosystems are typically influenced by a variety of pollutants, including agricultural contaminants that are the result of both biotic and abiotic byproducts of farming activities. The most prevalent agricultural sources of heavy metals include pesticides, fertilizers, bio-solids and sewage sludge [74].

6.2.2.3.1 Fertilizers

Fertilizers provide a variety of essential nutrients that boost plant growth and productivity while also enriching the organic content of the soil. Chemical fertilizers, especially inorganic fertilizers, comprise nitrogen (N), phosphorus (P) and potassium (K) and other mixed fertilizers. The excessive use of these fertilizers results in the nutrient leaching and pollution of various water sources, causing eutrophication [75]. This is widespread in developing countries where extensive agricultural practices are dominant. Phosphorus-based fertilizers have the highest concentration of heavy metal contaminants. For instance, superphosphate fertilizers may include contaminants such as Co, Cr, Cu, Pb, Ni, Cd and Zn [76, 77]. The consistent use of phosphorus-based fertilizers has shown a persistent increase in the Cd levels in the soil [78]. Cd is very harmful to plants because of its high solubility in soil solution. Besides phosphorus fertilizers, fertilizers comprising zinc sulphate, iron sulphate and copper sulphate also contain heavy metal contaminants like Pb [79, 80].

6.2.2.3.2 Pesticides

Pesticides are chemical compounds formulated to kill or inhibit the growth of various pests, herbicides are specially created for targeting weeds, fungicides for combating fungi and insecticides for controlling insects [81]. Pesticides hold a significant role in global agriculture, and it has been estimated that the absence of these chemicals can result in a nearly 40% reduction in the world's food production [82]. In recent times, the global consumption of pesticides has reached approximately 2 million tons, with herbicides accounting for 47.5%, insecticides for 29.5%, fungicides for 17.5% and the remaining 5.5% allocated to other types of pesticides [83, 84]. Until the late nineteenth century, copper-based fungicides, such as Bordeaux mixture ($CuSO_4 + Ca(OH)_2$), were employed in viticulture [85]. Ongoing use of copper compounds like $Cu(OH)_2$, Cu_2O, $CuCl_2$ and $CuSO_4$ has resulted in copper accumulation in the soil of grapevine plantation in Brazil, Spain, Croatia and France. Lead arsenate, an insecticide and phenylmercuric acetate, a fungicide, are among the numerous pesticides utilized in plantation of fruit in Australia and New Zealand [32]. These pesticides consist of several toxic heavy metals like Hg, As and Pb which leads to environmental contamination upon exposure. Pesticides contain several heavy metals, including Cu, Pb, Hg, Cr and Zn [86].

6.2.2.3.3 Bio-Solids and Manures

Bio-solids refer to organic materials found in sewage sludge, animal waste and industrial byproducts, including pulp sludge generated in paper production [87]. These solids are employed for soil improvement due to their nutrient content and their capacity to enhance soil properties such as water retention, fertility, porosity and pH [88]. Garden waste is also considered as a source of organic bio-solids, comprising additional materials such as sawdust. Bio-solids not only carry nutrients for enhancing soil fertility but also consist of various toxic heavy metals like Cr, Ni, Cu, Zn, Cd, Pb, Se, Sb, Mo and Hg [10]. This can lead to a gradual buildup of these metals in the longer period in the soil. It has been noticed that animal feeds are supplemented with organic and metallic products for growth promotion. Heavy metals like Zn, Cu, As and Cd are artificially added to commercial feeds with the aim of enhancing growth of animals and improve disease resistance [89, 90]. However, these animals are unable to digest these heavy metals and excrete them through manure [91]. Since heavy metals are non-degradable, they do not break down during the composting process [92]. Hence, prolonged and repeated usage of manures and compost may lead to the buildup of toxic heavy metals in agricultural soil, impacting crop health and productivity [93, 94].

6.2.2.4 Urbanization

Urbanization is the migration of individuals from rural areas to cities. The global urban population is expanding swiftly, with over half of the world's population (54%) currently residing in urban areas. The ongoing surge in urbanization is projected to reach 60% by the year 2050 [95]. From 2007 to 20250, the United Nations anticipates a global population increase from 6.7 billion to 9.2 billion. Concurrently, the global urban population is expected to almost double, growing from 3.3 billion to 6.4 billion [96]. This expansion will absorb not only the overall population growth but also the migration inflows from rural areas.

Figure 6.2 shows the population living in urban areas worldwide from 1950 to 2030.

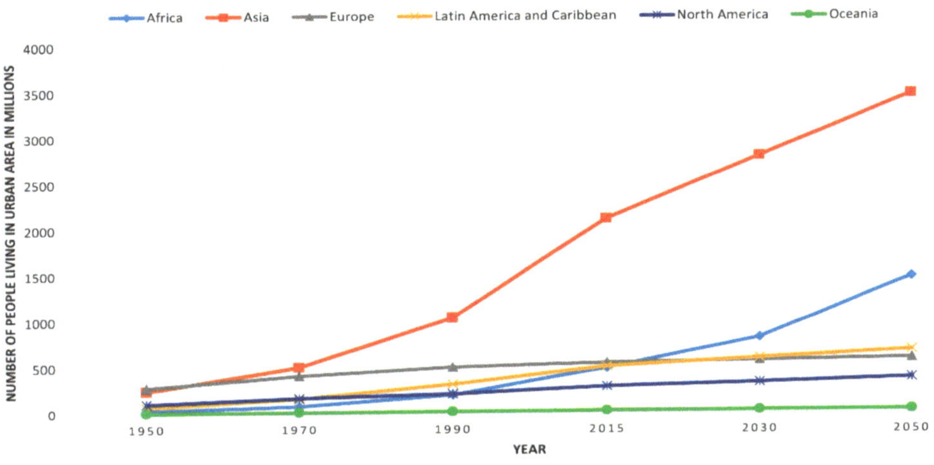

FIGURE 6.2 Population living in urban areas worldwide from 1950 to 2050. [See Ref. 97].

Urbanization is a multifaceted and continually changing process. The fast pace of urbanization leads to several environmental challenges and various forms of pollution that evolve with time and technology. The rise in urban population requires extra space for housing, businesses, roads, parking areas and vital infrastructure to meet the expanding requirements. Urban and economic growth significantly pollutes the environment by releasing several toxic heavy metals [98]. The overexploitation of natural resources to meet the needs of an unsustainable global development model has made the world more susceptible to shortages and deficiencies [99]. Urbanization results in the emission of heavy metals into the surroundings primarily due to vehicle exhausts, household septic tanks and manufacturing processes [100]. Progress in infrastructure, including buildings, roads and railways, has played a substantial role in releasing heavy metals into the environment.

6.2.2.4.1 *Municipal and Domestic Wastes*

Densely populated urban areas utilize large amounts of material and generate large amounts of solid waste like municipal wastes and domestic wastes [101]. The improper disposal of solid waste through burial, open dumps, landfills, incinerators is the main problems associated with urbanization [102]. The municipal solid wastes such as paper, metal scraps, food waste, ceramics, glass and ashes are usually disposed of either by incineration or landfilling. A common problem associated with the improper waste disposal practices is the occurrence of leachate from waste dumping sites. Open dump sites are the primary contributor to heavy metal contamination. Municipal wastewater contains heavy metals like Cd, Hg, Zn, Cr, Cu, Ni and Pb [103, 104]. The leachates carrying heavy metals seep from open dump sites to ground or surface water. When this interacts with the human food chain through vegetation or animals, it may lead to the phenomenon 'pollution magnification' [105, 106].

Domestic waste contains several metals which contribute to sewage sludge [107]. These metals originate from various sources such as faecal matter, detergents and cosmetics [108]. Shampoos and washing powder are found to contain Zn, Cd, Cr and Pb [108].

6.2.2.4.2 *Automobiles and Road Transport*

Urbanization of an area refers to the development of well-connected transportation routes. Roads have a major role in driving the social and economic development of a country. While road transportation accelerates a nation's development, it simultaneously introduces a substantial quantity of hazardous pollutants into the surroundings. Reports indicate that the automobiles contribute to over 40% of the overall air pollution, with this figure escalating to 70–80% in megacities of developing nations [109]. Roads and automobiles are currently recognized as major contributors to the release of heavy metals. Vehicles primarily release two types of pollutants: (i) gaseous pollutants and (ii) particulate pollutants. Gaseous pollutants are primarily emitted from the exhaust of fuels. Diesel and gasoline are the primary fossil fuels utilized to power most vehicles. The burning of diesel releases significant amounts of heavy metals such as Cd, As, Cu, Cr, Hg, Ni, Pb, Se and Zn. Gasoline consists of metalloids and metals such as Zn, Cu, Pb, Fe, As, Cd, Se, Ti, Pb and Ni [110]. Particulate matter (PM) emitted from vehicles in both solid and liquid form constitutes various inorganic and organic pollutants along with several toxic heavy metals [111]. Pb, Cu and Zn emerge as three prevalent heavy metals discharged during road journeys, constituting at least 90% of the total metals preset in road runoff. Various metals such as Ni and Cd are present in lower amounts in road runoff and automobile exhaust.

Brakes and tyres are important components of automobiles, where brakes are employed to halt vehicles, while tyres are used for acceleration. The friction between the road surface and vehicle components during braking and acceleration is responsible for the wear and tear of both brake pads and tyres. Tyres are primarily composed of rubber, which is fitted around metallic frame, while brake pads consist of a combination of metallic materials. Reports have shown that tyre dust contains various heavy metals like Fe, Cu, Cd, Co, Hg, Ni, Pb, Zn and Mo, while brake pads emit Fe, Sb, Sn, Ti, Cu, Ca, Cr, Mo, Zr, K and Co [112, 113]. Zn is the most abundant heavy metal released due to tyre wear [114].

6.2.2.4.3 Electronic Waste/E-Waste

Electronic devices comprise televisions, smartphones, computers, printers, refrigerators and various other electronic equipment. The rapid progress of technology and the rising consumer demand have led to global concern regarding E-waste. The uncontrolled disposal and improper handling of E-waste pose significant risk to humans and the environment [115]. E-wastes are widely recognized as a primary origin of heavy metals, dangerous substances and cancer-causing agents. E-wastes consist of various heavy metals like Hg, As, Ni, Cd and Pb [116]. Effective management and disposal of E-waste can help prevent various diseases linked to skin, digestive, respiratory, immune, endocrine and nervous system. To address the digital divide, there has been a significant increase in the utilization of electrical and electronic equipment (EEE) [117]. However, improper disposal of information and communication technology (ICT) waste can have severe repercussions for both environment and human beings. It is imperative to urgently adopt and enforce existing policies and guidelines in alignment with the international standards and policies to establish a proper E-management system for the well-being of both humans and the environment.

6.3 CONSEQUENCES OF HEAVY METAL CONTAMINATION ON HUMAN BEINGS AND THE ENVIRONMENT

Heavy metals can be highly toxic, sometimes carcinogenic and cause several health risks to living beings [133]. If the heavy metal concentration surpasses the levels set by the World Health Organization (WHO), it will lead to toxicity in both soil and aquatic systems [134]. Table 6.2 shows the permissible threshold of heavy metals concentration in drinking water as reported by various organizations. The heavy metal presence in soil degrades the quality of food, groundwater, growth of vegetation and agriculture friendly microorganisms [135]. These heavy metals circulate in the environment through food chain [136]. High concentrations of heavy metal contamination through food chain have the potential to induce significant stress on both human and the environment [137].

Arsenic (As) exists in inorganic forms arsenite and arsenate which are fatal to living beings. Short-term exposure to arsenic can lead to several health issues in the human body such as vomiting, nausea, confusion, hallucinations, diarrhoea and cardiac arrhythmia [138]. Conversely, long-term exposure to arsenic through food and water causes hyperpigmentation and cancer in skin,

TABLE 6.2
Permissible Threshold of Heavy Metals in Parts per Billion (ppb) in the Drinking Water as Reported by Various Foundations

Heavy Metals	WHO	BIS	USEPA	ECE	ADWG	NOM-127
Mercury (Hg)	6	–	2	1	1	1
Arsenic (As)	10	10	10	10	10	25
Chromium (Cr)	50	50	50	100	50	50
Lead (Pb)	10	10	15	10	10	10
Cadmium (Cd)	3	3	5	5	2	5
Zinc (Zn)	–	5000	500	–	3000	5000
Copper (Cu)	2000	1500	1500	–	–	–
Nickel (Ni)	702	20	20	–	20	–
Antimony (Sb)	20	–	6	5	3	–
Iron (Fe)	–	–	300	200	300	300
Silver (Ag)	–	–	100	–	100	–
Uranium (U)	30	–	30	–	17	–
Thallium (Tl)	–	–	2	–	–	–

Abbreviations: ADWG, Australian Drinking Water Guidelines; BIS, Bureau of Indian Standards; ECE, European Commission Environment; NOM-127, Norma Official Mexicana; USEPA, United States Environmental Protection Agency; WHO, World Health Organization

TABLE 6.3
Heavy Metals, Their Oxidation States, Sources and Effect on Human Beings

Metal	Oxidation States	Stable Form	Source	Effects	References
Mercury (Hg)	Hg^0 to Hg^{+2}	Hg(I), Hg(II)	Mining, thermal power plants, electrical appliances, fluorescent light lamps, metal coating, medical waste, fossil fuel combustion and fungicides	Autoimmune disease, nerve cell toxicity, vision loss, hair loss, dementia, decreased fertility, insanity, paralysis, deafness	[118, 119]
Arsenic (As)	As^{-3} to As^{+5}	As(III), As(V)	Thermal power plants, smelting operations, fossil fuel combustion and chemical pesticides	Cancer (liver, kidney, skin, bladder), cardiovascular diseases, kidney diseases	[120, 121]
Chromium (Cr)	Cr^{-2} to Cr^{+6}	Cr(III), Cr(VI)	Mining, tannery, fly ash, waste batteries, zinc smelting, dye and paint sludge, fertilizers and chromate synthesis process	Anaemia, allergies, gastrointestinal ulcers, hair loss, lung cancer, dermatitis, eczema, gingivitis, sinusitis, toxic nephritis, liver damage	[122, 123]
Lead (Pb)	Pb^{-4} to Pb^{+4}	Pb(II), Pb(IV)	Mining, fossil fuel combustion, smelting operations, ceramic industry, electroplating, thermal power plants, lead acid batteries, paints, ceramics and nuclear reactors	Inhibits brain development in children, memory loss, cardiovascular diseases, high blood pressure, growth retardation	[124]

(Continued)

TABLE 6.3 *(Continued)*
Heavy Metals, Their Oxidation States, Sources and Effect on Human Beings

Metal	Oxidation States	Stable Form	Source	Effects	References
Cadmium (Cd)	Cd^0 to Cd^{+2}	Cd(II)	Industrial waste disposal, dye industry electroplating, zinc smelting, paint sludge, fertilizers, battery and metal plating	Kidney and lung damage, cadmium pneumonitis, bone disease, infertility caused by reproductive system failure, DNA impairment, cancer, immune system deficiencies, central nervous system complications	[125, 126]
Zinc (Zn)	Zn^0 to Zn^{+2}	Zn(II)	Fossil fuel combustion, smelting, electroplating, industrial wastewater, mining, cosmetic products, pharmaceuticals	Memory loss, inhibits metabolic pathways, anaemia, prostate cancer, fatigue, stomach cramps, epigastric pain, impaired immune function, neutropenia	[127, 128]
Copper (Cu)	Cu^{-2} to Cu^{+4}	Cu(I), Cu(II)	Mining, smelting operations, electroplating and electronic manufacturing	Liver and kidney damage, anaemia, insomnia, restlessness, fatigue, jaundice, demyelination, hepatic cirrhosis, necrosis, insomnia, dizziness	[129, 130]
Nickel (Ni)	Ni^{-2} to Ni^{+4}	Ni(II)	Batteries, printing industry, fossil fuel combustion, smelting applications, power generation, waste incineration and car emission	Cancer, skin allergy, lung embolisms, reproductive disorder, hair loss, asthma, dizziness	[127, 128]
Molybdenum (Mo)	Mo^{-4} to Mo^{+6}	Mo(IV), Mo(VI)	Spent catalyst	Anorexia, headache, decrease in appetite, gout, testicular atrophy, hypochromic microcytic anaemia	[131]
Vanadium (V)	V^{+2} to V^{+5}	V(V)	Sulphuric acid plant, spent catalyst, oil fired power plants	Nose and eye irritations, throat pain, kidney haemorrhage, nausea, nervous system impairment, liver, cardiac disease, paralyses, behavioural changes and gastrointestinal organs inflammation	[132]

kidney, liver and lungs [139]. Additionally, arsenic also causes depression, sleeping disorders, headaches and numbness. Arsenic can also lead to reduced birth weight.

Antimony (Sb) closely resembles arsenic in terms of toxicity. Sb in +3 oxidation state is more carcinogenic than Sb in +5 state [140]. Inhaling Sb can cause respiratory system issues and several carcinogenic diseases. Higher intake of Sb leads to liver damage, miscarriage and premature delivery.

Cadmium (Cd) is found naturally in sedimentary rocks and marine phosphates. Exposure to low levels of cadmium causes headache, vomiting, cough and nausea. Conversely, high levels of cadmium can impact liver and kidney function and cause hypertension [141]. Cadmium has the ability to replace calcium in bones, leading to painful bone disorders. Prolonged exposure to Cd can result in chronic anaemia.

Mercury (Hg) is a silvery white colour metal that forms bond with both organic and inorganic compounds making it highly soluble in aquatic system. Mercury can enter the body through consumption of fish, food and water [142]. Persistent exposure to mercury can cause kidney disease, skin problems, damage to the circulatory system and nervous system. Mercury can also lead to rheumatoid arthritis.

Lead (Pb) is released in the atmosphere from automobiles and industries. The effects of lead exposure, whether short term or long term, have greater impact on children than on adults. Consuming lead through food and water can result in several health issues like kidney disease, damage to the brain, circulatory and nervous systems. Lead intake exceeding permissible limit is carcinogenic for humans. Pregnant individuals exposed to lead may face complications such as miscarriage, premature delivery, stillbirth and low birth weight [143].

Zinc (Zn) is considered a vital element and plays a crucial role in various metabolic processes, cell differentiation, embryonic development and cell proliferation. However, the consumption of excessive doses of zinc is perilous and can result in symptoms such as depression, increased thirst, lethargy and neurological issues [144].

Chromium (Cr) is a highly toxic element and exists in 0, +3 and +6 valence states. Both chromium in its Cr(III) and Cr(VI) forms are carcinogenic to humans. However, Cr(VI) is more dangerous than Cr(III) due to its oxidizing property. Prolonged exposure to Cr can harm the cardiovascular, respiratory, hepatic and gastrointestinal systems [145].

The effect of heavy metal contamination in human body is shown in Figure 6.3. The details of various heavy metals, their oxidation states, sources and their effect on human beings is shown in Tables 6.3 and the the various toxicity levels of heavy metals is shown in Table 6.4.

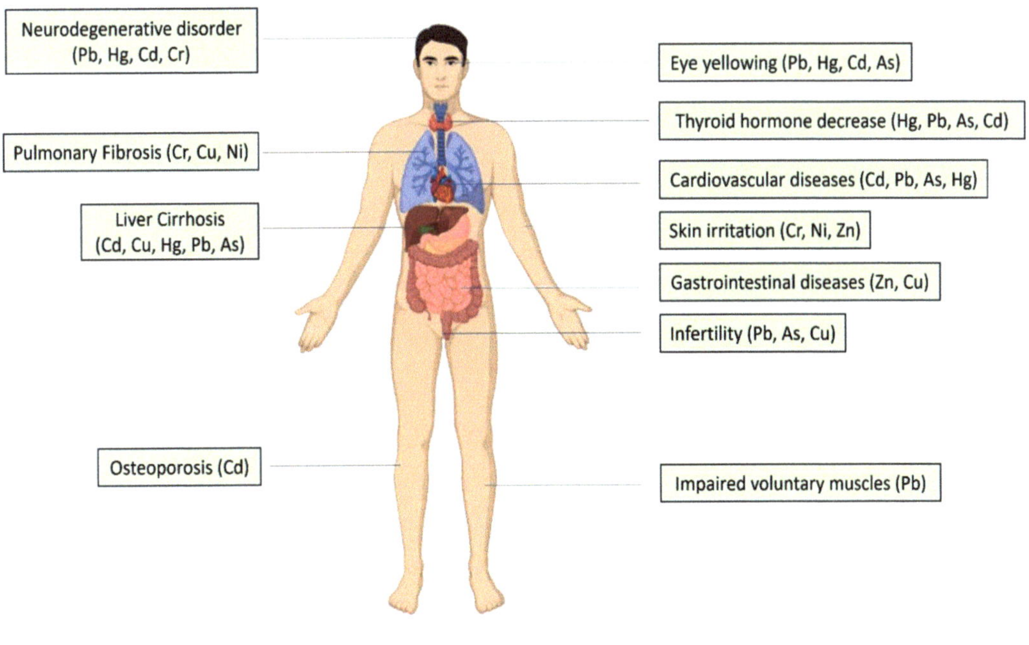

FIGURE 6.3 Heavy metal contamination in human body.

TABLE 6.4
Toxicity Levels of Heavy Metals

Sl. No.	Metal	Type	Toxicity Level
1.	Copper (Cu)	Metal	Environment hazard
2.	Zinc (Zn)	Metal	Environment hazard
3.	Silver (Ag)	Metal	Environment hazard
4.	Nickel (Ni)	Metal	Irritant and health hazard
5.	Arsenic (As)	Metalloid	Acute toxic and environment hazard
6.	Chromium (Cr)	Metal	Irritant, health hazard and environment hazard
7.	Iron (Fe)	Metal	Irritant
8.	Lead (Pb)	Metal	Health hazard and environment hazard
9.	Beryllium (Be)	Metalloid	Irritant, acute toxic and health hazard
10.	Antimony (Sb)	Metalloid	Irritant, acute toxic and health hazard
11.	Boron (B)	Metalloid	Irritant, acute toxic and health hazard
12.	Selenium (Se)	Metal	Health hazard, acute toxic and environment hazard
13.	Mercury (Hg)	Metal	Health hazard, acute toxic and environment hazard
14.	Cadmium (Cd)	Metal	Health hazard, acute toxic and environment hazard
15.	Thallium (Tl)	Metal	Health hazard and acute toxic
16.	Uranium (U)	Metal	Health hazard and acute toxic

6.4 ANALYSIS

The detection of heavy metals necessitates the use of advanced instruments that are both highly sensitive and sophisticated. These instruments should be capable of determining levels up to parts per million (ppm) and/or parts per billion (ppb) in air, soil, food and in various biological samples, including saliva, blood and serum [146, 147]. The techniques capable of detecting heavy metal levels up to ppb in various geological, environmental and biological samples are listed below:

- Inductively coupled plasma-atomic emission spectrometry (ICP-AES) or inductively coupled plasma-optical emission spectrometry (ICP-OES) or inductively coupled plasma-mass spectrometry (ICP-MS).
- Atomic absorption spectroscopy (AAS)
- Particle-induced X-ray emission (PIXE)
- X-ray fluorescence spectrometry (XRF) or bench-mounted energy-dispersive XRF (ED-XRF) or portable ED-XRF or synchrotron XRF or wavelength dispersive XRF (WD-XRF)
- Neutron activation analysis (NAA)

These techniques are extensively documented for their efficiency and wide-ranging capability in detecting heavy metals in the environment [148, 149].

6.5 MONITORING AND CONTROL TECHNIQUES FOR PARTICULATE MATTER (PM)

PM refers to the combination of solid particles and liquid droplets present in the air which vary in composition and size (PM2.5 and PM10). The existence of heavy metals in PM, especially in fine particles like PM2.5, represents a potential health risk for humans. PM consists of several heavy metals like Cd, Se, K, Ca, Ba, V and Sr. Various physical processes are employed for the removal of airborne particles from polluted air. Exposure to PM causes asthma, difficulty in breathing, coughing.

Long-term exposure to PM leads to chronic bronchitis, decreased lung capacity and irregular heartbeat. Typical equipment used for capturing fine particulates includes scrubbers, baghouse filters, electrostatic precipitator and cyclone separator.

6.5.1 Scrubbers

Scrubbers are systems employed to remove harmful pollutants from industrial exhaust before their release into the environment. There are two types of scrubbers: dry scrubbers and wet scrubbers.

6.5.1.1 Air Scrubbers

Air scrubbers are purification systems that filter and cool the incoming airstream within the scrubber to eliminate PM from air. The primary function of an air scrubber is to cleanse the air from pollutants, harmful gases, fumes and chemicals after contamination.

6.5.1.2 Wet Air Scrubbers

Wet air scrubbers function by soaking up contaminants with water or water-based solvent. The polluted gas enters the wet scrubber from bottom, ascends through packed bed and encounters downward moving solvent sprays. Prior to exiting the scrubber, the gas passes through a mist eliminator to capture any droplets, trapping contaminants within the solvent droplets. The effectiveness of impurity removal is influenced by the solvent composition with its electric charge being a critical factor. The solvent's capacity to form bonds with the inorganic contaminants is determined by its charge, which can be positive, negative or neutral.

6.5.1.3 Dry Air Scrubbers

In dry air scrubber, the contaminants are removed from the air stream by the interaction of the reagents and the pollutants. Dry scrubber is environment friendly; gathered pollutants are either incinerated in the heat of the air stream or captured in a filter. A dry scrubber eliminates the need for wastewater removal or storage, enhancing cost-effectiveness in operation. Dry scrubbers are generally employed to capture solvents and acidic vapours.

6.5.2 Baghouse Filters

The use of baghouse filters or fabric filter bags is an effect method for the elimination of suspended particulates. This system consists of an array of long, narrow bags, each approximately 25 cm in diameter, suspended upside down within a spacious enclosure. Fans blow dust-laden air upward through the enclosure, where particulates are captured inside filter bags. Clean air passed through the fabric, exiting at the top of the baghouse.

6.5.3 Electrostatic Precipitator

The electrostatic precipitator is a widely used control device which can remove solid, liquid or gaseous particles from air through the application of an electric charge. The precipitator captures smoke or gas emitted from a burner or furnace by passing the gas or smoke over wires or plates. In this procedure, the gas or smoke acquires a static charge and subsequently gathers on a second plate carrying negative charge which holds the pollutants. Here the dry gas enters the precipitator and circulates around the electrodes and plates. The electrode carries a negative charge, while the collection plate is positively charged, causing particles with high resistivity to be drawn towards the plate. The gas leaving the precipitator is purged with particulate and discharge into the atmosphere. Particles collected on the plates are released through various methods and descend to the bottom of the housing, where hoppers facilitate the removal of the accumulated particulate.

6.5.4 Cyclone Separator

Cyclone separators are devices for air pollution control designed to separate dry PM from gaseous pollutants. These devices use centrifugal force to gather and eliminate particulates. The gas streams flow into a cyclone and move through the cylindrical chamber in a spiral pattern. The swirling motion propels larger particles against the chamber wall, decreasing their inertia and leading them to descend into the collection hopper for further processing and disposal. The purified gas continues to move upward and exit the cyclone.

6.6 TECHNIQUES FOR REMOVAL OF HEAVY METAL

Heavy metal ion removal from point sources is crucial to safeguard the environment due to their toxic nature, capacity for bioaccumulation and persistence in the ecosystem [150]. Several methods and strategies have been employed to remediate heavy metals from soil and water. Some of these techniques are summarized in Table 6.5.

TABLE 6.5
Principles, Advantages, Disadvantages of Remedial Techniques from Removal of Heavy Metal Contamination from Soil and Water

Remedial Techniques	Principle	Advantages	Disadvantage	References
Ion-exchange	Ion-exchange resins are employed for the separation of metal ions from polluted water and soil	Wide range of heavy metal can be removed, low energy regeneration	Resin fouling, expensive	[151]
Chemical precipitation	Coagulants are utilized to induce the precipitation of heavy metals which are subsequently eliminated through filtration	Simple, low cost, less laborious	Sludge production, requirement of secondary treatment	[152]
Reverse osmosis	Removal of pollutants (heavy metal ions) from water using a semi-permeable membrane	Chemical free, easy operation	Costly, energy intensive, membrane fouling	[153]
Adsorption	The adsorbate (pollutant) in the liquid phase accumulates on the surface of the solid phase, which is the adsorbent	High removal efficiency, cost-effective, simple in design, can be used in wide pH range	Regeneration of adsorbent	[154]
Photocatalysis	Relies on the reactivity of electron-hole pairs produced within semiconductor particles when exposed to light	Degrades heavy metal complexes, no sludge production, high oxidizing potential	Expensive	[155]
Phytoremediation	Uses plants to remove heavy metals from air, soil and water	Can be used for large, contaminated area, non-toxic, easy implementation	Slow process, can be influenced by climate	[155]
Membrane filtration	Pollutants filtered from contaminated water through filters with various pore sizes	Efficient separation	Expensive, sludge production, membrane fouling	[156]

(Continued)

TABLE 6.5 (Continued)
Principles, Advantages, Disadvantages of Remedial Techniques from Removal of Heavy Metal Contamination from Soil and Water

Remedial Techniques	Principle	Advantages	Disadvantage	References
Coagulation/ flocculation	Coagulants are added to the contaminated water forming floc that has capability to remove heavy metals. Larger particles (floc) settle in water due to agglomeration	Easy separation	Secondary pollution generation	[157]
Electrochemical treatment	Removal of charged species from wastewater through application of electrochemical principles on the surface of electrodes	Minimum chemical usage, effective recycling	Costly, lacks efficiency and selectivity	[158]

6.7 CONCLUSION

Heavy metals have emerged as a significant environmental concern, particularly affecting plants, humans and the environment. Heavy metal contamination is increasing rapidly with changing lifestyle and increasing pressure on natural resources. Natural occurrence of heavy metals in environment has witnessed an incremental progression through human activities like vehicular emissions, various industrial and manufacturing processes.

The prolonged exposure of heavy metals in the environment presents a threat to both human health and the ecosystem. The monitoring and control of potential heavy metals and metalloids in different environmental segments hold considerable importance. The environmental contamination by heavy metals is a significant concern, and there is an urgent need for remediation from air, soil and water in urban and industrial areas worldwide. The best strategies may be employed to control the heavy metal contamination at the source of emission or release.

ACKNOWLEDGEMENTS

The authors acknowledge financial support from the Science and Engineering Research Board (SERB) core research grant (CRG/2021/007402). The authors are also thankful to BIT Mesra for necessary support for this work.

REFERENCES

1. C. Li, K. Zhou, W. Qin, C. Tian, M. Qi, X. Yan, W. Han, A review on heavy metals contamination in soil: effects, sources, and remediation techniques. Soil Sediment Contam. An Int. J. 28 (2019) 380–394.
2. S.S. Dhaliwal, J. Singh, P.K. Taneja, A. Mandal, Remediation techniques for removal of heavy metals from the soil contaminated through different sources: a review. Environ. Sci. Pollut. Res. 27 (2020) 1319–1333.
3. M. Jayakumar, U. Surendran, P. Raja, A. Kumar, V. Senapathi, A review of heavy metals accumulation pathways, sources and management in soils. Arab. J. Geosci. 14 (2021) 1–19.
4. G.L. Rowe, D.C. Reutter, D.L. Runkle, J.A. Hambrook, S.D. Janosy, L.H. Hwang, Water quality in the great and little Miami River basins. Ohio and Indiana, 1999–2001, 2004.
5. B.G. Lottermoser, Gold in municipal sewage sludges: a review on concentrations, sources and potential extraction. J. Solid Waste Technol. Manag. 27 (2001) 69–75.

6. I. Thornton, Impacts of mining on the environment; some local, regional and global issues. Appl. Geochem. 11 (1996) 355–361.
7. M. Kumar, A. Seth, A.K. Singh, M.S. Rajput, M. Sikandar, Remediation strategies for heavy metals contaminated ecosystem: a review. Environ. Sustain. Indic. 12 (2021) 100155.
8. T.A. Kirpichtchikova, A. Manceau, L. Spadini, F. Panfili, M.A. Marcus, T. Jacquet, Speciation and solubility of heavy metals in contaminated soil using X-ray microfluorescence, EXAFS spectroscopy, chemical extraction, and thermodynamic modeling. Geochim. Cosmochim. Acta. 70 (2006) 2163–2190.
9. J.P. Vareda, A.J.M. Valente, L. Durães, Heavy metals in Iberian soils: removal by current adsorbents/amendments and prospective for aerogels. Adv. Colloid Interface Sci. 237 (2016) 28–42.
10. K. Sardar, S. Ali, S. Hameed, S. Afzal, S. Fatima, M.B. Shakoor, S.A. Bharwana, H.M. Tauqeer, Heavy metals contamination and what are the impacts on living organisms. Greener J. Environ. Manag. Public Saf. 2 (2013) 172–179.
11. A.K. Zeraatkar, H. Ahmadzadeh, A.F. Talebi, N.R. Moheimani, M.P. McHenry, Potential use of algae for heavy metal bioremediation, a critical review. J. Environ. Manage. 181 (2016) 817–831.
12. S. Ferrando, M. Maisano, V. Parrino, T. Ferrando, L. Girosi, G. Tagliafierro, Gut morphology and metallothionein immunoreactivity in *Liza aurata* from different heavy metal polluted environments. Ital. J. Zool. 73 (2006) 7–14.
13. L.D. Benefield, J.F. Judkins, B.L. Weand, Process Chemistry for Water and Wastewater Treatment, Prentice-Hall, 1982.
14. Z. Hubicki, D. Kołodyńska, Selective removal of heavy metal ions from waters and waste waters using ion exchange methods. Ion Exch. Technol. 7 (2012) 193–240.
15. A. Selvi, D. Das, N. Das, Potentiality of yeast Candida sp. SMN04 for degradation of cefdinir, a cephalosporin antibiotic: kinetics, enzyme analysis and biodegradation pathway. Environ. Technol. 36 (2015) 3112–3124.
16. H.M. Selim, D.L. Sparks, Heavy Metals Release in Soils, CRC Press, 2001.
17. N. Sarwar, M. Imran, M.R. Shaheen, W. Ishaque, M.A. Kamran, A. Matloob, A. Rehim, S. Hussain, Phytoremediation strategies for soils contaminated with heavy metals: modifications and future perspectives. Chemosphere 171 (2017) 710–721.
18. V.R. Bagul, D.N. Shinde, R.P. Chavan, C.L. Patil, R.K. Pawar, New perspective on heavy metal pollution of water. J. Chem. Pharm. Res. 7 (2015) 700–705.
19. D. Langmuir, Aqueous Environmental, Geochemistry Prentice Hall Up, 1997: 600 p.
20. D.L. Sparks, Environmental soil chemistry: an overview. Environ. Soil Chem. 2 (2003) 1–42.
21. R.J. Blong, Volcanic Hazards: A Sourcebook on the Effects of Eruptions, Academic Press, 1984.
22. W.D. Huff, L.A. Owen, Volcanic landforms and hazards, in: Shroder, J., Owen, L.A. (Eds.), Treatise on Geomorphology, Vol. 5, Academic Press, 2015: pp. 148–192.
23. T. Komai, Y. Kawabe, J. Hara, Y. Sakamoto, M. Zhang, Geochemical survey of tsunami sediments and transport of toxic elements from offshore environment-Urgent investigation for earthquake March 11, 2011, in: ISOPE Int. Ocean Polar Eng. Conf., (2012): pp. ISOPE-I-12-376.
24. K. Simonen, M. Huang, C. Aicher, P. Morris, Embodied carbon as a proxy for the environmental impact of earthquake damage repair. Energy Build. 164 (2018) 131–139.
25. H. Leclère, D. Faulkner, S. Llana-Fúnez, J. Bedford, J. Wheeler, Reaction fronts, permeability and fluid pressure development during dehydration reactions. Earth Planet. Sci. Lett. 496 (2018) 227–237.
26. J. Hyks, T. Astrup, T.H. Christensen, Long-term leaching from MSWI air-pollution-control residues: leaching characterization and modeling. J. Hazard. Mater. 162 (2009) 80–91.
27. T. Ngulube, J.R. Gumbo, V. Masindi, A. Maity, An update on synthetic dyes adsorption onto clay based minerals: a state-of-art review. J. Environ. Manage. 191 (2017) 35–57.
28. G.D. Yuan, B.K.G. Theng, G.J. Churchman, W.P. Gates, Clays and clay minerals for pollution control, in: Developments in Clay Science, Elsevier, 2013: pp. 587–644.
29. L. Zhang, G. Zhu, X. Ge, G. Xu, Y. Guan, Novel insights into heavy metal pollution of farmland based on reactive heavy metals (RHMs): pollution characteristics, predictive models, and quantitative source apportionment. J. Hazard. Mater. 360 (2018) 32–42.
30. B. Clark, Thermal water pollution from nuclear power plants, Standford University. (2019).
31. E. Osann, B. Hayat, Protecting our waters from toxic power plant discharges and reducing water use in the process, NRDC Fact Sheet. (2014).
32. H.B. Bradl, Sources and origins of heavy metals, in: Interface Science and Technology, Elsevier, 2005: pp. 1–27.

33. T. Praharaj, M.A. Powell, B.R. Hart, S. Tripathy, Leachability of elements from sub-bituminous coal fly ash from India. Environ. Int. 27 (2002) 609–615.
34. S. Velusamy, A. Roy, S. Sundaram, T. Kumar Mallick, A review on heavy metal ions and containing dyes removal through graphene oxide-based adsorption strategies for textile wastewater treatment. Chem. Rec. 21 (2021) 1570–1610.
35. G. Moussavi, M. Mahmoudi, Removal of azo and anthraquinone reactive dyes from industrial wastewaters using MgO nanoparticles. J. Hazard. Mater. 168 (2009) 806–812.
36. N. Ristic, I. Ristic, Cationic modification of cotton fabrics and reactive dyeing characteristics. J. Eng. Fiber. Fabr. 7 (2012) 113–121.
37. A. Sharma, A. Ramotra, H.S. Sehgal, Physico–chemical analysis of paper industry effluents in Jammu city (J&K). Int. J. Sci. Res. Publ. 4 (2014) 1–4.
38. M. Kamali, Z. Khodaparast, Review on recent developments on pulp and paper mill wastewater treatment. Ecotoxicol. Environ. Saf. 114 (2015) 326–342.
39. P. Sharma, H.M.N. Iqbal, R. Chandra, Evaluation of pollution parameters and toxic elements in wastewater of pulp and paper industries in India: a case study. Case Stud. Chem. Environ. Eng. 5 (2022) 100163.
40. C.G. Anderson, R.C. Dunne, I.U. John, Mineral Processing and Extractive Metallurgy, Society for Mining, Metallurgy, and Exploration, 2014.
41. R. McLachlan, C. Clark, C. Harris, Micro Reform–Impacts on Firms: Aluminium Case Study, Industry Commission, 1998.
42. H. Sutar, S.C. Mishra, S.K. Sahoo, H.S. Maharana, Progress of red mud utilization: an overview. Am. Chem. Sci. J. 4 (2014) 255–279.
43. R. Islam, M.A. Hye, Metallurgical treatment processes of metals (Fe and steel, Al, Cu, Au) and their detrimental environmental issues – a mini review. Int. J. Sci. Res. Publ. 8 (2018) 677–691.
44. A. Boulamanti, J.A. Moya, Production costs of the non-ferrous metals in the EU and other countries: copper and zinc. Resour. Policy 49 (2016) 112–118.
45. G. Izydorczyk, K. Mikula, D. Skrzypczak, K. Moustakas, A. Witek-Krowiak, K. Chojnacka, Potential environmental pollution from copper metallurgy and methods of management. Environ. Res. 197 (2021) 111050.
46. I. Timofeev, N. Kosheleva, N. Kasimov, Contamination of soils by potentially toxic elements in the impact zone of tungsten-molybdenum ore mine in the Baikal region: a survey and risk assessment. Sci. Total Environ. 642 (2018) 63–76.
47. S.M. Serbula, J.S. Milosavljevic, A.A. Radojevic, J.V. Kalinovic, T.S. Kalinovic, Extreme air pollution with contaminants originating from the mining–metallurgical processes. Sci. Total Environ. 586 (2017) 1066–1075.
48. P.E. Rosenfeld, L. Feng, Risks of Hazardous Wastes, William Andrew, 2011.
49. B.G. Pagliari, M. de F.R. Moreira, C.F. Mannarino, G.B. dos Santos, Risk of exposure to metals in soil contaminated by steel industry waste for a population in Volta Redonda, RJ. Rev. Ambient. Água 16 (2021) 2696.
50. H. Motz, J. Geiseler, Products of steel slags an opportunity to save natural resources. Waste Manag. 21 (2001) 285–293.
51. G.V. Motuzova, T.M. Minkina, E.A. Karpova, N.U. Barsova, S.S. Mandzhieva, Soil contamination with heavy metals as a potential and real risk to the environment. J. Geochem. Explor. 144 (2014) 241–246.
52. P. Horký, R. Grabic, K. Grabicová, B.W. Brooks, K. Douda, O. Slavík, P. Hubená, E.M. Sancho Santos, T. Randák, Methamphetamine pollution elicits addiction in wild fish. J. Exp. Biol. 224 (2021) jeb242145.
53. C. Sirtori, A. Zapata, I. Oller, W. Gernjak, A. Agüera, S. Malato, Decontamination industrial pharmaceutical wastewater by combining solar photo-Fenton and biological treatment. Water Res. 43 (2009) 661–668.
54. B. Ramola, A. Singh, Heavy metal concentrations in pharmaceutical effluents of industrial area of Dehradun (Uttarakhand), India. J. Environ. Anal. Toxicol. 3 (2013) 525–2161.
55. S. Randhawa, L. Chopra, Pestilential impacts of battery industry discharged metal waste on human health. Mater. Today Proc. 52 (2022) 434–438.
56. R. Xiao, S. Wang, R. Li, J.J. Wang, Z. Zhang, Soil heavy metal contamination and health risks associated with artisanal gold mining in Tongguan, Shaanxi, China. Ecotoxicol. Environ. Saf. 141 (2017) 17–24.
57. I. Park, C.B. Tabelin, S. Jeon, X. Li, K. Seno, M. Ito, N. Hiroyoshi, A review of recent strategies for acid mine drainage prevention and mine tailings recycling. Chemosphere 219 (2019) 588–606.

58. L. Sumi, S. Thomsen, Mining in remote areas, issues and impacts, Minind Watch Canada/Mines Alerte. (2001) 1–33.
59. D.C. Jhariya, R. Khan, G.S. Thakur, Impact of mining activity on water resource: an overview study, in: Proceeding of Recent Practices and Innovations in Mining Industry, Raipur, India, 2016: pp. 271–277.
60. V. Sheoran, A.S. Sheoran, R.P. Choudhary, Biogeochemistry of acid mine drainage formation: a review, Mine Drain. Relat. Probl. 4(2011) 119–154.
61. B. Dhir, Biotechnological tools for remediation of acid mine drainage (removal of metals from wastewater and leachate), in: Bio-Geotechnologies for Mine Site Rehabilitation, Elsevier, 2018: pp. 67–82.
62. H. Zhao, B. Xia, J. Qin, J. Zhang, Hydrogeochemical and mineralogical characteristics related to heavy metal attenuation in a stream polluted by acid mine drainage: a case study in Dabaoshan Mine, China. J. Environ. Sci. 24(2012) 979–989.
63. D. Katoria, D. Sehgal, S. Kumar, Environment impact assessment of coal mining, Int. J. Environ. Eng. Manag. 4 (2013) 245–250.
64. C.H.U. Daozhong, Z.H.U. Qingli, W. Jie, Z. Xiaozhi, Comparative analysis of ecological rucksack between open-pit and underground coal mine, Energy Procedia 5 (2011) 1116–1120.
65. R.E. Masto, S. Sheik, G. Nehru, V.A. Selvi, J. George, L.C. Ram, Assessment of environmental soil quality around Sonepur Bazari mine of Raniganj coalfield, India. Solid Earth 6 (2015) 811–821.
66. S. De, A.K. Mitra, Mobilization of heavy metals from mine spoils in a part of Raniganj coalfield, India: causes and effects. Environ. Geosci. 11 (2004) 65–76.
67. E.F. Da Silva, C. Zhang, L.S. Pinto, C. Patinha, P. Reis, Hazard assessment on arsenic and lead in soils of Castromil gold mining area, Portugal. Appl. Geochem. 19 (2004) 887–898.
68. T.V. Durkin, J.G. Herrmann, Focusing on the problem of mining wastes: an introduction to acid mine drainage, in: Managing Environmental Problems at Inactive and Abandoned Mine Sites, US EPA Seminar Publication, 1994: pp. 1–3.
69. K.H. Telmer, M.M. Veiga, World emissions of mercury from artisanal and small scale gold mining, in: Mercury Fate and Transport in the Global Atmosphere. Springer, 2009: pp. 131–172.
70. D. Shang, B. Geissler, M. Mew, L. Satalkina, L. Zenk, H. Tulsidas, L. Barker, A. El-Yahyaoui, A. Hussein, M. Taha, Unconventional uranium in China's phosphate rock: review and outlook. Renew. Sustain. Energy Rev. 140 (2021) 110740.
71. M. Yin, J. Sun, Y. Chen, J. Wang, J. Shang, N. Belshaw, C. Shen, J. Liu, H. Li, W. Linghu, Mechanism of uranium release from uranium mill tailings under long-term exposure to simulated acid rain: geochemical evidence and environmental implication. Environ. Pollut. 244 (2019) 174–181.
72. B. Liu, T. Peng, H. Sun, H. Yue, Release behavior of uranium in uranium mill tailings under environmental conditions. J. Environ. Radioact. 171 (2017) 160–168.
73. B.L. Jagetiya, P. Purohit, Effects of various concentrations of uranium tailings on certain growth and biochemical parameters in sunflower. Biologia (Bratisl) 61 (2006) 103–107.
74. B.J. Alloway, Sources of heavy metals and metalloids in soils. Heavy Metal in Soils: Trace Metals and Metalloids in Soils and Their Bioavailability. Springer, 2013: pp. 11–50.
75. A.C. Gonçalves Jr, H. Nacke, D. Schwantes, G.F. Coelho, Heavy metal contamination in Brazilian agricultural soils due to application of fertilizers. Environ. Risk Assess. Soil Contam. 4 (2014) 105–135.
76. B. Wei, J. Yu, Z. Cao, M. Meng, L. Yang, Q. Chen, The availability and accumulation of heavy metals in greenhouse soils associated with intensive fertilizer application. Int. J. Environ. Res. Public Health 17 (2020) 5359.
77. S.S. Mar, M. Okazaki, T. Motobayashi, The influence of phosphate fertilizer application levels and cultivars on cadmium uptake by Komatsuna (*Brassica rapa* L. var. perviridis), Soil Sci. Plant Nutr. 58 (2012) 492–502.
78. Z. Zhuang, H. Mu, P. Fu, Y. Wan, Y. Yu, Q. Wang, H. Li, Accumulation of potentially toxic elements in agricultural soil and scenario analysis of cadmium inputs by fertilization: a case study in Quzhou county. J. Environ. Manage. 269 (2020) 110797.
79. Z. Atafar, A. Mesdaghinia, J. Nouri, M. Homaee, M. Yunesian, M. Ahmadimoghaddam, A.H. Mahvi, Effect of fertilizer application on soil heavy metal concentration. Environ. Monit. Assess. 160 (2010) 83–89.
80. S. Satarug, J.R. Baker, S. Urbenjapol, M. Haswell-Elkins, P.E.B. Reilly, D.J. Williams, M.R. Moore, A global perspective on cadmium pollution and toxicity in non-occupationally exposed population. Toxicol. Lett. 137 (2003) 65–83.
81. D.R. Wallace, Environmental pesticides and heavy metals—role in breast cancer. in: Toxicity and Hazard of Agrochemicals (2015): pp. 39–70.
82. E.-C. Oerke, Crop losses to pests. J. Agric. Sci. 144 (2006) 31–43.

83. A. Sharma, V. Kumar, B. Shahzad, M. Tanveer, G.P.S. Sidhu, N. Handa, S.K. Kohli, P. Yadav, A.S. Bali, R.D. Parihar, Worldwide pesticide usage and its impacts on ecosystem. SN Appl. Sci. 1 (2019) 1–16.
84. A. De, R. Bose, A. Kumar, S. Mozumdar, Targeted Delivery of Pesticides Using Biodegradable Polymeric Nanoparticles, Springer, 2014.
85. A.C. Zolnowski, M.K. Busse, P.K. Zajac, Response of maize (*Zea mays* L.) to soil contamination with copper depending on applied contamination neutralizing substances. J. Elem. 18 (2013) 507–520.
86. K.A. Lewis, J. Tzilivakis, D.J. Warner, A. Green, An international database for pesticide risk assessments and management. Hum. Ecol. Risk Assess. An Int. J. 22 (2016) 1050–1064.
87. D.C. Adriano, D.C. Adriano, Nickel, Springer, 2001.
88. M.L.A. Silveira, L.R.F. Alleoni, L.R.G. Guilherme, Biosolids and heavy metals in soils. Sci. Agric. 60 (2003) 793–806.
89. F. Zhang, Y. Li, M. Yang, W. Li, Content of heavy metals in animal feeds and manures from farms of different scales in Northeast China. Int. J. Environ. Res. Public Health 9 (2012) 2658–2668.
90. Y. Hu, W. Zhang, G. Chen, H. Cheng, S. Tao, Public health risk of trace metals in fresh chicken meat products on the food markets of a major production region in Southern China. Environ. Pollut. 234 (2018) 667–676.
91. J. Jensen, M.M. Larsen, J. Bak, National monitoring study in Denmark finds increased and critical levels of copper and zinc in arable soils fertilized with pig slurry. Environ. Pollut. 214 (2016) 334–340.
92. C. Lopes, M. Herva, A. Franco-Uría, E. Roca, Inventory of heavy metal content in organic waste applied as fertilizer in agriculture: evaluating the risk of transfer into the food chain. Environ. Sci. Pollut. Res. 18 (2011) 918–939.
93. Y. Zhao, Z. Yan, J. Qin, Z. Xiao, Effects of long-term cattle manure application on soil properties and soil heavy metals in corn seed production in Northwest China. Environ. Sci. Pollut. Res. 21 (2014) 7586–7595.
94. X. Yang, Q. Li, Z. Tang, W. Zhang, G. Yu, Q. Shen, F.-J. Zhao, Heavy metal concentrations and arsenic speciation in animal manure composts in China. Waste Manag. 64 (2017) 333–339.
95. B. Shaharoona, S. Al-Ismaily, A. Al-Mayahi, N. Al-Harrasi, R. Al-Kindi, A. Al-Sulaimi, H. Al-Busaidi, M. Al-Abri, The role of urbanization in soil and groundwater contamination by heavy metals and pathogenic bacteria: a case study from Oman. Heliyon 5 (2019) e01771.
96. L. Jiang, M.H. Young, K. Hardee, Population, urbanization and the environment. World Watch 9 (2008) 34–39.
97. C. CHANGE, According to the United Nations (World Urbanization Prospects Revised, WUP, 2014), the global urban population in 2014 was close to 3.9 billion, approximately 54% of the total, and it is estimated to grow to 6.6 billion by 2050, 66% of the global population. On the other hand, the rural population is expected to remain 3.4 billion up to 2020, and it is estimated it will decrease to approximately 3.2 billion by 2050. (2014).
98. J. Peng, Y. Song, P. Yuan, X. Cui, G. Qiu, The remediation of heavy metals contaminated sediment. J. Hazard. Mater. 161 (2009) 633–640.
99. K.P. Shimod, V. Vineethkumar, T.K. Prasad, G. Jayapal, Effect of urbanization on heavy metal contamination: a study on major townships of Kannur District in Kerala, India. Bull. Natl. Res. Cent. 46 (2022) 1–14.
100. K.H. Vardhan, P.S. Kumar, R.C. Panda, A review on heavy metal pollution, toxicity and remedial measures: current trends and future perspectives. J. Mol. Liq. 290 (2019) 111197.
101. A. Lagerkvist, L. Dahlén, Solid waste generation and characterization, In: Meyers, R.A. (Eds.), Encyclopedia of Sustainability Science and Technology, Springer, 2012.
102. M. Azarafza, A. Ghazifard, Urban geology of Tabriz City: environmental and geological constraints. Adv. Environ. Res. 5 (2016) 95–108.
103. S. Kanmani, R. Gandhimathi, Assessment of heavy metal contamination in soil due to leachate migration from an open dumping site. Appl. Water Sci. 3 (2013) 193–205.
104. M. Jaishankar, T. Tseten, N. Anbalagan, B.B. Mathew, K.N. Beeregowda, Toxicity, mechanism and health effects of some heavy metals. Interdiscip. Toxicol. 7 (2014) 60.
105. G.T. Müller, A. Giacobbo, E.A. dos Santos Chiaramonte, M.A.S. Rodrigues, A. Meneguzzi, A.M. Bernardes, The effect of sanitary landfill leachate aging on the biological treatment and assessment of photoelectrooxidation as a pre-treatment process. Waste Manag. 36 (2015) 177–183.
106. A.R. de Pauli, F.R. Espinoza-Quinones, D.E.G. Trigueros, A.N. Modenes, A.R.C. de Souza, F.H. Borba, A.D. Kroumov, Integrated two-phase purification procedure for abatement of pollutants from sanitary landfill leachates. Chem. Eng. J. 334 (2018) 19–29.

107. J. Drozdova, H. Raclavska, K. Raclavsky, H. Skrobankova, Heavy metals in domestic wastewater with respect to urban population in Ostrava, Czech Republic. Water Environ. J. 33 (2019) 77–85.
108. A. Kalinowska, M. Szopińska, S. Chmiel, M. Kończak, Ż. Polkowska, W. Artichowicz, K. Jankowska, A. Nowak, A. Łuczkiewicz, Heavy metals in a high arctic fiord and their introduction with the wastewater: a case study of Adventfjorden-Longyearbyen system, Svalbard. Water 12 (2020) 794.
109. S.K. Bharti, D. Kumar, S. Anand, S.C. Barman, N. Kumar, Characterization and morphological analysis of individual aerosol of PM10 in urban area of Lucknow, India. Micron 103 (2017) 90–98.
110. H. Ozaki, I. Watanabe, K. Kuno, Investigation of the heavy metal sources in relation to automobiles. Water. Air. Soil Pollut. 157 (2004) 209–223.
111. V. Gupta, Vehicle-generated heavy metal pollution in an urban environment and its distribution into various environmental components, in: Environmental Concerns and Sustainable Development, Vol. 1, Springer, 2020: pp. 113–127.
112. E. Apeagyei, M.S. Bank, J.D. Spengler, Distribution of heavy metals in road dust along an urban-rural gradient in Massachusetts. Atmos. Environ. 45 (2011) 2310–2323.
113. S.T. Hjortenkrans, B.G. Bergbäck, A.V. Häggerud, Metal emissions from brake linings and tires: case studies of Stockholm, Sweden 1995/1998 and 2005. Environ. Sci. Technol. 41 (2007) 5224–5230.
114. E. Smolders, F. Degryse, Fate and effect of zinc from tire debris in soil. Environ. Sci. Technol. 36 (2002) 3706–3710.
115. S. Prakash, A. Manhart, Y. Amoyaw-Osei, O.O. Agyekum, Socio-economic assessment and feasibility study on sustainable e-waste management in Ghana, Öko-Institut EV Coop. with Ghana Environ. Prot. Agency Green Advocacy Ghana, Minist. Housing, Spat. Plan. Environ. VROM-Inspectorate. 118 (2010).
116. S.S. Sonone, S. Jadhav, M.S. Sankhla, R. Kumar, Water contamination by heavy metals and their toxic effect on aquaculture and human health through food chain. Lett. Appl. NanoBioScience. 10 (2020) 2148–2166.
117. G.P. Singh, B. Chaturvedi, R. Kumar, Effect of electronic waste on environmental & human health – a review. IOSR J. Environ. Sci. Toxicol. Food Technol. 10 (2016) 98–104.
118. C. Johansson, A.F. Castoldi, N. Onishchenko, L. Manzo, M. Vahter, S. Ceccatelli, Neurobehavioural and molecular changes induced by methylmercury exposure during development. Neurotox. Res. 11 (2007) 241–260.
119. J.-D. Park, W. Zheng, Human exposure and health effects of inorganic and elemental mercury. J. Prev. Med. Public Heal. 45 (2012) 344.
120. S.A. Bhat, T. Hassan, S. Majid, Heavy metal toxicity and their harmful effects on living organisms – a review. Int. J. Med. Sci. Diagnosis Res. 3 (2019) 106–122.
121. U. ur Rehman, S. Khan, S. Muhammad, Ingestion of arsenic-contaminated drinking water leads to health risk and traces in human biomarkers (hair, nails, blood, and urine), Pakistan. Expo. Heal. 12 (2020) 243–254.
122. J.R. Peralta-Videa, M.L. Lopez, M. Narayan, G. Saupe, J. Gardea-Torresdey, The biochemistry of environmental heavy metal uptake by plants: implications for the food chain. Int. J. Biochem. Cell Biol. 41 (2009) 1665–1677.
123. J. Zhang, R. Yang, Y.C. Li, Y. Peng, X. Wen, X. Ni, Distribution, accumulation, and potential risks of heavy metals in soil and tea leaves from geologically different plantations. Ecotoxicol. Environ. Saf. 195 (2020) 110475.
124. T.V. Lam, P. Agovino, X. Niu, L. Roché, Linkage study of cancer risk among lead-exposed workers in New Jersey. Sci. Total Environ. 372 (2007) 455–462.
125. P. Koedrith, H. Kim, J.-I. Weon, Y.R. Seo, Toxicogenomic approaches for understanding molecular mechanisms of heavy metal mutagenicity and carcinogenicity. Int. J. Hyg. Environ. Health 216 (2013) 587–598.
126. P.K. Rai, S.S. Lee, M. Zhang, Y.F. Tsang, K.-H. Kim, Heavy metals in food crops: health risks, fate, mechanisms, and management. Environ. Int. 125(2019) 365–385.
127. M. Lal, B.L. Sau, J. Patidar, A. Patidar, Climate change and groundwater: impact, adaptation and sustainable. Int. J. Bio-Resource Stress Manag. 9 (2018) 408–415.
128. T.K. Marella, A. Saxena, A. Tiwari, Diatom mediated heavy metal remediation: a review. Bioresour. Technol. 305 (2020) 123068.
129. R.L. Hough, N. Breward, S.D. Young, N.M.J. Crout, A.M. Tye, A.M. Moir, I. Thornton, Assessing potential risk of heavy metal exposure from consumption of home-produced vegetables by urban populations. Environ. Health Perspect. 112 (2004) 215–221.
130. R.A. Wuana, F.E. Okieimen, Heavy metals in contaminated soils: a review of sources, chemistry, risks and best available strategies for remediation. Int. Sch. Res. Netw. 2011 (2011) 1–20.

131. K. Kot, D. Kosik-Bogacka, P. Ziętek, M. Karaczun, Ż. Ciosek, N. Łanocha-Arendarczyk, Impact of varied factors on iron, nickel, molybdenum and vanadium concentrations in the knee joint. Int. J. Environ. Res. Public Health 17 (2020) 813.
132. P. Zhang, K. Inoue, K. Yoshizuka, H. Tsuyama, Extraction and selective stripping of molybdenum (VI) and vanadium (IV) from sulfuric acid solution containing aluminum (III), cobalt (II), nickel (II) and iron (III) by LIX 63 in Exxsol D80. Hydrometallurgy 41 (1996) 45–53.
133. Y.K. Leong, J.-S. Chang, Bioremediation of heavy metals using microalgae: recent advances and mechanisms. Bioresour. Technol. 303 (2020) 122886.
134. H. Ali, E. Khan, I. Ilahi, Environmental chemistry and ecotoxicology of hazardous heavy metals: environmental persistence, toxicity, and bioaccumulation. J. Chem. 2019 (2019) 1–14.
135. I.V. Popescu, C. Stihi, G.V. Cimpoca, G. Dima, G. Vlaicu, A. Gheboianu, I. Bancuta, V. Ghisa, G. State, Environmental samples analysis by atomic absorption spectrometry (AAS) and inductively coupled plasma-optical emission spectroscopy (ICP-AES). Rom J Phys. 54 (2009) 731–741.
136. I. Ali, X.Y. Mbianda, A. Burakov, E. Galunin, I. Burakova, E. Mkrtchyan, A. Tkachev, V. Grachev, Graphene based adsorbents for remediation of noxious pollutants from wastewater, Environ. Int. 127 (2019) 160–180.
137. L. Zhao, D. Gong, W. Zhao, L. Lin, W. Yang, W. Guo, X. Tang, Q. Li, Spatial-temporal distribution characteristics and health risk assessment of heavy metals in surface water of the Three Gorges Reservoir, China. Sci. Total Environ. 704 (2020) 134883.
138. B.K. Mandal, K.T. Suzuki, Arsenic round the world: a review. Talanta 58 (2002) 201–235.
139. J.L. Barringer, P.A. Reilly, Arsenic in groundwater: a summary of sources and the biogeochemical and hydrogeologic factors affecting arsenic occurrence and mobility. Curr. Perspect. Contam. Hydrol. Water Resour. Sustain. 34 (2013) 83–116.
140. S. Sundar, J. Chakravarty, Antimony toxicity. Int. J. Environ. Res. Public Health 7 (2010) 4267–4277.
141. H. Sharma, N. Rawal, B.B. Mathew, The characteristics, toxicity and effects of cadmium. Int. J. Nanotechnol. Nanosci. 3 (2015) 1–9.
142. R.A. Bernhoft, Mercury toxicity and treatment: a review of the literature. J. Environ. Public Health 2012 (2012) 460508.
143. A. Raviraja, G. Babu, A. Bijoor, G. Menezes, T. Venkatesh, Lead toxicity in a family as a result of occupational exposure. Arch. Ind. Hyg. Toxicol. 59 (2008) 127–133.
144. M.A. Hashim, S. Mukhopadhyay, J.N. Sahu, B. Sengupta, Remediation technologies for heavy metal contaminated groundwater. J. Environ. Manage. 92 (2011) 2355–2388.
145. A. Zhitkovich, Chromium in drinking water: sources, metabolism, and cancer risks. Chem. Res. Toxicol. 24 (2011) 1617–1629.
146. B. Bansod, T. Kumar, R. Thakur, S. Rana, I. Singh, A review on various electrochemical techniques for heavy metal ions detection with different sensing platforms. Biosens. Bioelectron. 94 (2017) 443–455.
147. S. Lamine, G. Petropoulos, P. Brewer, P.K. Srivastava, K. Manevski, C. Kalaitzidis, M. Macklin, Heavy metals soil contamination detection using combined geochemistry and ASD field spectrometry over a highly contaminated floodplain site in the United Kingdom. Sensors 19 (2019) 762.
148. L. Zhang, M. Fang, Nanomaterials in pollution trace detection and environmental improvement. Nano Today 5 (2010) 128–142.
149. P.K. Srungaram, K.K. Ayyalasomayajula, F. Yu-Yueh, J.P. Singh, Comparison of laser induced breakdown spectroscopy and spark induced breakdown spectroscopy for determination of mercury in soils. Spectrochim. Acta Part B At. Spectrosc. 87 (2013) 108–113.
150. B. Geng, Z. Xu, P. Liang, J. Zhang, P. Christie, H. Liu, S. Wu, X. Liu, Three-dimensional macroscopic aminosilylated nanocellulose aerogels as sustainable bio-adsorbents for the effective removal of heavy metal ions. Int. J. Biol. Macromol. 190 (2021) 170–177.
151. N. Kansara, L. Bhati, M. Narang, R. Vaishnavi, Wastewater treatment by ion exchange method: a review of past and recent researches. ESAIJ (Environ. Sci. An Indian J.). 12 (2016) 143–150.
152. G. Crini, E. Lichtfouse, Advantages and disadvantages of techniques used for wastewater treatment. Environ. Chem. Lett. 17 (2019) 145–155.
153. N. Abdullah, N. Yusof, W.J. Lau, J. Jaafar, A.F. Ismail, Recent trends of heavy metal removal from water/wastewater by membrane technologies. J. Ind. Eng. Chem. 76 (2019) 17–38.
154. R. Chakraborty, A. Asthana, A.K. Singh, B. Jain, A.B.H. Susan, Adsorption of heavy metal ions by various low-cost adsorbents: a review. Int. J. Environ. Anal. Chem. 102 (2022) 342–379.
155. Y. Deng, C. Feng, L. Tang, G. Zeng, Z. Chen, M. Zhang, Nanohybrid photocatalysts for heavy metal pollutant control, in: Nanohybrid and Nanoporous Materials for Aquatic Pollution Control, Elsevier, 2019: pp. 125–153.

156. S. Imdad, R.K. Dohare, A critical review on heavy metals removal using ionic liquid membranes from the industrial wastewater. Chem. Eng. Process. Intensif. 173 (2022) 108812.
157. D. Sakhi, Y. Rakhila, A. Elmchaouri, M. Abouri, S. Souabi, A. Jada, Optimization of coagulation flocculation process for the removal of heavy metals from real textile wastewater, in: Proceedings of the International Conference on Advanced Intelligent Systems for Sustainable Development, Tangier, Morocco, Springer, 2019: pp. 257–266.
158. H.I. Maarof, W.M.A.W. Daud, M.K. Aroua, Recent trends in removal and recovery of heavy metals from wastewater by electrochemical technologies. Rev. Chem. Eng. 33 (2017) 359–386.

7 Distribution of Heavy Metals Contamination in the Environment and Their Toxicological Impacts

Manish Pant and Jabrinder Singh*
Land, Air and Water – Centre of Excellence,
DIT University, Dehradun, Uttarakhand, India

7.1 INTRODUCTION

Natural ecosystems comprise all living things, including people, plants, animals and microorganisms along with abiotic entities. All components are in the harmonious coexistence by maintaining the equilibrium of the environment through biogeochemical cycles and food webs. Heavy metals are metallic elements found naturally in the environment (lithogenic source) with relatively high densities (more than 5 g/cm^3) and atomic weights. Many of them, among which are Cu, Co, Zn, Fe, Mo, Mn and others, are biologically necessary in trace amounts but harmful at higher concentrations. However, numerous toxic heavy metal/metalloids, which include arsenic – As, lead – Pb, cadmium – Cd and mercury – Hg are hazardous to the environs and human health. Chemicals that are in excess concentration in the environment comparative to natural are known as environmental contaminants or toxicants [1–3]. Heavy metal(loid)s pollution in soil has recently been found as a global environmental problem, with approximately 10 million polluted sites including more than 50% of land contaminated with harmful metal(loid)s [4]. Although traces of these elements are present naturally in rocks and aquifers, high concentrations of these elements have been found in several places, both naturally and through anthropogenic sources such as industrial wastes, urban and agricultural runoff, domestic wastewater effluents, medical, atmospheric deposition, mining activities and the transformation in technology resulting in a major concern for heavy metal toxicity to the human and his environment. Due to the reckless exploitation of the planet's natural resources, environmental pollution has become a major global problem [5]; and pollutants such as inorganic ions, organic composites, gaseous poisons, radioactive variants, organometallic compounds and nano-particles have all wreaked havoc on the environment. High concentrations of these toxicants in air, water, soil and food adversely impact not just ecosystem flora and wildlife, but also human health via bioaccumulation/biomagnification in living tissues [2]. Because it comes mostly from a single source, localized contamination may have little impacts; nevertheless, nonspecific deposition or release of heavy metals will cause severe damage. Because of their poisonous nature, long-term persistence and bioaccumulation, high metal concentrations in the environment are harmful to ecosystems [6]. Vegetation (cereals, vegetables, etc.) cultivated in contaminated soils or irrigated with polluted water accrue substantial quantity of heavy metal/metalloids in various plant tissues, which reach humans via drinking water and contaminated food, resulting in many times the prescribed limits set by various regulatory bodies. Because of their great toxicity potential, they are known as systematic toxicants. Their effects vary depending on numerous parameters, such as dose and

* Corresponding author: manishpant66@gmail.com

DOI: 10.1201/9781032685793-7

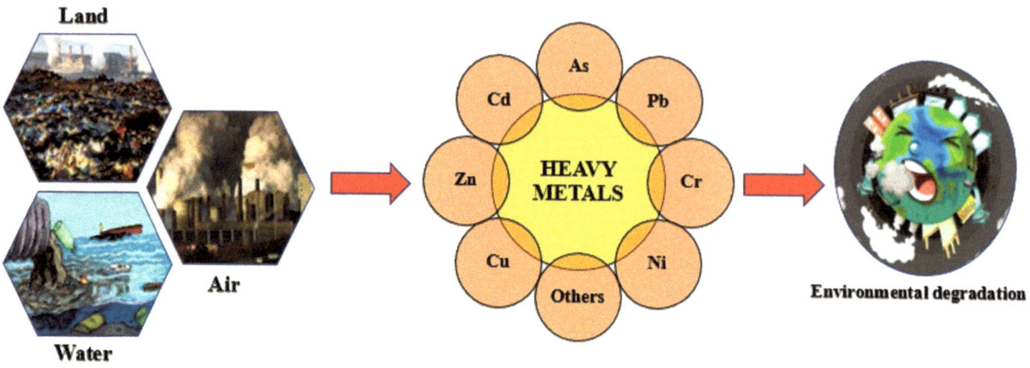

FIGURE 7.1 Anthropogenic sources of heavy metals and environmental impacts.

exposure, and they can induce systemic organ failure. Metals have disrupted numerous ecosystems and impaired the ecological balance due to the significant expansion in their use in several industries such as industry, agriculture, domestic, and technology [1], neurological and other chronic disorders. They also have the potential for genotoxicity and are carcinogenic even at low levels of exposure. Figure 7.1 depicts outline graph of heavy metal fate. Globally all nations are working to combat the pollutants consequences and reduce the menace of heavy metal toxicity by implementing the various regulations and trying to achieve the 18 SDGs of UN. In this part of the book, an attempt has been made to discuss the dispersal of heavy metallic ions in ecosystems and impacts of numerous heavy metals on soils, water and human health (Figure 7.1).

7.2 HEAVY METALS: SOURCES AND DISTRIBUTION

7.2.1 Zinc and Zinc Compounds

Among the predominant elements in the Earth's crust is zinc, usually found in water, soil, air and all foods. Zinc is a blue-white and glossy metal that is brittle at room temperature and malleable around 100–150°C; commonly used in electroplating industry to prevent rust and corrosion by coating steel, iron and other metals with zinc (galvanization). Metallic Zn forms, when it is combined with other metals then it forms alloys such as bronze and brass. In the United States for making pennies zinc and copper alloy is used. Dry cell batteries are also made from metallic zinc [7]. Zinc may react with other metals like oxygen, sulphur and chlorine to generate zinc compounds like zinc oxide, zinc sulphide, zinc sulphate and zinc chloride. The majority of zinc ore is found in the nature in the form of zinc sulphide. The majority of zinc ore found in nature is in the form of zinc sulphide. Ceramics, white paints and other items contain zinc sulphide and zinc oxide. Zinc enters the environment through natural processes as well as human activity.

Most zinc enters the environment through mining, lead and cadmium ores, purification of zinc, steel manufacture, coal and waste combustion. These activities have potential to raise zinc levels in the environment. Zinc can enter aquifers through wastewater channels and other discharge from metallurgical and chemical industries, runoff from zinc-enriched soils and domestic wastewater. Zinc soil levels rise mostly as a result of zinc waste discharged by metal manufacturing business companies and coal ash discharged by power utilities [7]. Zinc is widely distributed in the atmosphere as fine dust particles that settle over land and water. Zinc in lakes and rivers can settle to the bottom, dissolve in water or remain in the form of small floating particles. Zinc can be ingested by fish through the water and their feeding patterns. Some zinc from hazardous waste sites may percolate into the soil and thereby contaminate groundwater, depending on the kind of soil. Animals may consume zinc through feeding or drinking contaminated water. China was recently named one among the world's leading metal manufacturers and consumers. Nonferrous metals are vital fundamental commodities of strategic importance for China's national economy, driving growing demand for metal ores and increasing mining and smelting. In 2018, China's nonferrous metal productivity was around 58 million tonnes, with 5.11 million tonnes producing lead (Pb) and 9.03 million tonnes producing zinc (Zn) [8].

7.2.2 Arsenic and Arsenic Compounds

Inorganic arsenic – As is exists in the Earth's crust naturally with a molar wt. −74.92 g/mol and is a grey-coloured element. Parallel, inorganic arsenic, coexists frequently with the various elements such as O_2, Cl and S. Organic arsenic is formed when carbon and hydrogen are combined with arsenic. Arsenobetaine, dimethylarsinic acid and arsanilic acid are organic arsenic compounds. The pure form of arsenic is scarce in nature and is not extremely poisonous, but it may be quickly oxidized in moist air and converted to arsenic anhydride, which is toxic. In nature, arsenic has four distinct oxidation processes and is mainly found in the +3 and +5 valencies [9–11].

Arsenic compounds have numerous industrial applications. Some toxic compounds such as arsenic trichloride ($AsCl_3$), arsenic anhydride (As_2O_5) and arsenic triiodide (AsI_3), copper arsenide [$Cu_3(AsO_3)_2$] and very poisonous chemicals such as lead arsenide [$Pb_3(AsO_4)_2$], sodium arsenate (Na_2HAsO_4) and calcium arsenide [$Ca_3(AsO_4)_5$] are used to make paint (green, yellow), semiconductors, glass and ceramics.

Insecticides can also be made from arsenic compounds. In viticulture, potato producing and fruit cultivation, sodium arsenate – Na_2HAsO_4 is used to eradicate parasites. Arsenic compounds have also been employed in the production of chemical weapons. Arsenic trioxide (As_2O_3), often known as arsenious anhydride, is one of the deadliest arsenic compounds. The substance is a white precipitate which dissolves easily in water, bases and acids, and it includes arsenic acid (H_3AsO_3). Being volatile, it quickly enters the respiratory tract; nonetheless, there is a danger of loss during toxicological searches. As_2O_3 is commonly used as a rat repellent since mediaeval times. Other arsenic compounds are also being utilized as a rodenticide in small concentrations (10%). Arsenic compound toxicity is generally determined by their water solubility; pure arsenic metal and inorganic compounds (sulphides) are water insoluble and have no hazardous effects in theory, but it can be oxidized to poisonous chemicals (As_2O_3) in a humid environment. The inorganic compounds (arsenous anhydride, alkaline arsenates, arsenide,) which are water soluble are extremely toxic, and arsine (AsH_3) or arsenic hydrogen has a different toxicity as a haemolytic poison than the other forms of inorganic arsenic. Organic arsenic compounds are less harmful and are removed from organisms more quickly [9, 12–14].

In accordance for the IARC – International Agency for Research on Cancer, inorganic arsenic represents a Group 1 carcinogen to humans [15]. However, the EPA – Environmental Protection Agency classed it as a Group A human carcinogen. The EPA determined the cancer gradient factor for inorganic arsenic oral exposure to be 1.5 mg/kg/day. Arsenic has not been studied for cancer inhalation in people or animals. The EPA does not consider arsine to be carcinogenic. The EPA has established restrictions on the amount of arsenic that can be discharged into the nature from industrialized wastes. The arsenic limit in potable water was determined to be 0.01 mg/L [1].

7.2.3 Lead and Lead Compounds

Lead and their compounds are industrial hazardous chemicals that have been used extensively by humans for ages and hence contribute significantly to environmental contamination. Lead is a metal that occur naturally having a blue-grey metal texture that's uncommon in the Earth's crust. It is found mostly as lead ore – lead sulphide (galena). It can be found in many different forms, including lead chloride, lead acetate, lead chromate and oxides of lead. While pure lead is unsolvable in water, lead complexes may vary in their solubility and its chemical symbol is Pb, atomic weight is 207.2 g/mol.

At 1000°C, its vapour pressure is 1.77 mm Hg [9]. Rarely found in pure form, oxides form and inorganic salts such as sulphate, carbonate, acetate and nitrate and organic molecules – lead tetraethyl can also observed in nature. The principal application of Pb is in the production of batteries and frequently employed in the manufacture of paint, electronic conductors, alloys, metal products (lead-tin and lead-antimony) in industry, sound insulation and pipes and puppets or waterproof plates (roofing). Lead acetate is also used in the manufacturing of several ointments. Diluted lead acetate is used as an abortive in treatment under the brand name "white water," while lead arsenate is utilized as a pesticide in agriculture. Lead concentration in fuel is officially limited in several countries (0.5 g/L), and the unleaded petrol is only permitted in vehicles like in India. These organic lead compounds added to petrol contribute significantly to air pollution (particularly in metro cities) [9, 16]. Through ingestion (as food, drinks, vapour, liquids or particles), Pb enters the body mostly and the respiratory tract and a very poisonous metal which enters the biological systems through the interaction of respired air and nutrients from the air, soil and water. The lungs absorb both solid and gaseous up to 90% of the lead in the atmosphere. Small amounts of vapour or lead dust, particularly in the air, induce lead compounds (salts or oxides) to enter in the body through breathing.

7.2.4 Chromium and Chromium Compounds

Chromium – Cr (atomic weight is 51.996 g/mol) is a solid with grey steel texture having a high melting point and goes through various oxidation processes from Cr-II to Cr-VI. It may be used to make a wide range of compounds commonly in Cr-III and Cr-VI forms as trivalent form of chromium compounds are the most stable, and second most stable is the hexavalent form. While Cr-VI composites are easily water soluble, very minimally soluble is chromium(III) compounds, with the water-soluble portion present as Cr-III complexes. Chromium is an element which naturally occurred found in rocks, soil, animals, plants, gases and volcanic dust. Cr is usually found naturally occurring in the form of occurring Cr-III and Cr-VI which is most commonly manufactured by industrial processes, as well as the lesser known metallic Cr(0). Cr-III is an essential dietary element since it is required for appropriate glucose, protein and fat metabolism. Adults should take 50–200 µg/day of Cr-III, which is an important element. The body has several processes in place to convert Cr-VI to Cr-III [17]. The most significant sources in the atmosphere of industrial chromium are related to the manufacturing of ferro-chromes. Metal chrome is mostly utilized in the production of steel and other alloys. It is utilized in copiers as toner and in lesser amounts in other industries e.g., textiles and drilling muds. The whole public is exposed to chromium (Cr-III in general) through drinking water, eating and breathing contaminated air. For both acute and chronic exposure, chromium-VI is far more hazardous than chromium-III. In humans, the respiratory tract is the principal target organ for Cr-VI after exposed during inhalation [1, 18, 19]. Only Cr-VI was determined by the EPA to be a human carcinogen. Various studies on animal suggest that the Cr-VI tempts lung tumours after prolonged inhalation.

7.2.5 Cadmium and Cadmium Compounds

Cadmium (Cd) (atomic weight is 112.41 g/mol) is hard, lustrous and brittle metal which is silver-grey and it contaminates soil and water by transferring waste from industries and residues into the soil and water. In many countries, agricultural soils are exposed to low-moderate cadmium concentration as a result of long-term usage of phosphor-fertilizers and STP sludge. Cd have high mobility in water resulting quickly enter the food chain and the soil-plant system, making it hazardous to human health, plant, and animal. In air, Cd is accumulated by the combustion of fossil fuels (coal or petroleum), as well as garbage. Cd can also be released into the atmosphere as a result of smelting ores like copper, zinc and lead. Cd in the environment is further boosted through the usage of cadmium dyes and the combustion of nickel-cadmium batteries. Cadmium salts include, iodide, acetate, bromide, fluoride, chloride, carbonate, chloride, salicylate, nitrate, cyanide, oxide and tungstate, which are either free or in alloys with nickel, zinc silver and lead. Cd is utilized in cells, nuclear material, porcelain, accumulators, dental amalgam, vacuum tubes, metal coating, plastic and other applications. Cadmium oxide and anthranilate are anthelmintic drugs used in the production and veterinary care of pigs and poultry [1, 18]. The poisoning of Cd occurs through inhalation of dust or vapours at low concentrations over time, as well as ingestion of dissolved Cd and its salts mixed with food. Cigarette smoking is a significant cause of Cd exposure. The smokers of cigarette have twice the amount of cadmium in their bodies as nonsmokers. Food is the most common cause of cadmium exposure. The utilization of phosphate fertilizer or sewage sludge to the agricultural areas can hike the levels of cadmium in some foods. Poisoning is caused by the dark brown vapours produced when cadmium is heated outside. Occupational illnesses are caused by the presence of cadmium's major mineral components in the environment or working air [1]. In the most visible case of Cd poisoning today, 200 people in the Japanese city of Toyama were found to have abnormalities in the bones and kidney as a result of severe Cd poisoning caused by excessive intake of Cd – contaminated rice [20]. The poisoning of Cd is on the list of occupational illnesses in the EU. Cd poisoning was dubbed Itai Itai (pain pain) illness by the Japanese. It is designated as a Group B1 carcinogen by the EPA because it may impacts human physiology [21].

7.2.6 Nickel and Nickel Compounds

The chemical representation of nickel is Ni has an atomic weight of 58.71 g/mol and is a hard metal which is silvery-white that occurs naturally as a component of sulphide, arsenide ores and silicate. Ni is mostly found in the environment as oxides or sulphides with oxygen or sulphur. The Ni^{+2} state is the most prevalent in nature. The physical properties of nickel vary depending on its form. Nickel sulphate, nickel chloride and nickel nitrate are examples of soluble nickel salts. Nickel carbonyl is a very unstable state and not found in the nature and decomposes quickly. In the Earth's crust, nickel is a naturally occurring element because it can be found in trace levels in water, food, soil and air. The most common source of nickel exposure is food. The daily consumption of adults ranges between 100 and 300 µg. Furthermore, those who work in industries that involve the processing, production and use of nickel, as well as those who come into contact with nickel-containing routine products such as jewellery, dining utensils – stainless steel cooking, and who smoke cigarettes, may be exposed to Ni. A compound Nickel carbonyl – $Ni(CO)_4$ is the most toxic nickel compound. The reaction of metallic nickel and CO_2 produces nickel carbonyl, which is employed in the metallurgy and refining sectors. If 30 ppm nickel carbonyl is inhaled for 20 minutes, it can result in lethal toxic consequences [1, 22, 23]. At very low concentrations nickel is found in ambient air as a result of discharges from coal and oil combustion, sewage sludge burning, nickel processing, manufacturing facilities and other sources. Nickel carbonyl exposure is relatively uncommon due to its high instability. It's utilized in electroplating, industrial installations, coinage, stainless steel, batteries (cadmium-nickel mixture, nickel-chromium resistance wires), machine parts, spark-plug and catalysts [1, 9]. Nickel salts, nickel sulphate inhaled and consumed, and nickel acetate are not carcinogenic in rat/mice studies. The EPA has not investigated the possible carcinogenicity of nickel soluble salts in humans. In rats exposed to nickel carbonyl via inhalation, lung tumours were reported to grow. Nickel carbonyl was designated by the EPA as Group B2, a potential human carcinogen. Several studies on animals have also revealed lung tumours caused by nickel subsulphide exposure and nickel refinery dust. Nickel subsulphide and nickel refinery powder have been documented as Group-A human carcinogens by the EPA [24].

7.2.7 Copper and Copper Compounds

Copper is a malleable metallic element that is pale reddish-brown in colour. On the periodic table of elements, it is denoted by the symbol "Cu" and has an atomic number of 29 and an atomic weight of 64 [25, 26]. Copper occurs naturally in rock as a variety of mineral formations, either pure or compounded. It's also been discovered in soils, water and sediments [27, 28]. It can also be found in municipal incinerator zones, metal smelting sites [28], foundries and power plants as a byproduct. Copper sulphate and copper oxide are the most common naturally occurring copper compounds; nevertheless, they mix with other metals such as zine and tin to make alloys such as brass and bronze [25, 26]. At room temperature, copper is a solid metal with good electrical and thermal conductivity. It does not react with water, but rather with the oxygen in the air to generate a thin layer of dark-brown copper oxide. It does not react with sulphide, ammonia or chloride. Copper has 29 known isotopes ranging from 52 to 80. Only two of these are stable and occur naturally, "^{63}Cu" and "^{65}Cu." Cu is the most abundant element (69%). At room temperature, copper is a solid metal with good electrical and thermal conductivity. Copper is widely utilized in the manufacture of wires electrical conductors, pipe and plumbing fixtures, sheet metals, coinage, cooking utensils and other metal items, either in pure form or as an alloy [27, 29]. Copper compounds have found widespread use in agriculture, where they are utilized as fungicides. It is also used in water treatment, specifically to eradicate algae. Other uses include the manufacture of preservative timber, tannery and textiles. Copper is also an important component in pottery, glaze and glass crafts (Figure 7.2).

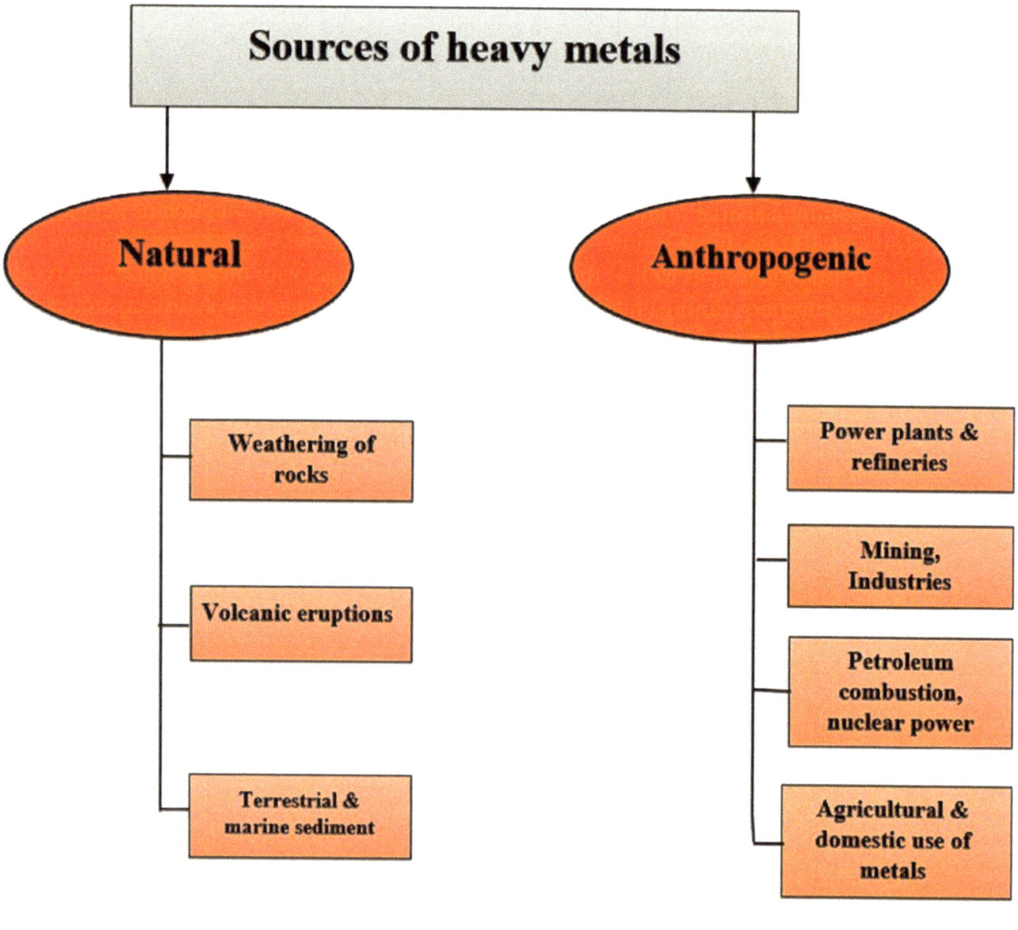

FIGURE 7.2 Heavy metal sources – Natural and anthropogenic.

7.3 HEAVY METAL POLLUTION

Heavy metals naturally occur in various forms in the Earth's crust. The tremendous growth in heavy metal use by human developmental activities has resulted in an influx of metallic compounds in both terrestrial and aquatic ecosystems [5]. Heavy metal pollution – excess concentration in nature – has resulted primarily from anthropogenic activity viz., smelting, metal mining, foundries and other metallurgical industries as well as metal leaching from a variety of sources including landfills, runoffs, automobiles, waste dumps, excretion, chicken manure and roadworks. Figure 7.2 define contamination in aquatic ecosystem. Agriculture has been a secondary source of heavy metal contamination because of the use of heavy metals in pesticides fertilizers, and other chemicals, further accumulated in food chains (Figure 7.3).

7.3.1 Heavy Metal Pollution: Water

Heavy metals (HMs) poisoning of water is a severe environmental issue that impacts humans, plants and animals [30]. Heavy metals are also extremely hazardous to marine animals at very low concentrations [31]. These components have the ability to cause severe physiological changes in the body as well as histological abnormalities in aquatic organisms such as fish tissues [32]. Water pollutants (heavy metal) concentrated by a multitude of factors. One of the major primary causes of rising concentration of heavy metals in water is increasing industrialization and urbanization. Heavy metals are transported via runoff from agri and mine fields, industrial estates, towns and cities [33]. Unprocessed industrial waste discharge into water bodies is a major source of ground and surface water contamination [34]. Contamination of water bodies with heavy metals is a global issue due to the bioaccumulation, environmental permanence and bio-amplification of food chains as well as the toxicity of these elements impacts human health [35].

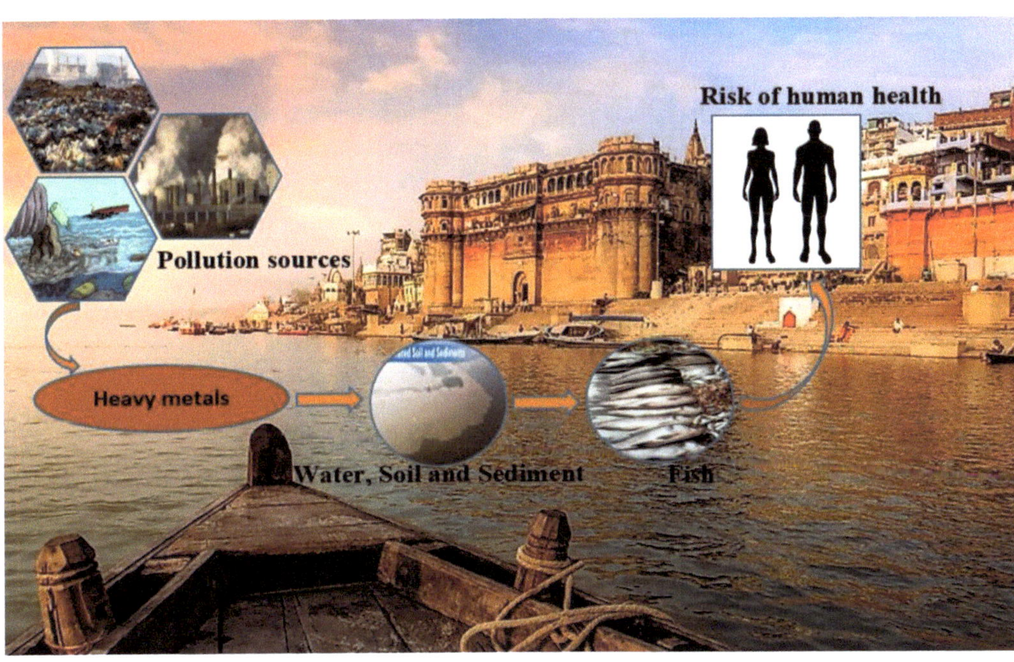

FIGURE 7.3 Heavy metal contamination in aquatic ecosystem and risk on health.

7.3.2 Heavy Metal Pollution: Soil/Sediments

Heavy metal contamination of soil is caused by industrial manufacturing processes, mining runoff, disposal of untreated metallurgical waste, leaded fuel, paints/varnishes, agri application of fertilizers and insecticides, sewage sludge/animal excreta, wastewater irrigation, coal combustion residues and petrochemical leakages in coastal region. Soils have been identified as main sinks for heavy metal ions released into the environment by the aforementioned anthropogenic activities. Because most heavy metals are non-degradable and do not disintegrate through microbial or chemical degradation, their total amounts stay in the environment for a long time after they are discharged [36, 37]. Heavy metals in soils are a major issue because they amplify in food chains through biomagnification, wreaking havoc on the entire ecosystem. While organic contaminants are biodegradable, the heavy metals existence in the environment reduces the decomposition rate and enhances the environmental contamination (organic pollutants with heavy metals). Heavy metals represent a multitude of dangers to animals including humans, plants and entire ecosystems. Soil quality is affected by direct ingestion, ingesting contaminated water, absorption by plants, food chains, changes in soil pH, colour, porosity and natural ecology [38].

In the aquatic environment, sediments are the principal source of HMs. Their quality can provide information about the level of contamination in the water [39]. Sediments act in the water column as both a washbasin and a source of heavy metals [40]. Heavy metal buildup in sediments can potentially lead to pollution of groundwater [41]. Hydro-dynamic circumstances, organic matter and salinity, microbial content, temperature, redox state and particle size are the physicochemical variables that impact heavy metal adsorption – desorption and further sediments contamination via its concentration [42]. The chemical makeup, particle size and total organic matter concentration of sediments all influence heavy metal dispersion. The pH of sediments is critical in determining metal bioavailability. A fall in pH promotes competition between metallic ions and hydroxyl ions for sediment binding positions which may contribute to the breakdown of metal complexes and the release of free metal ions into the water column [43].

7.3.3 Heavy Metal Pollution: Organisms (Food Chain) – Terrestrial and Aquatic

Water, soil, sediments and food all expose aquatic/terrestrial biota to heavy metals in different ways [44]. Freshwater fish are exposed to a variety of harmful HMs released into bodies of freshwater from natural and humanmade sources. Heavy metal pollution in fish has become an important issue worldwide because it endangers fish and endangers the health of seafood consumers. It is crucial to examine HM bioaccumulation in fish species from various aquatic settings [45]. Minamata disease was the first case study of biomagnification of heavy metal in food chain. The analysis of heavy metal concentrations in fish tissues is crucial for marine environment protection and human consumption of fish [46]. In terrestrial food chains also heavy metal contamination is observed. Through soils, wastewater irrigation, pesticides and other anthropogenic causes, they enter into living tissues of crops and further magnify in food chain. Toxicity by heavy metals in aquatic organisms/terrestrial food chains (grains, fruits, vegetables, etc.) is considered a health risk to vulnerable groups of people such as children, pregnant/feeding mothers and those at risk of other chronic diseases.

7.4 HEAVY METAL TOXICITY AND HEALTH HAZARDS

7.4.1 Plants

7.4.1.1 Zinc

Zn is a micronutrient which is essential for all the living species including phytons [47]. The second most abundant transition metal after iron in organisms is Zn and with all enzymatic processes has a strong relationship with Zn. Typically, Zn^{2+} is ecstatic from the soil to the plant via the roots [48]. Zn is important in photosynthetic action (redox reactions) [49]. Zn buildup in the plant roots or shoots causes serious harm. Excess of Zn in the plant tissues causes instability in physiological processes, ultimately resulting in plant death [50]. A prior work by Ebbs et al. [51] found that Zn poisoning caused chlorosis in budding leaves priory, then progressed to older leaves. Hammerschmitt and associates [52] conducted another investigation on different types of plants to explore Zn toxicity in young peach trees. Zn buildup in the root tissues prevents components from being transported to the leaves, according to the findings. Furthermore, productivity of dry matter suffered significantly. According to Song et al. [47], Zn exposure dramatically reduced root length and photosynthesis.

7.4.1.2 Cadmium

Cd is classified in the category of non-essential harmful heavy metal because it enters the environment from anthropogenic sources, posing dangers to the entire ecosystem [53, 54]. Because of mobility, bioavailability and concentration, plant roots absorb the majority of Cd ions. The remainder, on the other hand, may be directly absorbed from the air [55]. Cadmium takes entry into plant cells through various carriers, such as Calcium – Ca channels, and accumulates in shoots, roots and edible sections [54]. Many physiological and biochemical changes occur in plants when Cd levels are high [56]. Moreover, Cd buildup in plants has harmful effects such as the suppression of some activities (mineral transport, nutrient uptake and photosynthetic apparatus). Furthermore, it can impede Fe transport into plant shoots [53, 57]. Cd has a negative impact on plant phenotypic (for example, lowering plant weight and shoot and root length), cytotoxicity (for example, limiting photosynthetic effectiveness and reducing chlorophyll content) and metabolic processes (for example, cell damage and chlorosis) [58]. Seregin et al. [59] discovered that heavy metal translocation into conducting tissues was mediated through the xylem parenchyma; still Cd transport across the xylem was restricted. Cd poisoning, according to Gratão et al. [60], could reduce the dry mass and length of the roots. Vardhan et al. [61] used Pichia sp. and *Azotobacter chroococcum* to assess the ecological impact of Cd ions on *Lactuca sativa* seedlings. According to the findings, Cd(II) inhibited seed germination, impeded root growth and reduced biomass and development. Furthermore, Cd(II) had a negative impact on microbial development.

7.4.1.3 Lead

Pb like Hg and Cd is not required for the plant growth [62]. Pb is regarded as both a helpful and a hazardous metal [63]. Because of its extreme toxicity, it has been designated as a significant pollutant [64]. Pb ions are gradually normally transported via the soil to the plant through the xylem via the roots [65]. Even at low concentrations, Pb poisoning is harmful to plants, reducing healthy plant development and decreasing agricultural productivity and yield [66]. Pb accumulation in plants causes physiological problems involving DNA damage and the destruction of root and shoot systems, as well as a decrease in enzyme activity [67, 68]. The toxic effects of Pb in plants was investigated by Nas and Ali [69]. The findings revealed that a high concentration of Pb had an impact on fresh biomass and growth in plants. The results were consistent with the same as Cimrin et al. [70]. Pb, in accordance with Kushwaha et al. [71], reduced seed germination and protein content.

7.4.1.4 Copper

Third most consumed metal in the world is Cu, after only steel and aluminium. Seed development and plant nutrition require a tiny quantity of Cu. On the other hand, Cu is a very toxic metal at high concentrations [72, 73]. A range of variables impact plant Cu absorption, including soil physico-chemical qualities and other physiological elements [74]. Although the root system is in charge of absorbing Cu ions from the soil, the concentration of Cu in the roots is normally higher than in the shoots. According to Kopittke et al. [75], the root epidermis contained the highest concentration of Cu. According to Adrees et al. [76], Cu toxicity reduced crop production, chlorophyll biosynthesis and plant productivity through photosynthesis and nutrient alteration. Cu concentrations of 5 mg/kg are enough to injure the plant, lowering plant development and productivity. Barbosa et al. [77] investigated the influence of high Cu concentration on the maize plant's productivity and morphology. Their findings showed that increasing the applied Cu dose dramatically reduced plant height. Aly et al. [78] investigated the adverse effects of high concentrations on maize plants. The data revealed a significant decrease in length of stem, demonstrating that Cu's harmful effect reaches all regions of the plant.

7.4.1.5 Arsenic

The adverse effects of arsenic on the fruit yield (dropped) and leaf fresh weight (decreased) in tomato (*Lycopersicon esculentum*) [79] were observed. Also causes, stunted development, chlorosis and wilting in canola – *Brassica napus* [80]. Arsenic also lowers seed germination, leaf area, seedling height as well as dry matter production in rice – *Oryza sativa* [81, 82].

7.4.1.6 Chromium

Chromium is a toxic metal that can affect both animals and plants. Chromium-induced oxidative stress in plants induces lipid peroxidation, which damages cell membranes severely. The degradation of photosynthetic pigments caused by chromium-induced oxidative stress leads to a decline in growth. High chromium concentrations can disrupt the photosynthetic process by altering the ultrastructure of chloroplasts. Because first physiological activity affected by Cr is the seed germination, the capacity of a germinate seed in Cr-containing medium reveals its tolerance to this metal [83]. With 200 M Cr, *Echinochloa colona* seed germination was reduced to 25% [84]. The sequence of metal toxicity to new root primordia in *Salix viminalis*, according to Prasad [85], is Cd > Cr > Pb; with Chromium having a higher effect on root length than the other heavy metals studied. In respect to CO_2 fixation, photophosphorylation, enzyme activity electron transport, chromium stress is an important factor impacting photosynthesis [86].

7.4.1.7 Nickel

Nickel is a crucial nutrient for the development of plants. The amount of Ni needed for typical plant growth is really little. It is essential to understand the beneficial roles of Ni in plants and its harmful consequences as the concentration of Ni pollution in the environmental matrix rises. Growing plants in soil with elevated Ni^{2+} levels revealed nutritional imbalances that interfered with cell membrane function. Ni^{2+} therefore had an impact on the plasma membrane's lipid composition and H-ATPase activity in *O. sativa* shoots [87]. Changes in water balance were associated with additional symptoms observed in Ni^{2+}-treated plants. Monocot and dicot plant species have lower water content as a result of the greater Ni^{2+} absorption. The course of Ni^{2+} toxicity is monitored by the decline in water absorption in plants.

7.4.2 Humans

7.4.2.1 Effect of Heavy Metal Toxicity on Human Health

Heavy metal pollution has prompted significant worry, long-term accumulation, even at low concentrations due to its harmful consequences and biomagnification qualities. Heavy metals are among the most hazardous toxicants found in multilayered soils and environmental pollutants [88, 89]. The heavy metals presence in the environment increases the possibility of such dangerous components being consumed by living beings and accumulating in various body organs such as the bone, kidney, liver and so on. Furthermore, heavy metal buildup harms a range of physiological systems, including the endocrine, nervous, skeletal, immunological and circulatory systems [90, 91]. Humans are exposed to heavy metals all throughout the world through ingestion (drinking or eating) or inhalation (breathing). Individuals are endangered by working in or near facilities that employ these metals and their compounds, in addition to being nearby a location in which such metals have been illegally discarded of. Fishing and hunting practices may also provide greater risks of exposure and harmful health impacts to subsistence hunters and fishermen. Because of the widespread exposure, the impact of these hazardous substances on the health of people has recently become an important issue. Environmental hazardous metal emissions reached alarming levels as a result of the increased usage of a varied range of metals in industries and in our daily lives via novel applications [92].

7.4.2.2 Effect of Heavy Metal Toxicity on Children's Health

Toxicity of heavy metals enter the body via a variety of routes, including ingestion (food, water), dermal absorption and inhalation (air), results in health consequences from heavy metal exposure. The impacts of heavy metals on the children's health have become more serious than those on the adults. Heavy metals should be based on more thought due to their significant toxicity risk, prevalence and widespread use [93]. Even during the embryo stage the heavy metals are impacting the living tissues.

As per Chunhabundit [94], Cd poisoning can cause renal impairment due to damaged PCT – proximal convoluted tubules, which are linked to mitochondrial dysfunctioning. Furthermore, Cd exposure has been linked to osteoporosis paediatric cancer and delayed development in children [95, 96]. Gardner et al. [97] reported that Cd exposure was inversely related to newborn baby growth (height and weight). Lead poisoning is one of the most common avoidable childhood poisonings. Children are especially vulnerable to Pb poisoning, which produces long-term brain abnormalities that impair learning and behaviour [98]. Furthermore, exceptionally high Pb exposure levels have been reported to induce an increase in irritability, dullness and a shorter attention span in the CNS – central nervous system, which can lead to headache, epilepsy, seizures, coma, neurological problems and even death [99].

Cu is essential for brain functioning; however, if the concentration in the cells above the metabolic requirement, it can be harmful [100]. Serum Cu levels that are elevated have been associated to impaired working memory in children. According to Zhou et al. [101], elevated Cu concentrations can impair working memory via influencing attention. Wilson disease (WD) is a genetic condition characterized by hepatocellular Cu disposition that is caused by Wilson ATPase enzyme failing, a P1B-ATPase represented by the ATP7B gene. Although hepatic dysfunction is common in children and adolescents, clinical symptoms of behavioural, neurological and haemorrhagic diseases may develop. As a result of WD, very young children may acquire clinically apparent liver disease [102].

Zn poisoning in youngsters has been observed seldom. Low appetite, headaches, diarrhoea, nausea and vomiting have all been connected to acute consumption. Arsenault and Brown [103] discovered in 2003 that pre-schoolers reported Zn intakes that above the dietary recommendation consumption, and that 36% of the children had meals containing Zn that exceeded their recommended higher intake range. Combining a greater availability of Zn-fortified foods with a rise in use of Zn supplements in the United States may result in excessive consumption among children [104]. Cu deficiency owing to Zn was discovered as a result of Cu competition on the same absorption location [105]. Excessive Zn supplementation has been related to a decreased immune response [106] and may interact with a number of drugs, antibiotics (quinolone and tetracycline) and diuretics (e.g., thiazides) [107].

7.4.2.3 Effect of Heavy Metal Toxicity on Adult's Health

Certain heavy metals have the potential to harm nearly all human cells and tissues. Cadmium and its derivatives can disrupt calcium metabolism, osteoporosis, bone disorders, renal tubular failure and lung cancer [108]. Cd poisoning has been linked to breast and prostate cancer, neurodegenerative disorders and diabetes [109]. Cd exposure has an impact on male reproductive systems, sperm quality as well as spermatogenesis, namely hormone synthesis/release and sperm motility. Cd also decreases the fertility, menstrual cycles, reproductive hormonal balance according to clinical and human investigations [110]. Scientists found that Cd may be harmful to expecting women. According to a study, Cd may cause placental harm and diminish newborn baby weight [111]. Cd inhalation can cause considerable lung damage and may possibly result in death [112].

Pb has a wide range of harmful effects on biological systems. Symptoms are mostly nonspecific such as decreased infertility in men, miscarriage in females, cognitive function in adults and behavioural problems in infants. Common symptoms include renal failure, stunted development, anaemia, hypertension and abdominal colic [113]. Pb has been shown to be hazardous by replacing Zn in haem production and decreasing the function of haem synthesizing enzymes. It is carcinogenic and excessive amounts of exposure can result in death [114].

Cu is a hazardous metal that is abundant in the liver, brain and kidneys [115]. Cu poisoning is commonly associated with gastrointestinal (GI) side symptoms such as stomach discomfort, anorexia, vomiting and jaundice, haematemesis, melena as well as erosive gastropathy [116]. Excess Cu causes oxidative stress, damage to DNA and a reduction in cell development [117].

Zn is recognized to be nontoxic, especially when given orally. With increased Zn intake, however, indications of overt poisoning symptoms (epigastric discomfort, nausea, exhaustion and vomiting) will ensue. Although evident symptoms of poisoning necessitate the administration of rather large amounts of Zn, there is evidence that utilizing Zn supplements by people can have negative effects under specific conditions [118]. Neuronal damage and mortality have been associated to Zn-induced neurotoxicity in stroke, epilepsy, neurodegenerative illnesses and traumatic brain injury. Inorganic arsenic causes acute, subacute and chronic toxic effects that might be local or systemic. Abdominal cramps, extremities (hyperesthesia), abdominal patellar reflexes and abdominal electrocardiograms are examples of acute toxic consequences [119]. Such effects are common at doses of exposure comparable to ~50 glkg weight/day. Chronic As poisoning, on the other hand, causes anaemia, liver and kidney damage. The majority of the negative effects of Cr on humans are related with its hexavalent form. Chromium poisoning is characterized by hepatic necrosis, nephrites and gastrointestinal discomfort [120]. Cr^{+6} can also induce nasal, and mucous membrane ulcers, as well as dermatitis through skin contact.

For adults, the World Health Organization (WHO) recommends a preliminary tolerated weekly intake (PTWI) is Ni approximately 0.035 mg/kg body weight [121]. Nickel causes embryo toxicity and nephron toxicity, as well as allergic responses and contact dermatitis [122]. In the general population, nickel sensitization occurs as a result of contact to coinage, jewellery, fasteners, clothing and watchcases. Nickel causes asthma, conjunctivitis and local or systemic reactions to Ni-containing prostheses such as cardiac valve replacements, cardiac pacemaker wires, joint replacements, pins and dental inlays [120]. Nickel being a possible lung carcinogen can induce lung fibrosis, skin allergies and respiratory tract cancer in occupationally exposed people [123].

7.5 CONCLUSION AND FUTURE PERSPECTIVES

Heavy metals, i.e. arsenic – Ar, lead – Pb, zinc – Zn, cadmium – Cd, chromium – Cr, copper – Cu and nickel – Ni, have an adverse impact on the living organisms, environment and mainly human beings. This needs to be addressed at local and global level. The legislation, standard operating procedures and guidelines by authorities are necessary to be executed. Failure for controlling the exposures serious consequences in the forthcoming times due to the adverse effects of the heavy metals. Scientific, technological and legislative solutions can reduce occupational health risks and community consequences caused by heavy metal exposure. Addressing this issue requires understanding the major sources and consequences thereto, implementing proper heavy metal waste management, strict regulations and responsible handling. Advanced remediation techniques like phytoremediation, microbial treatment, etc., can also help alleviate contamination effects. Immediate attention and strenuous efforts from govt. authorities, industries, communities and individuals are necessary to prevent further contamination and ensure a cleaner sustainable environment for forthcoming generations. Monitoring exposure and potential interventions to shrink further exposure to heavy metals in the environment and humans can be a significant step towards prevention. Local and worldwide collaboration is crucial for developing adequate heavy metal prevention methods. Extensive studies regarding monitoring, remediation and control of heavy metal toxicity is need of the hour.

ACKNOWLEDGEMENT

Authors extend their heartfelt gratitude to Department of Chemistry, The School of Physical Sciences (SoPS), DIT University, Dehradun, for necessary support during this work.

REFERENCES

1. Masindi, Vhahangwele, and Khathutshelo L. Muedi. "Environmental contamination by heavy metals." *Heavy Metals* 10 (2018): 115–132.
2. Walker, Colin Harold, R. M. Sibly, Stephen P. Hopkin, and David B. Peakall. *Principles of ecotoxicology*. CRC Press, 2012.
3. Martin, Yvonne E., and Edward A. Johnson. "Biogeosciences survey: Studying interactions of the biosphere with the lithosphere, hydrosphere and atmosphere." *Progress in Physical Geography* 36, no. 6 (2012): 833–852.
4. Li, Hao, Jun Yao, Ning Min, Zhihui Chen, Miaomiao Li, Wancheng Pang, Bang Liu, Ying Cao, Dongyang Men, and Robert Duran. "Comprehensive evaluation of metal (loid)s pollution risk and microbial activity characteristics in non-ferrous metal smelting contaminated site." *Journal of Cleaner Production* 344 (2022): 130999. https://doi.org/10.1016/j.jclepro.2022.130999
5. Gautam, Pavan Kumar, Ravindra Kumar Gautam, Sushmita Banerjee, M. C. Chattopadhyaya, and J. D. Pandey. "Heavy metals in the environment: Fate, transport, toxicity and remediation technologies." In Deepak Pathania (ed.), *Heavy metals: Sources toxicity and remediation techniques*, vol. 60. Nava Science Publishers, Inc., pp. 101–130. 2016.
6. Vieira, Maria Conceiçao, Rafael Torronteras, Francisco Córdoba, and Antonio Canalejo. "Acute toxicity of manganese in goldfish *Carassius auratus* is associated with oxidative stress and organ specific antioxidant responses." *Ecotoxicology and Environmental Safety* 78 (2012): 212–217.
7. Roney, Nickolette. *Toxicological profile for zinc*. Agency for Toxic Substances and Disease Registry, 2005.
8. Song, Le, Jiazhong Qian, Fawang Zhang, Xiangke Kong, Hui Li, Song Luan, Qinjun Zhang, Zhiqiang Kang, Zhantao Han, and Zhaoji Zhang. "An ecological remediation model combining optimal substrate amelioration and native hyperaccumulator colonization in non-ferrous metal tailings pond." *Journal of Environmental Management* 322 (2022): 116141. https://doi.org/10.1016/j.jenvman.2022.116141
9. Dökmeci, Ayşe Handan. "Environmental impacts of heavy metals and their bioremediation." In Mazen Khaled Nazal and Hongbo Zhao (eds.), *Heavy metals-their environmental impacts and mitigation*. IntechOpen, 2020.
10. Misra, Sandhya, and Krishna G. Misra. "Phytoremediation: An alternative tool towards clean and green environment." In Sachi Shah, V. Venkatramanan and Ram Prasad (eds.), Sustainable green technologies for environmental management. Springer, pp. 87–109. 2019. https://doi.org/10.1007/978-981-13-2772-8_5
11. Dökmeci, Ayşe Handan, T. Yildiz, A. Ongen, and N. Sivri. "Heavy metal concentration in deepwater rose shrimp species (*Parapenaeus longirostris*, Lucas, 1846) collected from the Marmara Sea Coast in Tekirdağ." *Environmental Monitoring and Assessment* 186 (2014): 2449–2454.
12. Loprieno, N. "International Agency for Research on Cancer (IARC) monographs on the evaluation of carcinogenic risk of chemicals to man:" Relevance of data on mutagenicity"." *Mutation Research* 31, no. 3 (1975): 210.
13. California Environmental Protection Agency (CalEPA). *Technical support document for the determination of noncancer chronic reference exposure levels*. Office of Environmental Health Hazard Assessment, 2008.
14. Mohammed, Ali Seid, Anil Kapri, and Reeta Goel. "Heavy metal pollution: Source, impact, and remedies." In Mohammad Saghir Khan, Almas Zaidi, Reeta Goel and Javed Musarrat (eds.), Biomanagement of metal-contaminated soils. Springer, pp. 1–28. 2011.
15. U.S. Environmental Protection Agency (USEPA). Toxics Release Inventory Guidance for Reporting the Lead and Lead Compounds Category. USA, 2020: 20–34.
16. Ara, Anjum, and Jawed Ahmad Usmani. "Lead toxicity: A review." *Interdisciplinary Toxicology* 8, no. 2 (2015): 55–64.
17. Khan, Alam, and Mahpara Safdar. "Role of diet, nutrients, spices and natural products in diabetes mellitus." *Pakistan Journal of Nutrition* 2 (2003): 1–12.
18. Tabelin, Carlito Baltazar, Toshifumi Igarashi, Mylah Villacorte-Tabelin, Ilhwan Park, Einstine M. Opiso, Mayumi Ito, and Naoki Hiroyoshi. "Arsenic, selenium, boron, lead, cadmium, copper, and zinc in naturally contaminated rocks: A review of their sources, modes of enrichment, mechanisms of release, and mitigation strategies." *Science of the Total Environment* 645 (2018): 1522–1553. https://doi.org/10.1016/j.scitotenv.2018.07.103

19. Huvinen, M., J. Uitti, P. Oksa, P. Palmroos, and P. Laippala. "Respiratory health effects of long-term exposure to different chromium species in stainless steel production." *Occupational Medicine* 52, no. 4 (2002): 203–212. https://doi.org/10.1093/occmed/52.4.203
20. Genchi, Giuseppe, Maria Stefania Sinicropi, Graziantonio Lauria, Alessia Carocci, and Alessia Catalano. "The effects of cadmium toxicity." *International Journal of Environmental Research and Public Health* 17, no. 11 (2020): 3782.
21. Huang, Yingying, Chuntao He, Chuang Shen, Jingjie Guo, Samavia Mubeen, Jiangang Yuan, and Zhongyi Yang. "Toxicity of cadmium and its health risks from leafy vegetable consumption." *Food & Function* 8, no. 4 (2017): 1373–1401. https://doi.org/10.1039/C6FO01580H
22. Barceloux, Donald G., and Donald Barceloux. "Nickel." *Journal of Toxicology: Clinical Toxicology* 37, no. 2 (1999): 239–258.
23. Roychowdhury, Tarit, Hiroshi Tokunaga, and Masanori Ando. "Survey of arsenic and other heavy metals in food composites and drinking water and estimation of dietary intake by the villagers from an arsenic-affected area of West Bengal, India." *Science of the Total Environment* 308, no. 1–3 (2003): 15–35. https://doi.org/10.1016/S0048-9697(02)00612-5
24. Zhao, Jinshun, Xianglin Shi, Vincent Castranova, and Min Ding. "Occupational toxicology of nickel and nickel compounds." *Journal of Environmental Pathology, Toxicology and Oncology* 28, no. 3 (2009). https://doi.org/10.1615/JEnvironPatholToxicolOncol.v28.i3.10
25. The facts on copper, Dartmouth College, Hanover, NH, 2015. http://www.dartmouth.edu/
26. Canadian soil quality guidelines for copper environmental and human health, Canadian Council of Environmental Ministers, Winnipeg, Manitoba, 1997.
27. US EPA. Toxics Release Inventory. U.S. Environmental Protection Agency, Washington, DC, 1999. www.epa.gov
28. Copper fact sheet. Bureau of Reclamation, Technical Service Center. Water Treatment Engineering and Research Group. D-8230, Denver, CO, 2001.
29. Bettelle. "Environmental Technology Verification Report Environmental Bio-detection Products Inc. Toxi-Chromotest". US Environment Protection Agency. 2006.
30. Rezania, Shahabaldin, Shazwin Mat Taib, Mohd Fadhil Md Din, Farrah Aini Dahalan, and Hesam Kamyab. "Comprehensive review on phytotechnology: Heavy metals removal by diverse aquatic plants species from wastewater." *Journal of Hazardous Materials* 318 (2016): 587–599. https://doi.org/10.1016/j.jhazmat.2016.07.053
31. Abrar, Maal, Z. Hussain, M. Akif, K. Sok, A. Muhammad, A. R. Khan, and M. Khan. "Textile effluents and their contribution towards aquatic pollution in the Kabul River (Pakistan)." *Journal of the Chemical Society of Pakistan* 24, no. 4 (2011): 106.
32. Ahmed, Md Kawser, Elora Parvin, Md Monirul Islam, Mosammat Salma Akter, Shahneawz Khan, and Md Habibullah Al-Mamun. "Lead-and cadmium-induced histopathological changes in gill, kidney and liver tissue of freshwater climbing perch *Anabas testudineus* (Bloch, 1792)." *Chemistry and Ecology* 30, no. 6 (2014): 532–540.
33. Zhuang, Ping, Zhi-an Li, Murray B. McBride, Bi Zou, and Gang Wang. "Health risk assessment for consumption of fish originating from ponds near Dabaoshan mine, South China." *Environmental Science and Pollution Research* 20 (2013): 5844–5854. https://doi.org/10.1007/s11356-013-1606-0
34. Afzal, Muhammad Salman, Adil Ashraf, and Mohammad Nabeel. "Characterization of industrial effluents and groundwater of Hattar industrial estate, Haripur." *Advances in Agriculture and Environmental Science: Open Access (AAEOA)* 1, no. 2 (2018): 70–77.
35. Rajaei, Ghasem, Borhan Mansouri, Hamid Jahantigh, and Amir Hossein Hamidian. "Metal concentrations in the water of Chah Nimeh reservoirs in Zabol, Iran." *Bulletin of Environmental Contamination and Toxicology* 89 (2012): 495–500.
36. Lepp, Nicholas W. *Effect of heavy metal pollution on plants: Metals in the environment.* Pollution Monitoring Series. Springer Dordrecht, 2012.
37. Athar, M., and S. B. Vohora. "*Heavy metals and environment.* New Age International (P) Limited, 2001.
38. Musilova, Janette, Julius Arvay, Alena Vollmannova, Tomas Toth, and Jan Tomas. "Environmental contamination by heavy metals in region with previous mining activity." *Bulletin of Environmental Contamination and Toxicology* 97 (2016): 569–575.
39. Zahra, Azmat, Muhammad Zaffar Hashmi, Riffat Naseem Malik, and Zulkifl Ahmed. "Enrichment and geo-accumulation of heavy metals and risk assessment of sediments of the Kurang Nallah—Feeding tributary of the Rawal Lake Reservoir, Pakistan." *Science of the Total Environment* 470 (2014): 925–933.

40. Fernandes, Lina L., and G. N. Nayak. "Heavy metals contamination in mudflat and mangrove sediments (Mumbai, India)." *Chemistry and Ecology* 28, no. 5 (2012): 435–455.
41. Sanyal, Tanmay, Anilava Kaviraj, and Subrata Saha. "Deposition of chromium in aquatic ecosystem from effluents of handloom textile industries in Ranaghat–Fulia region of West Bengal, India." *Journal of Advanced Research* 6, no. 6 (2015): 995–1002. https://doi.org/10.1016/j.jare.2014.12.002
42. Zhao, Shengnan, Xiaohong Shi, Changyou Li, Hanmeng Zhang, and Yong Wu. "Seasonal variation of heavy metals in sediment of Lake Ulansuhai, China." *Chemistry and Ecology* 30, no. 1 (2014): 1–14. https://doi.org/10.1080/02757540.2013.841894
43. Ali Azadi, Nammam, Borhan Mansouri, Lucia Spada, Mohammad Hosein Sinkakarimi, Younes Hamesadeghi, and Ali Mansouri. "Contamination of lead (Pb) in the coastal sediments of north and south of Iran: A review study." *Chemistry and Ecology* 34, no. 9 (2018): 884–900.
44. Ali, Mir Mohammad, Delower Hossain, A. Al-Imran, Md. Suzan Khan, Maksuda Begum, and Mahadi Hasan Osman. "Environmental pollution with heavy metals: A public health concern." In Heavy *metals-their environmental impacts and mitigation* (Eds. Mazen Khaled Nazal and Hongbo Zhao). Intech Open, pp. 771–783. 2021. https://doi.org/10.5772/intechopen.96805
45. Russo, Rosario, A. Lo Voi, A. De Simone, F. P. Serpe, Aniello Anastasio, Tiziana Pepe, D. Cacace, and Lorella Severino. "Heavy metals in canned tuna from Italian markets." *Journal of Food Protection* 76, no. 2 (2013): 355–359. https://doi.org/10.4315/0362-028X.JFP-12-346
46. Ali, Hazrat, and Ezzat Khan. "Trophic transfer, bioaccumulation, and biomagnification of non-essential hazardous heavy metals and metalloids in food chains/webs—Concepts and implications for wildlife and human health." *Human and Ecological Risk Assessment: An International Journal* 25, no. 6 (2019): 1353–1376. https://doi.org/10.1080/10807039.2018.1469398
47. Song, Changzheng, Yifan Yan, Abel Rosado, Zhenwen Zhang, and Simone Diego Castellarin. "ABA alleviates uptake and accumulation of zinc in grapevine (*Vitis vinifera* L.) by inducing expression of ZIP and detoxification-related genes." *Frontiers in Plant Science* 10 (2019): 872.
48. Broadley, Martin R., Philip J. White, John P. Hammond, Ivan Zelko, and Alexander Lux. "Zinc in plants." *New Phytologist* 173, no. 4 (2007): 677–702.
49. Hafeez, Babar, Mohd Khanif Yusop, and Muhammad Saleem. "Role of zinc in plant nutrition – A review." *American Journal of Experimental Agriculture* 3, no. 2 (2013): 374–391.
50. Liang, Jian, and Weifei Yang. "Effects of zinc and copper stress on antioxidant system of olive leaves." *IOP Conference Series: Earth and Environmental Science* 300, no. 5 (2019), p. 052058.
51. Ebbs, Stephen D., and Leon V. Kochian. Toxicity of zinc and copper to Brassica species: Implications for phytoremediation. *Journal of Environmental Quality* 26, no. 3. (1997): 776–781. https://doi.org/10.2134/jeq1997.00472425002600030026x
52. Hammerschmitt, Rodrigo K., Tadeu L. Tiecher, Daniela B. Facco, Lincon OS Silva, Raissa Schwalbert, Gerson L. Drescher, Edicarla Trentin et al. "Copper and zinc distribution and toxicity in 'Jade'/'Genovesa' young peach tree." *Scientia Horticulturae* 259 (2020): 108763. https://doi.org/10.1016/j.scienta.2019.108763
53. Jibril, Sani Ahmad, Siti Aishah Hassan, Che Fauziah Ishak, and Puteri Edaroyati Megat Wahab. "Cadmium toxicity affects phytochemicals and nutrient elements composition of lettuce (*Lactuca sativa* L.)." *Advances in Agriculture* 2017 (2017). https://doi.org/10.1155/2017/1236830
54. Zhou, Cheng, Ninggao Ge, Jiansheng Guo, Lin Zhu, Zhongyou Ma, Shiyong Cheng, and Jianfei Wang. "Enterobacter asburiae reduces cadmium toxicity in maize plants by repressing iron uptake-associated pathways." *Journal of Agricultural and Food Chemistry* 67, no. 36 (2019): 10126–10136. https://pubs.acs.org/doi/abs/10.1021/acs.jafc.9b03293
55. Ismael, Marwa A., Ali Mohamed Elyamine, Mohamed G. Moussa, Miaomiao Cai, Xiaohu Zhao, and Chengxiao Hu. "Cadmium in plants: Uptake, toxicity, and its interactions with selenium fertilizers." *Metallomics* 11, no. 2 (2019): 255–277. https://doi.org/10.1039/c8mt00247a
56. Feng, Jianpeng, Qinghua Shi, Xiufeng Wang, Min Wei, Fengjuan Yang, and Huini Xu. "Silicon supplementation ameliorated the inhibition of photosynthesis and nitrate metabolism by cadmium (Cd) toxicity in *Cucumis sativus* L." *Scientia Horticulturae* 123, no. 4 (2010): 521–530. https://doi.org/10.1016/j.scienta.2009.10.013
57. Schützendübel, Andres, Petia Nikolova, Claudia Rudolf, and Andrea Polle. "Cadmium and H_2O_2-induced oxidative stress in *Populus × canescens* roots." *Plant Physiology and Biochemistry* 40, no. 6–8 (2002): 577–584. https://doi.org/10.1016/S0981-9428(02)01411-0
58. Hayat, Malik Tahir, Muhammad Nauman, Nida Nazir, Shafaqat Ali, and Nazneen Bangash. "Environmental hazards of cadmium: Past, present, and future." In Mirza Hasanuzzaman, Majeti Narasimha Vara Prasad and Masayuki Fujita (eds.), *Cadmium toxicity and tolerance in plants*. Academic Press, pp. 163–183. 2019.

59. Seregin, I. V., and A. D. Kozhevnikova. "Roles of root and shoot tissues in transport and accumulation of cadmium, lead, nickel, and strontium." *Russian Journal of Plant Physiology* 55 (2008): 1–22. https://doi.org/10.1134/S1021443708010019
60. Gratão, Priscila L., Carolina C. Monteiro, Mônica L. Rossi, Adriana P. Martinelli, Lázaro EP Peres, Leonardo O. Medici, Peter J. Lea, and Ricardo A. Azevedo. "Differential ultrastructural changes in tomato hormonal mutants exposed to cadmium." *Environmental and Experimental Botany* 67, no. 2 (2009): 387–394. https://doi.org/10.1016/j.envexpbot.2009.06.017
61. Vardhan, Kilaru Harsha, Ponnusamy Senthil Kumar, and Rames C. Panda. "A review on heavy metal pollution, toxicity and remedial measures: Current trends and future perspectives." *Journal of Molecular Liquids* 290 (2019): 111197. https://doi.org/10.1016/j.molliq.2019.111197
62. Diaconu, Mariana, Lucian Vasile Pavel, Raluca-Maria Hlihor, Mihaela Rosca, Daniela Ionela Fertu, Markus Lenz, Philippe Xavier Corvini, and Maria Gavrilescu. "Characterization of heavy metal toxicity in some plants and microorganisms—A preliminary approach for environmental bioremediation." *New Biotechnology* 56 (2020): 130–139. https://doi.org/10.1016/j.nbt.2020.01.003
63. Uzu, Gaëlle, Sophie Sobanska, Yassine Aliouane, Philippe Pradere, and Camille Dumat. "Study of lead phytoavailability for atmospheric industrial micronic and sub-micronic particles in relation with lead speciation." *Environmental Pollution* 157, no. 4 (2009): 1178–1185. https://doi.org/10.1016/j.envpol.2008.09.053
64. Chauhan, Priti, A. B. Rajguru, M. Y. Dudhe, and Jyoti Mathur. "Efficacy of lead (Pb) phytoextraction of five varieties of *Helianthus annuus* L. from contaminated soil." *Environmental Technology & Innovation* 18 (2020): 100718. https://doi.org/10.1016/j.eti.2020.100718
65. Gupta, Neha, Krishna Kumar Yadav, Vinit Kumar, Sandeep Kumar, Richard P. Chadd, and Amit Kumar. "Trace elements in soil-vegetables interface: Translocation, bioaccumulation, toxicity and amelioration – A review." *Science of the Total Environment* 651 (2019): 2927–2942. https://doi.org/10.1016/j.scitotenv.2018.10.047
66. Ashraf, Umair, Adam S. Kanu, Quanquan Deng, Zhaowen Mo, Shenggang Pan, Hua Tian, and Xiangru Tang. "Lead (Pb) toxicity; physio-biochemical mechanisms, grain yield, quality, and Pb distribution proportions in scented rice." *Frontiers in Plant Science* 8 (2017): 259. https://doi.org/10.3389/fpls.2017.00259
67. Gichner, Tomáš, Irena Žnidar, and Jiřina Száková. "Evaluation of DNA damage and mutagenicity induced by lead in tobacco plants." *Mutation Research/Genetic Toxicology and Environmental Mutagenesis* 652, no. 2 (2008): 186–190. https://doi.org/10.1016/j.mrgentox.2008.02.009
68. Reddy, Alavala Matta, Surabhi Giridara Kumar, Gottimukkala Jyothsnakumari, S. Thimmanaik, and Chinta Sudhakar. "Lead induced changes in antioxidant metabolism of horsegram (*Macrotyloma uniflorum* (Lam.) Verdc.) and bengalgram (*Cicer arietinum* L.)." *Chemosphere* 60, no. 1 (2005): 97–104. https://doi.org/10.1016/j.chemosphere.2004.11.092
69. Nas, Farouk S., and Muhammad Ali. "The effect of lead on plants in terms of growing and biochemical parameters: A review." *MOJ Ecology & Environmental Sciences* 3, no. 4 (2018): 265–268. https://doi.org/10.15406/mojes.2018.03.00098
70. Cimrin, K., Metin Turan, and Burcak Kapur. "Effect of elemental sulphur on heavy metals solubility and remediation by plants in calcareous soils." *Fresenius Environmental Bulletin* 16 (2007): 1113–1120.
71. Kushwaha, Anamika, Nidhi Hans, Sanjay Kumar, and Radha Rani. "A critical review on speciation, mobilization and toxicity of lead in soil-microbe-plant system and bioremediation strategies." *Ecotoxicology and Environmental Safety* 147 (2018): 1035–1045. https://doi.org/10.1016/j.ecoenv.2017.09.049
72. Wuana, Raymond A., and Felix E. Okieimen. "Heavy metals in contaminated soils: A review of sources, chemistry, risks and best available strategies for remediation." *International Scholarly Research Notices* 2011 (2011): 402647.
73. Chiou, Wei-Yang, and Fu-Chiun Hsu. "Copper toxicity and prediction models of copper content in leafy vegetables." *Sustainability* 11, no. 22 (2019): 6215. https://doi.org/10.3390/su11226215
74. Bjuhr, Josef. "Trace metals in soils irrigated with wastewater in a periurban area downstream Hanoi City, Vietnam." UNSPECIFIED, Uppsala (2007). https://stud.epsilon.slu.se/11698/1/bjuhr_j_171123.pdf
75. Kopittke, Peter M., Neal W. Menzies, Martin D. de Jonge, Brigid A. McKenna, Erica Donner, Richard I. Webb, David J. Paterson et al. "In situ distribution and speciation of toxic copper, nickel, and zinc in hydrated roots of cowpea." *Plant Physiology* 156, no. 2 (2011): 663–673. https://doi.org/10.1104/pp.111.173716
76. Adrees, Muhammad, Shafaqat Ali, Muhammad Rizwan, Muhammad Ibrahim, Farhat Abbas, Mujahid Farid, Muhammad Zia-ur-Rehman, Muhammad Kashif Irshad, and Saima Aslam Bharwana. "The effect of excess copper on growth and physiology of important food crops: A review." *Environmental Science and Pollution Research* 22 (2015): 8148–8162. https://doi.org/10.1007/s11356-015-4496-5

77. Barbosa, Rogério Hidalgo, Luciane Almeri Tabaldi, Fábio Rodrigo Miyazaki, Márcio Pilecco, Samir Oliveira Kassab, and Daísa Bigaton. "Foliar copper uptake by maize plants: Effects on growth and yield." *Ciência Rural* 43 (2013): 1561–1568. https://doi.org/10.1590/S0103-84782013000900005
78. Aly, Amina A., and Amal A. Mohamed. "The impact of copper ion on growth, thiol compounds and lipid peroxidation in two maize cultivars ('*Zea mays*' L.) grown' in vitro'." *Australian Journal of Crop Science* 6, no. 3 (2012): 541–549.
79. Barrachina, A. Carbonell, F. Burlo Carbonell, and J. Mataix Beneyto. "Arsenic uptake, distribution, and accumulation in tomato plants: Effect of arsenite on plant growth and yield." *Journal of Plant Nutrition* 18, no. 6 (1995): 1237–1250.
80. Cox, M. S., P. F. Bell, and J. L. Kovar. "Differential tolerance of canola to arsenic when grown hydroponically or in soil." *Journal of Plant Nutrition* 19, no. 12 (1996): 1599–1610.
81. Marin, A. R., S. R. Pezeshki, P. H. Masschelen, and H. S. Choi. "Effect of dimethylarsenic acid (DMAA) on growth, tissue arsenic, and photosynthesis of rice plants." *Journal of Plant Nutrition* 16, no. 5 (1993): 865–880. https://doi.org/10.1080/01904169309364580
82. Abedin, Md Joinal, J. Cotter-Howells, and Andy A. Meharg. "Arsenic uptake and accumulation in rice (*Oryza sativa* L.) irrigated with contaminated water." *Plant and Soil* 240 (2002): 311–319. https://doi.org/10.1023/A:1015792723288
83. Peralta, J. R., Jorge L. Gardea-Torresdey, K. J. Tiemann, E. Gomez, S. Arteaga, E. Rascon, and J. G. Parsons. "Uptake and effects of five heavy metals on seed germination and plant growth in alfalfa (*Medicago sativa* L.)." *Bulletin of Environmental Contamination and Toxicology* 66 (2001): 727–734. https://doi.org/10.1007/s001280069
84. Rout, Gyana R., Sanghamitra Samantaray, and Premananda Das. "Effects of chromium and nickel on germination and growth in tolerant and non-tolerant populations of *Echinochloa colona* (L.) Link." *Chemosphere* 40, no. 8 (2000): 855–859. https://doi.org/10.1016/S0045-6535(99)00303-3
85. Prasad, M. N. V., Maria Greger, and Tommy Landberg. "*Acacia nilotica* L. bark removes toxic elements from solution: Corroboration from toxicity bioassay using *Salix viminalis* L. in hydroponic system." *International Journal of Phytoremediation* 3, no. 3 (2001): 289–300. https://doi.org/10.1080/15226510108500060
86. Clijsters, H., and F. Van Assche. "Inhibition of photosynthesis by heavy metals." *Photosynthesis Research* 7 (1985): 31–40. https://doi.org/10.1007/BF00032920
87. Ros, Roc, David T. Cook, Carmen Martinez-Cortina, and Isabel Picazo. "Nickel and cadmium-related changes in growth, plasma membrane lipid composition, ATPase hydrolytic activity and proton-pumping of rice (*Oryza sativa* L. cv. Bahia) shoots." *Journal of Experimental Botany* 43, no. 11 (1992): 1475–1481.
88. Jiang, Xiaolu, Bin Zou, Huihui Feng, Jingwen Tang, Yulong Tu, and Xiuge Zhao. "Spatial distribution mapping of Hg contamination in subclass agricultural soils using GIS enhanced multiple linear regression." *Journal of Geochemical Exploration* 196 (2019): 1–7. https://doi.org/10.1016/j.gexplo.2018.10.002
89. Xiao, Ran, Di Guo, Amjad Ali, Shenshen Mi, Tao Liu, Chunyan Ren, Ronghua Li, and Zengqiang Zhang. "Accumulation, ecological-health risks assessment, and source apportionment of heavy metals in paddy soils: A case study in Hanzhong, Shaanxi, China." *Environmental Pollution* 248 (2019): 349–357. https://doi.org/10.1016/j.envpol.2019.02.045
90. Lamas, Gervasio A., Ana Navas-Acien, Daniel B. Mark, and Kerry L. Lee. "Heavy metals, cardiovascular disease, and the unexpected benefits of chelation therapy." *Journal of the American College of Cardiology* 67, no. 20 (2016): 2411–2418. https://doi.org/10.1016/j.jacc.2016.02.066
91. Ma, Yukun, Prasanna Egodawatta, James McGree, An Liu, and Ashantha Goonetilleke. "Human health risk assessment of heavy metals in urban stormwater." *Science of the Total Environment* 557 (2016): 764–772. https://doi.org/10.1016/j.scitotenv.2016.03.067
92. Wang, Feifei, Qingyu Guan, Jing Tian, Jinkuo Lin, Yanyan Yang, Liqin Yang, and Ninghui Pan. "Contamination characteristics, source apportionment, and health risk assessment of heavy metals in agricultural soil in the Hexi Corridor." *Catena* 191 (2020): 104573. https://doi.org/10.1016/j.catena.2020.104573
93. Al Osman, Muwaffak, Fei Yang, and Isaac Yaw Massey. "Exposure routes and health effects of heavy metals on children." *Biometals* 32 (2019): 563–573. https://doi.org/10.1007/s10534-019-00193-5
94. Chunhabundit, Rodjana. "Cadmium exposure and potential health risk from foods in contaminated area, Thailand." *Toxicological Research* 32 (2016): 65–72. https://doi.org/10.5487/TR.2016.32.1.065
95. Schoeters, Greet, Elly Den Hond, Moniek Zuurbier, Rima Naginiene, Peter Van Den Hazel, Nikolaos Stilianakis, Roberto Ronchetti, and Janna G. Koppe. "Cadmium and children: Exposure and health effects." *Acta Paediatrica* 95 (2006): 50–54. https://doi.org/10.1080/08035320600886232

96. Sherief, Laila M., Elhamy R. Abdelkhalek, Amal F. Gharieb, Hanan S. Sherbiny, Doaa M. Usef, Mohamed AA Almalky, Naglaa M. Kamal, Mostafa A. Salama, and Wafaa Gohar. "Cadmium status among pediatric cancer patients in Egypt." *Medicine* 94, no. 20 (2015): e740.
97. Gardner, Renee M., Maria Kippler, Fahmida Tofail, Matteo Bottai, Jena Hamadani, Margaretha Grandér, Barbro Nermell, Brita Palm, Kathleen M. Rasmussen, and Marie Vahter. "Environmental exposure to metals and children's growth to age 5 years: A prospective cohort study." *American Journal of Epidemiology* 177, no. 12 (2013): 1356–1367. https://doi.org/10.1093/aje/kws437
98. Maret, Wolfgang. "The bioinorganic chemistry of lead in the context of its toxicity." *Lead: Its Effects on Environment and Health* 17 (2017): 1–20.
99. Pfadenhauer, Lisa Maria, Jacob Burns, Anke Rohwer, and Eva Annette Rehfuess. "A protocol for a systematic review of the effectiveness of interventions to reduce exposure to lead through consumer products and drinking water." *Systematic Reviews* 3, no. 1 (2014): 1–13.
100. Lieberman, Harris R., Robin B. Kanarek, and Chandan Prasad, eds. *Nutritional neuroscience*. CRC Press, 2005.
101. Zhou, Guoping, Xiaopeng Ji, Naixue Cui, Siyuan Cao, Chang Liu, and Jianghong Liu. "Association between serum copper status and working memory in schoolchildren." *Nutrients* 7, no. 9 (2015): 7185–7196. https://doi.org/10.3390/nu7095331
102. Roberts, Eve A., and Piotr Socha. "Wilson disease in children." In Anna Członkowska and Michael L. Schilsky (eds.), *Wilson disease*. Handbook of clinical neurology 142, Elsevier, pp. 141–156. 2017. https://doi.org/10.1016/B978-0-444-63625-6.00012-4
103. Arsenault, Joanne E., and Kenneth H. Brown. "Zinc intake of US preschool children exceeds new dietary reference intakes." *The American Journal of Clinical Nutrition* 78, no. 5 (2003): 1011–1017.
104. Black, Jennifer L., Domingo J. Piñero, and Niyati Parekh. "Zinc and cognitive development in children: Perspectives from international studies." *Topics in Clinical Nutrition* 24, no. 2 (2009): 130–138. https://doi.org/10.1097/TIN.0b013e3181a6b947
105. Shaikhkhalil, Ala K., Jennifer Curtiss, Teresa D. Puthoff, and Christina J. Valentine. "Enteral zinc supplementation and growth in extremely-low-birth-weight infants with chronic lung disease." *Journal of Pediatric Gastroenterology and Nutrition* 58, no. 2 (2014): 183. https://doi.org/10.1097%2FMPG.0000000000000145
106. Chandra, Ranjit Kumar. "Excessive intake of zinc impairs immune responses." *JAMA* 252, no. 11 (1984): 1443–1446.
107. Willoughby, Jennifer L., and Christine N. Bowen. "Zinc deficiency and toxicity in pediatric practice." *Current Opinion in Pediatrics* 26, no. 5 (2014): 579–584.
108. Avenant-Oldewage, A., and H. M. Marx. "Bioaccumulation of chromium, copper and iron in the organs and tissues of *Clarias gariepinus* in the Olifants River, Kruger National Park." *Water SA* 26, no. 4 (2000): 569–582.
109. Jiang, Ling-Feng, Tian-Ming Yao, Zhi-Liang Zhu, Chong Wang, and Liang-Nian Ji. "Impacts of Cd(II) on the conformation and self-aggregation of Alzheimer's tau fragment corresponding to the third repeat of microtubule-binding domain." *Biochimica et Biophysica Acta (BBA)-Proteins and Proteomics* 1774, no. 11 (2007): 1414–1421. https://doi.org/10.1016/j.bbapap.2007.08.014
110. Kumar, Sunil, and Anupama Sharma. "Cadmium toxicity: Effects on human reproduction and fertility." *Reviews on Environmental Health* 34, no. 4 (2019): 327–338. https://doi.org/10.1515/reveh-2019-0016
111. Esteban-Vasallo, María D., Nuria Aragones, Marina Pollan, Gonzalo López-Abente, and Beatriz Perez-Gomez. "Mercury, cadmium, and lead levels in human placenta: A systematic review." *Environmental Health Perspectives* 120, no. 10 (2012): 1369–1377. https://doi.org/10.1289/ehp.1204952
112. Fatima, Ghizal, Ammar Mehdi Raza, Najah Hadi, Nitu Nigam, and Abbas Ali Mahdi. "Cadmium in human diseases: It's more than just a mere metal." *Indian Journal of Clinical Biochemistry* 34 (2019): 371–378. https://doi.org/10.1007/s12291-019-00839-8
113. Wani, A. L., A. Ara, and J. A. Usmani. "Lead toxicity: A review." *Interdisciplinary Toxicology* 8 (2015): 55–64.
114. Ogwuegbu, M. O., and M. A. Ijioma. "Effects of certain heavy metals on the population due to mineral exploitation." In *International conference on scientific and environmental issues in the population, environment and sustainable development in Nigeria*. University of Ado Ekiti, pp. 8–10. 2003.
115. Uriu-Adams, Janet Y., and Carl L. Keen. "Copper, oxidative stress, and human health." *Molecular aspects of medicine* 26, no. 4–5 (2005): 268–298. https://doi.org/10.1016/j.mam.2005.07.015
116. Gamakaranage, Champika SSK, Chaturaka Rodrigo, Sajitha Weerasinghe, Ariaranee Gnanathasan, Visvalingam Puvanaraj, and Harshani Fernando. "Complications and management of acute copper sulphate poisoning; a case discussion." *Journal of Occupational Medicine and Toxicology* 6 (2011): 1–5. https://doi.org/10.1186/1745-6673-6-34

117. Oe, Shinji, Koichiro Miyagawa, Yuichi Honma, and Masaru Harada. "Copper induces hepatocyte injury due to the endoplasmic reticulum stress in cultured cells and patients with Wilson disease." *Experimental Cell Research* 347, no. 1 (2016): 192–200. https://doi.org/10.1016/j.yexcr.2016.08.003
118. Hambidge, K. M., C. E. Casey and N. F. Krebs. Zinc. In *Trace elements in human and animal nutrition*. Ed. W Mertz. 5th Edition. pp 1137. Academic Press, Inc., New York (1986).
119. Franzblau, Alfred, and Ruth Lilis. "Acute arsenic intoxication from environmental arsenic exposure." *Archives of Environmental Health: An International Journal* 44, no. 6 (1989): 385–390. https://doi.org/10.1080/00039896.1989.9935912
120. Athar, Mohammad, and Shashi B. Vohora. *Heavy metals and environment*. New Age International, 1995.
121. WHO (World Health Organization), "Quality Directive of Potable Water." Geneva, second ed., p. 197. (1994).
122. US EPA. Nickel and Nickel Compound Pollution Prevention Fact Sheet, 96, 1–2, US EPA's Integrated Risk Information System (IRIS) (2002).
123. Kasprazak, K. S., F. W. Sunderman, and K. Salnikow. "Nickel carcinogenesis." *Mutation Research/Fundamental and Molecular Mechanisms of Mutagenesis* 533, no. 1–2 (2003): 67–97.

8 Routes of Exposure (Inhalation, Ingestion, Dermal Contact) of Heavy Metals and Their Implications for Human Health

*Priya Shukla, Pradeep Srivastava, and Abha Mishra**
School of Biochemical Engineering, Indian Institute of Technology (Banaras Hindu University), Varanasi, India

8.1 INTRODUCTION

Heavy metals are defined as metals having a comparatively higher atomic number, weight, and density. Significant worries exist over these elements' environmental consequences and potential adverse effects on human health due to their broad residential, industrial, medical, agricultural, and technological usage.[1] Because heavy metals are so common, exposure to humans can happen by ingestion, absorption through the skin, or inhalation. Numerous heavy metals have undergone active toxicological investigations due to their high level of toxicity, and the metals are handled through legislation. Diverse cancers were also brought on by heavy metal exposure. Arsenic, chromium, nickel, and cadmium have all been categorized as Group 1 carcinogens.[2]

Although heavy metals exist naturally in the atmosphere, human endeavors, including farming, smelting, mining, and manufacturing, contribute considerably to their release.[3] Additional possible causes of contamination comprise atmospheric deposition, corrosion of the metals, resurfacing of the sediment, leaching, and erosion of their ions.[4] They also influence the environmental pollution caused by such natural occurrences as volcanic eruptions and weathering.[5] High heavy metal concentrations in soil can result from many agricultural techniques and have hazardous effects. Some heavy metals are harmful at low concentrations and are persistent and non-biodegradable. They represent a serious risk to humans (Figure 8.1) and soil health in areas where they are bioavailable. Plants can absorb heavy metals accumulated in soil, allowing them to enter the body of humans and food web.[6]

Researches show that minerals like chromium and copper are crucial for a variety of biochemical and physical processes, and that deficiencies in these minerals can lead to diseases or syndromes. As a result, heavy metals are either necessary or not. In plants and animals, essential metals are used for physical and metabolic processes. To give an instance, copper is mandatory for the hemoglobin synthesis and the metabolism of carbohydrates. But in excess, it harms cellular structures. Metals including lead, barium, aluminum, cadmium, and mercury are non-essential.[3]

A number of cellular structures and functions, for instance the mitochondria, lysosomes, nuclei, endoplasmic reticulum, cell membrane, and several enzymes associated with metabolism, bodily cleansing, and damage repair, are known to be inhibited by heavy metals. Furthermore, they engage with proteins and DNA, resulting in damage to DNA and structural alterations that could result in apoptosis, cancer, or cycle regulation.[3]

* **Corresponding Author:** abham.bce@itbhu.ac.in

DOI: 10.1201/9781032685793-8

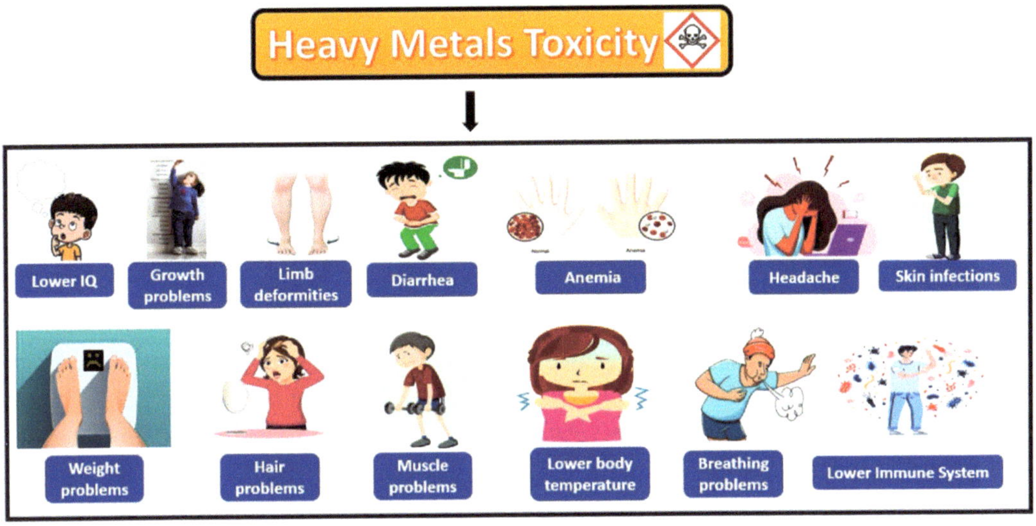

FIGURE 8.1 Symptoms of heavy metals on public health.

Mercury, lead, chromium, cadmium, and barium present a substantial threat to public health since they are highly poisonous. They are categorized as systemic toxicants as a result. The chapter concentrates on these five heavy metals despite the fact that there are many more due to their (1) extensive usage, (2) danger associated with few of their forms, and (3) predominance in the surroundings. It is important to be aware that even at lower exposure levels, these metals can harm different human organs. These are either categorized as "confirmed" or "probable" human carcinogens, depending on the mechanism(s) through which they cause poisoning. It should be mentioned that each of them has unique traits and chemical compositions that determine how hazardous it is. Some of these components' taxological mechanisms and activities are not adequately explained. To address the health problems in public health caused by exposure to heavy metals, there exists a requirement for a coherent branch of knowledge that could direct the deployment of countermeasures. The pathways of exposure to chromium, mercury, cadmium, lead, and barium, as well as the detrimental effects on health and toxicological mechanisms, are discussed in this work (Figure 8.1).

8.2 PATHWAYS OF HEAVY METAL EXPOSURE, HEALTH IMPACTS, AND TOXICOLOGICAL MECHANISMS

8.2.1 Mercury (Hg)

8.2.1.1 Routes of Exposure

Different types of mercury (inorganic, organic, and elemental) differ in their characteristics, applications, and levels of toxicity. The natural degassing of the Earth's crust causes mercury to be released into the environment.[7] Human actions might potentially result in mercury pollution. The main route of the pollutant's global transportation, the atmosphere, contains it as mercury vapor. It doesn't change while in the atmosphere for nearly a year. Then it transforms into a form that is soluble in water and reinstates to the Earth's surface. Microorganisms, primarily bacteria, then transform it back into mercury vapor. Planktons and fish are just two examples of the aquatic creatures that can introduce mono-methyl mercury into the aquatic food chain. Consequently, mercury exposure in humans is mostly caused through eating seafoods.[8,9]

Contact to mercury comes through multiple sources to humans. For instance, although industrial contact with mercury is typically brought on by breathing elemental mercury and consuming seafood from polluted water sources, non-occupational communities get their exposure by means of dietary and dental amalgam. As a result, the crucial form of mercury exposure for ASGM miners is elemental vapor, which can be precisely detected in urine.[10] Urine levels of mercury indicate chronic, long-term exposure; however, blood mercury values point to more recent contact. In urine the half-time of inorganic mercury has been calculated to be between 40 and 80 days, and that of methamphetamine in the blood and throughout the body has been calculated to be between 70 and 50 days, respectively.[11] Chronic mercury markers include hair exposure because mercury rarely separates and is consistently present once integrated with hair. It is possible to compute historical mercury exposure amounts using the supposition that hair grows 1 centimeter per month. Methylmercury excretes slowly from the human system, as evidenced by its half-life of 35–100 days in hair. Pre-analytical variance is associated with growth rates, and elemental mercury can also be assimilated into hair through synthetic hair-waving, coloring, and dyeing.[12]

8.2.1.2 Health Effects

Children's exposure to mercury can have negative effects in two different ways. First, maternal consumption of contaminated seafood during pregnancy has been linked to unfavorable developmental consequences.[13] The neurotoxicity of mercury is also connected to impaired neurological and cognitive development.[14] The verbal IQ scores of kids subjected to mercury from their moms remained lower, and they performed poorly in social growth, hand-eye coordination, and social conduct.[15] Mental impairment has also been linked to mercury exposure in young children.[16, 17]

Growth problems, epilepsy, excessive salivation, limb deformities, damaged cerebellum, misaligned eyes, chorea and athetosis, and primitive reflexes are some of the ways that mercury toxicity can manifest itself, according to Liu et al.[18] Additionally, according to the World Health Organization,[7] breathing in mercury vapor can damage the function of important bodily functions like the kidneys, lungs, digestive, immunological, and brain systems. In addition to corroding the skin and eyes, mercury exposure has been linked to neurocognitive issues such as tremors, sleeplessness, and memory loss.[19–21]

A rise in type 2 diabetes occurrences has also been linked to mercury toxicity.[22,23] Methylmercury has negative effects on the growth and operation of pancreatic beta (β) cells, according to Schumacher and Abbott. Therefore, impaired pancreatic function may result in hyperglycemia, insulin resistance, and finally diabetes.[22] According to Jeon et al.,[24] oxidative stress brought on by mercury poisoning causes the pancreatic beta cells to die and become dysfunctional. Additionally, it is being shown that oxidative stress brought on by mercury can disrupt the insulin signaling pathway. Mercury poisoning can also alter the mitochondria's normal structure and operation, according to Jia et al.,[25] which means it can lead to cardiorenal metabolic syndrome once more. Consequently, persons exposed to mercury have been reported to have metabolic, renal, and cardiovascular problems.[25]

8.2.1.3 Mechanism of Toxicity

As ions of mercury enter the body, they cause widespread corrosion, enzyme inhibition, and protein precipitation, all of which have harmful effects. Mercury ions can bind proteins having amine, amide, carboxyl, and phosphoryl groups, among others. These groups are highly susceptible to the mercury reaction in proteins. These proteins bind to mercury and stop functioning. The chemical makeup and oxidative state of mercury affect its toxicity. For instance, elemental Hg quickly crosses plasma membranes because it is extremely soluble in lipids. Mercury is more hazardous when it is divalently oxidized than when it is monovalently oxidized. Furthermore, biological forms of mercury absorb more energy (90%) than inorganic forms (only 10%). Therefore, it has been suggested that organic mercury compounds have more corrosive effects on the gastrointestinal mucosa.[26]

8.2.2 Lead (Pb)

8.2.2.1 Routes of Exposure

Actual exposure to lead represents combined ancient and modern contaminants considering lead (Pb) can accumulate in a number of mediums and is a widely recognized environmental pollutant. The two main problems, therefore, are locating up-to-date data to obtain an accurate depiction of the causes of pollution and calculating the human body levels of exposure and outcomes that occur under protracted exposure settings. Air, sediments, water, and soil are just a few of the environmental media that contain lead. Lead is found in drinking water,[27] edibles,[28] and even breast milk[29] due to this contamination, assemblage, and subsequent redistribution.

Road traffic, industry, energy generation, and municipal waste combustion facilities are all sources of lead emissions in the air compartment.[30] The most frequent method of entry is through fumes, which are the main means of lead transmission. Although lead is rarely absorbed via the skin, since they regularly use their mouths and gnaw on things that are not edible such as paint chips, lead-coated household items, and toys, kids tend to be more susceptible to lead toxicity.[31] Additionally, the manufacture of asphalt uses slag remnants from municipal solid waste incineration amenities, which leaches lead into the Earth's crust and freshwater. The biggest contributors to water pollution with regard to water and sediment are cables and lost fishing equipment abandoned on the ocean bed.[32]

8.2.2.2 Health Effects

The impact of exposure of lead on people varies according to the quantity or intensity of exposure. Kids subjected to very high proportion of lead have been shown to have shorter attention spans, increased irritability, and dullness in their central nervous systems, which can lead to seizures, headaches, comas, and even death.[33] According to Lanphear et al., lead's detrimental effects on human health start to manifest when blood lead levels approach 10 g/dL.[34] Children's intellect has been linked to lead exposure above this threshold. Similar conclusions have been drawn by Hong et al. who discovered school-age children's impulsivity and IQ are significantly impacted by modest levels of lead exposure.[35] Once more, lead levels in blood and academic performance have been inversely correlated, and lead exposure in childhood has been linked to children's failure in reading and math.[36]

According to the Institute for Health Metrics and Evaluation (IHME), lead poisoning was responsible for an estimated 540,000 deaths and 13.9 million healthy life years (DALYs) lost globally in 2016. In addition, IHME anticipated that there was a risk of intellectual disability in 63.8% of lead exposure patients. The most frequently impacted nations were those with low and moderate incomes. It continuously and significantly affects public health. Additionally, this increases the strain on these nations, who are already dealing with other well-known public health issues. High numbers of DALYs related to lead exposure might be utilized for the remediation of toxic waste in general to increase sustenance and capital.[37–39]

8.2.2.3 Mechanism of Toxicity

The term "neurotoxicity" refers to both an imbalance in the neurophysiological function and disclosure of the whole body to lead-toxic substances. Neurocognitive symptoms and psychiatric abnormalities are the two main symptoms brought on by such exposures. It primarily affects the CNS, often the growing brain, and has an impact on nearly every organ system.[40] Blood vessels and hypertension are both impacted by lead. Blood arteries clogged with lead can result in a fatal heart attack right away. Recent evidence suggests that lowered levels of lead in the blood may be associated with high blood pressure, and scientists have found a fascinating correlation between lower lead amounts in blood and cardiovascular disease.[41]

Lead is capable of blocking the functioning of enzymes, accelerate the creation of oxygen species, and intensify oxidative stress when it is present in the blood in low amounts.

According to several reports, the crucial factor in the development of lead-associated poisoning is oxidative stress. The central nervous system was identified by Nemsadze et al. as the primary organ affected by lead toxicity. Lead poisoning is connected with several neural system mechanisms, together with the initiation of oxidative stress, inhibition of calcium ion-dependent enzymes, and an escalation in nerve cell death.[42] Additionally, it is said that lead is mostly deposited in the teeth and bones, where it builds up over time and affects the bones, kidneys, and liver.[43]

8.2.3 Chromium

8.2.3.1 Routes of Exposure

Chromium is an organically formed heavy metal which is typically found in hexavalent (Cr(VI)) and trivalent (Cr(III)) forms in the environment. Reactive intermediates are created during the reduction of Cr(VI) to Cr(III), which adds to the cytotoxicity, genotoxicity, and carcinogenicity of compounds containing Cr(VI). Humans obtain most of their chromium from diet, including of their chromium from diet, which includes meat, vegetables, urban air, cigarettes, and hip or knee replacements.[44,45] Industrial chemicals like Cr(VI) are frequently employed in the production of paints, metal coatings, stainless steel, chrome, alloy cast irons, and wood treatments. Contrarily, Cr(III) salts including chromium picolinate (CrP), chromium polynicotinate, and chromium chloride are utilized as nutritional supplements and micronutrients and have been shown to have a variety of positive health impacts in both humans and animals.[46]

Important to note, infertility, cardiovascular disease, and diabetes are all linked to a lack of the necessary vitamin Cr(III), while Cr(VI) causes cancer. Chromium enters the body through skin ingestion, absorption, and inhalation, according to the CDC.[7] Inhalation and skin contact are the two important routes that people are exposed to chromium at work. However, consumption of tainted food and water accounts for a significant part of chromium subjection in the general population.

8.2.3.2 Health Impacts

Being exposed to chromium has been correlated with adverse impacts on respiratory health. Scientists looked into the respiratory consequences of prolonged exposure in the industries related to electroplating. They discovered that employees subjected to professional chromium toxicity had more instances of the harmful pulmonary effects of chromium, comprising wheezing, phlegm, shortness of breath, and productive coughing.[47] Neghab et al. also reported that compared to non-exposed individuals, workers who had professional subjection to the chromium had lower parameters of lung health.[47] Chromium exposure is characterized by perforations of the irritation of the larynx, mucous membranes of the nasal septum, and edema, throat, co-occurrence of acute bronchitis and asthma, and restriction of the bronchial airways.[48] Other respiratory impacts of hexavalent chromium exposure comprise bronchitis, pneumonia, discomfort and itching in the nose, and reduced pulmonary function.[49] Thus, the verification recommends that occupational exposure may result in compromised respiratory function.

Once more, chromium exposure has been connected with negative impacts on the gastrointestinal system, hematology, immunological system, kidneys, and liver.[49] Furthermore, the metal has been linked to dermatitis, heightened skin sensitivity, and skin ulcers.[50] Ray's most current research on the hematological effects of chromium on blood cells.[51] Scientists showed that chromium exposure can reduce the volume of packed red blood cells, the number of immature erythrocytes after ingestion, the total number of red blood cells, the levels of leucocytes, illuminating intravascular hemolysis, and the mean corpuscular hemoglobin concentration.[52,53]

8.2.3.3 Mechanism of Toxicity

Hexavalent chromium is linked to chromium poisoning, according to laboratory and clinical findings. Chromium-induced molecular damage linked to intracellular decrease can also be used to explain a large portion of its toxicity. Numerous kinds of point mutations in deoxyribonucleic acid have been documented to result from subjection to hexavalent chromium, in addition to harm to the chromosome. Additionally, its toxicity has been shown to cause oxidative protein modifications and the creation of adducts.[54] The toxicity of chromium compounds is associated with the conversion of hexavalent chromium to trivalent chromium inside of cells. The breakdown of cellular components when Cr(VI) is being reduced to Cr(III) has also been linked with intracellular chromium toxicity. Deoxyribonucleic acid damage, among others, results from the creation of free radicals in this damage.

8.2.4 Cadmium

8.2.4.1 Routes of Exposure

Eight naturally occurring isotopes make up cadmium, which doesn't exist in its basic form in nature but rather as a compound.[55] Carbon oxide, cadmium carbonate, and cadmium sulfide have been named the three most prevalent cadmium compounds. It can enter the body through ingesting contaminated foods like shellfish, smoking, drinking water, and inappropriate handling of the metal itself that has been contaminated with cadmium. Notably, exposure to cadmium can also happen indirectly, particularly through plants. They take up cadmium from Earth's crust that has been polluted by cadmium-infected irrigation water, cadmium-enriched fertilizers, and cadmium fallout from the atmosphere.[56]

8.2.4.2 Health Effects

The effects of cadmium on the menstrual cycle and reproductive hormones, as well as low birth weight, early birth, and pregnancy loss, have been documented.[57] In postmenopausal individuals, cadmium was also inversely correlated with bone mineral density. Women have a greater body burden of cadmium than men due to having larger amounts of the metal in their urine, blood, and kidneys. Intestinal cadmium absorption rises with overt iron shortage and a reduced bodily iron storage, circumstances that are common in fertile-age women.[58] Recent research has looked at the relationship between the vascular pathology of cadmium and the risk of subarachnoid hemorrhage.[59] Being an endocrine disruptor, cadmium is thought to adhere to estrogen receptor alpha and change signal transduction via the MAPK and estrogen signaling pathways. Cadmium may interact with several hormonal signaling pathways. The cardiovascular system is impacted by low levels of cadmium.[60] According to epidemiological research, cadmium exposure may encourage the onset of musculoskeletal conditions such as rheumatoid arthritis (RA), osteoporosis, and osteoarthritis (OA).[61]

Cadmium has strong teratogenic and mutagenic effects, but it additionally exhibits adverse impacts at low concentrations on both male and female reproductive systems, which may disrupt pregnancy or its outcome. Small amounts of cadmium exposure can harm the cardiovascular system, liver, skeletal system, kidneys, sight, hearing, and kidney function.[62] This is brought on by alterations in how many genes are expressed in the embryo, which causes anomalous methylation of genes responsible in both the placenta and the embryo. The propensity of cadmium to readily bind to thiols in the absence of the methyl donor S-adenosyl methionine has been connected to the epigenetic modification patterns it causes, which in turn affects the methylome and, ultimately, the activity of DNA methyltransferases. Disorders of placental and fetal development may result from this.[63]

8.2.4.3 Mechanism of Toxicity

In the words of Rani et al.,[64] the ability of cadmium to cause genomic instability through multiple and multifactorial mechanisms is related to its ability to cause toxicity. Its interactions with DNA repair pathways, creation of reactive oxygen species, and induction of cell death are some of the most important mechanisms. Cadmium toxicity is also linked to its induction of pro-inflammatory cytokines and suppression of nitric oxide production, which plays a protective role.[65] Cadmium-induced damage has additionally been associated with a rise in oxidative stress because of its catalytic involvement in producing reactive oxygen species, accelerated lipid peroxidation, the depletion of protein-bound glutathione, and the oxidization of sulfhydryl groups.[65]

8.2.5 Barium

8.2.5.1 Routes of Exposure

The two main forms of barium that are found in nature are barium carbonate and barium sulfate.[7] Inhaling polluted air and consuming impure food and drink are the important ways that people are exposed to barium. According to the CDC,[7] minute amounts of barium may enter the body when it comes into close proximity to the skin. The solubility of the substance affects how easily it can be absorbed into the bloodstream. Strongly soluble substances like barium nitrate and chloride into the bloodstream are more rapidly absorbed than insoluble substances like barium carbonate and sulfate.

8.2.5.2 Health Effects

Depending on how much compounds of barium dissolve in the stomach or water, different barium exposures have different negative effects on health. According to the CDC,[7] soluble barium compounds like hydroxide, barium sulfide, chloride, acetate, and nitrate have negative effects on health when consumed through tainted food and water, while insoluble barium compounds like barium carbonate and sulfate have little to no negative effects on health. If medical help is not immediately sought after, this may ultimately lead to an abnormal cardiac rhythm, paralysis, or death. Additionally, gastrointestinal issues like diarrhea, cramping in the abdomen, and vomiting may also be brought on. Additionally, the CDC[7] reported that a small amount of water-soluble barium being subjected may result in harmful effects on the heart and blood vessels, including low or high blood pressure that causes hypokalemia, and even breathing difficulties.

8.2.5.3 Mechanism of Toxicity

Although the exact mechanisms underlying toxicity of barium are yet not completely understood, it appears to be linked to ability of barium to serve as an antagonistic K^+ channel blocker and obstruct passive intracellular K^+ efflux. This results in a transfer of K^+ to intracellular compartments from extracellular, which lowers the ability of neuronal cells to respond to stimulation and also lowers resting membrane potential. Additionally, barium prevents the metabolic control of K^+ levels.

The two synergistic mechanisms that cause hypokalemia to be stimulated are as follows. Barium ions are initially preferentially blocked by the potassium rectifier channel before it can actively pump intracellular K^+ out of the cell. Barium's capacity to raise the permeability of plasma membrane to Na^+ is also correlated with its toxicity. Following this, extracellular K^+ is transported into the cell, which results in an increase in the sodium-potassium pump bioelectrogenesis. Both paralysis and intracellular K^+ trapping have been linked to depolarized membranes.

8.3 CONCLUSION

The goal of the current chapter was to look into how humans are exposed to heavy metals and their injurious impacts on their health. According to the study's findings, these metals enter the body via four different pathways: cutaneous absorption, contaminated food consumption, and polluted water consumption. Public health suffers when they are exposed to heavy metals. According to reports, different heavy metals can have adverse health effects through different processes. Lead, mercury, cadmium, chromium, and barium pose particularly serious health risks. This occurs as a result of their pervasive usage, increased levels of toxicity, and prevalence.

Depending on the metal in question, the harmful effects of heavy metal poisoning present themselves in various ways. However, the detrimental effects on health that the majority of heavy metals exhibit are consistent. For instance, it has been discovered that the majority of these metals can lead to neurocognitive impairments in children, including memory loss and reduced IQ. In turn, the neurological impairments hinder a child's intellectual development in addition to resulting in behavioral problems. Moreover, children's growth is hampered by mercury, lead, and cadmium. Additionally, mercury and chromium both cause diabetes, while chromium and cadmium cause cancer, particularly malignancies of the nose, lungs, and sinuses. In addition to irritating airways and obstructing them, heavy metal toxicity can lead to respiratory issues. They also contribute to the development of cardiovascular conditions such as stroke, low or high blood pressure, heart failure, and coronary heart disease. Endeavors should be implemented to reduce heavy metal toxicity and environmental spreading due to the harmful health impacts of heavy metal poisoning.

ACKNOWLEDGMENT

Authors are grateful to IIT (BHU) for providing necessary support for this work.

REFERENCES

1. Pereira SC, Oliveira PF, Oliveira SR, et al. Impact of environmental and lifestyle use of chromium on male fertility: Focus on antioxidant activity and oxidative stress. Antioxidants. 2021; 10(9):1365.
2. Forte IM, Indovina P, Costa A, et al. Blood screening for heavy metals and organic pollutants in cancer patients exposed to toxic waste in southern Italy: A pilot study. Journal of Cellular Physiology. 2020; 235(6):5213–5222.
3. Tchounwou PB, Yedjou CG, Patlolla AK, et al. Heavy Metal Toxicity and the Environment. In: A Luch, editor. Molecular, Clinical and Environmental Toxicology, Volume 3. Basel: Springer; 2012. pp. 133–164.
4. Cai L-M, Wang Q-S, Luo J, et al. Heavy metal contamination and health risk assessment for children near a large Cu-smelter in central China. Science of the Total Environment. 2019; 650:725–733.
5. Wu W, Wu P, Yang F, et al. Assessment of heavy metal pollution and human health risks in urban soils around an electronics manufacturing facility. Science of the Total Environment. 2018; 630:53–61.
6. Demková L, Árvay J, Bobuľská L, et al. Accumulation and environmental risk assessment of heavy metals in soil and plants of four different ecosystems in a former polymetallic ores mining and smelting area (Slovakia). Journal of Environmental Science and Health, Part A. 2017; 52(5):479–490.
7. Al Osman M, Yang F, Massey IY. Exposure routes and health effects of heavy metals on children. Biometals. 2019; 32:563–573.
8. Habiba G, Abebe G, Bravo AG, et al. Mercury human exposure in populations living around Lake Tana (Ethiopia). Biological Trace Element Research. 2017; 175:237–243.
9. Sagiv SK, Thurston SW, Bellinger DC, et al. Prenatal exposure to mercury and fish consumption during pregnancy and attention-deficit/hyperactivity disorder–related behavior in children. Archives of Pediatrics & Adolescent Medicine. 2012; 166(12):1123–1131.
10. WHO. Elemental Mercury and Inorganic Mercury Compounds: Human Health Aspects, Concise International Chemical Assessment Document 50. World Health Organization, Geneva. 2003.
11. Akerstrom M, Barregard L, Lundh T, et al. Relationship between mercury in kidney, blood, and urine in environmentally exposed individuals, and implications for biomonitoring. Toxicology and Applied Pharmacology. 2017; 320:17–25.
12. Dakcishi M, Nakai K, Sakamoto M, et al. Effects of hair treatment on hair mercury—The best biomarker of methylmercury exposure? Environmental Health and Preventive Medicine. 2005; 10:208–212.
13. Rice KM, Walker EM Jr, Wu M, et al. Environmental mercury and its toxic effects. Journal of Preventive Medicine and Public Health. 2014; 47(2):74.
14. Myers GJ, Thurston SW, Pearson AT, et al. Postnatal exposure to methyl mercury from fish consumption: A review and new data from the Seychelles Child Development Study. Neurotoxicology. 2009; 30(3):338–349.
15. Oken E, Bellinger DC. Fish consumption, methylmercury and child neurodevelopment. Current Opinion in Pediatrics. 2008; 20(2):178.

16. Ye BS, Leung AOW, Wong MH. The association of environmental toxicants and autism spectrum disorders in children. Environmental Pollution. 2017; 227:234–242.
17. Mahmud HNME, Huq AO, binti Yahya R. The removal of heavy metal ions from wastewater/aqueous solution using polypyrrole-based adsorbents: A review. RSC Advances. 2016; 6(18):14778–14791.
18. Liu Y, McDermott S, Lawson A, et al. The relationship between mental retardation and developmental delays in children and the levels of arsenic, mercury and lead in soil samples taken near their mother's residence during pregnancy. International Journal of Hygiene and Environmental Health. 2010; 213(2):116–123.
19. Kaur S, Kaur A, Singh G, et al. Mercurius solubilis attenuates scopolamine-induced memory deficits and enhances the motor coordination in mice. International Journal of Neuroscience. 2018; 128(3): 219–230.
20. Do SY, Lee CG, Kim JY, et al. Cases of acute mercury poisoning by mercury vapor exposure during the demolition of a fluorescent lamp factory. Annals of Occupational and Environmental Medicine. 2017; 29(1):1–8.
21. Calabrese EJ, Iavicoli I, Calabrese V, et al. Elemental mercury neurotoxicity and clinical recovery of function: A review of findings, and implications for occupational health. Environmental Research. 2018; 163:134–148.
22. Schumacher L, Abbott LC. Effects of methyl mercury exposure on pancreatic beta cell development and function. Journal of Applied Toxicology. 2017; 37(1):4–12.
23. Wallin A, Di Giuseppe D, Orsini N, et al. Fish consumption and frying of fish in relation to type 2 diabetes incidence: A prospective cohort study of Swedish men. European Journal of Nutrition. 2017; 56:843–852.
24. Jeon JY, Ha KH, Kim DJ. New Risk Factors for Obesity and Diabetes: Environmental Chemicals. Copenhagen: Wiley Online Library; 2015. pp. 109–111.
25. Jia G, Aroor AR, Martinez-Lemus LA, et al. Mitochondrial functional impairment in response to environmental toxins in the cardiorenal metabolic syndrome. Archives of Toxicology. 2015; 89:147–153.
26. Broussard LA, Hammett-Stabler CA, Winecker RE, et al. The toxicology of mercury. Laboratory Medicine. 2002; 33(8):614–625.
27. Corfitzen CB, Albrechtsen H-J, editors. Sampling for drinking water quality in drinking water installations regarding metal concentrations. DRITA-Drinkable Water from Tap; 2008: Citeseer.
28. Donadini G, Spalla S, Beone G. Arsenic, cadmium and lead in beers from the Italian market. Journal of the Institute of Brewing. 2008; 114(4):283–288.
29. Dorea JG. Mercury and lead during breast-feeding. British Journal of Nutrition. 2004; 92(1):21–40.
30. Schröder W, Holy M, Pesch R, et al. Are cadmium, lead and mercury concentrations in mosses across Europe primarily determined by atmospheric deposition of these metals? Journal of Soils and Sediments. 2010; 10:1572–1584.
31. He K, Wang S, Zhang J. Blood lead levels of children and its trend in China. Science of the Total Environment. 2009; 407(13):3986–3993.
32. Pizzol M, Thomsen M, Andersen MS. Long-term human exposure to lead from different media and intake pathways. Science of the Total Environment. 2010; 408(22):5478–5488.
33. Pfadenhauer LM, Burns J, Rohwer A, et al. A protocol for a systematic review of the effectiveness of interventions to reduce exposure to lead through consumer products and drinking water. Systematic Reviews. 2014; 3(1):1–13.
34. Lanphear BP, Hornung R, Khoury J, et al. Low-level environmental lead exposure and children's intellectual function: An international pooled analysis. Environmental Health Perspectives. 2005; 113(7):894–899.
35. Hong S-B, Im M-H, Kim J-W, et al. Environmental lead exposure and attention deficit/hyperactivity disorder symptom domains in a community sample of South Korean school-age children. Environmental Health Perspectives. 2015; 123(3):271–276.
36. Evens A, Hryhorczuk D, Lanphear BP, et al. The impact of low-level lead toxicity on school performance among children in the Chicago Public Schools: A population-based retrospective cohort study. Environmental Health. 2015; 14(1):1–9.
37. Debnath B, Singh WS, Manna K. Sources and toxicological effects of lead on human health. Indian Journal of Medical Specialities. 2019; 10(2):66–71.
38. Cao J, Li M, Wang Y, et al. Environmental lead exposure among preschool children in Shanghai, China: Blood lead levels and risk factors. PLoS One. 2014; 9(12):e113297.
39. Caravanos J, Carrelli J, Dowling R, et al. Burden of disease resulting from lead exposure at toxic waste sites in Argentina, Mexico and Uruguay. Environmental Health. 2016; 15:1–9.

40. Rehman K, Fatima F, Waheed I, et al. Prevalence of exposure of heavy metals and their impact on health consequences. Journal of Cellular Biochemistry. 2018; 119(1):157–184.
41. Kim H-C, Jang T-W, Chae H-J, et al. Evaluation and management of lead exposure. Annals of Occupational and Environmental Medicine. 2015; 27(1):1–9.
42. Nemsadze K, Sanikidze T, Ratiani L, et al. Mechanisms of lead-induced poisoning. Georgian Medical News. 2009; (172–173):92–96.
43. Hanna-Attisha M, Lanphear B, Landrigan P. Lead Poisoning in the 21st Century: The Silent Epidemic Continues. Washington, DC: American Public Health Association; 2018. pp. 1430–1430.
44. Hertel R. Sources of exposure and biological effects of chromium. IARC Scientific Publications. 1986; (71):63–77.
45. Das AP, Singh S. Occupational health assessment of chromite toxicity among Indian miners. Indian Journal of Occupational and Environmental Medicine. 2011; 15(1):6.
46. Anderson RA. Chromium in the prevention and control of diabetes. Diabetes and Metabolism. 2000; 26(1):22–28.
47. Neghab M, Azad P, Honarbakhsh M, et al. Acute and chronic respiratory effects of chromium mists. Journal of Health Sciences & Surveillance System. 2015; 3(3):119–124.
48. Rasoul GMA, Abou Salem ME, Allam HK, et al. Health-related disorders on occupational exposure to chromium in a leather tanning factory (Menoufia, Egypt). Menoufia Medical Journal. 2017; 30(1):92.
49. Saha R, Nandi R, Saha B. Sources and toxicity of hexavalent chromium. Journal of Coordination Chemistry. 2011; 64(10):1782–1806.
50. Buters J, Biedermann T. Chromium (VI) contact dermatitis: Getting closer to understanding the underlying mechanisms of toxicity and sensitization! Journal of Investigative Dermatology. 2017; 137(2):274–277.
51. Ray RR. Adverse hematological effects of hexavalent chromium: An overview. Interdisciplinary Toxicology. 2016; 9(2):55.
52. Sharma B, Singhal P, Chugh K. Intravascular haemolysis and acute renal failure following potassium dichromate poisoning. Postgraduate Medical Journal. 1978; 54(632):414–415.
53. Ramzan M, Malik MA, Iqbal Z, et al. Study of hematological indices in tannery workers exposed to chromium in Sheikhupura (Pakistan). Toxicology and Industrial Health. 2011; 27(9):857–864.
54. Dayan A, Paine A. Mechanisms of chromium toxicity, carcinogenicity and allergenicity: Review of the literature from 1985 to 2000. Human & Experimental Toxicology. 2001; 20(9):439–451.
55. Prozialeck WC, Edwards JR. Mechanisms of cadmium-induced proximal tubule injury: New insights with implications for biomonitoring and therapeutic interventions. Journal of Pharmacology and Experimental Therapeutics. 2012; 343(1):2–12.
56. Olympio KPK, da Rocha Silva JP, da Silva AS, et al. Blood lead and cadmium levels in preschool children and associated risk factors in São Paulo, Brazil. Environmental Pollution. 2018; 240:831–838.
57. Thompson J, Bannigan J. Cadmium: Toxic effects on the reproductive system and the embryo. Reproductive Toxicology. 2008; 25(3):304–315.
58. Kippler M, Ekström E-C, Lönnerdal B, et al. Influence of iron and zinc status on cadmium accumulation in Bangladeshi women. Toxicology and Applied Pharmacology. 2007; 222(2):221–226.
59. Söderholm M, Borné Y, Hedblad B, et al. Blood cadmium concentration and risk of subarachnoid haemorrhage. Environmental Research. 2020; 180:108826.
60. Fatima G, Raza AM, Hadi N, et al. Cadmium in human diseases: It's more than just a mere metal. Indian Journal of Clinical Biochemistry. 2019; 34:371–378.
61. Reyes-Hinojosa D, Lozada-Pérez C, Cuevas YZ, et al. Toxicity of cadmium in musculoskeletal diseases. Environmental Toxicology and Pharmacology. 2019; 72:103219.
62. Kumar S, Sharma A. Cadmium toxicity: Effects on human reproduction and fertility. Reviews on Environmental Health. 2019; 34(4):327–338.
63. Geng H-X, Wang L. Cadmium: Toxic effects on placental and embryonic development. Environmental Toxicology and Pharmacology. 2019; 67:102–107.
64. Rani A, Kumar A, Lal A, et al. Cellular mechanisms of cadmium-induced toxicity: A review. International Journal of Environmental Health Research. 2014; 24(4):378–399.
65. Navas-Acien A, Selvin E, Sharrett AR, et al. Lead, cadmium, smoking, and increased risk of peripheral arterial disease. Circulation. 2004; 109(25):3196–3201.

9 Chronic and Acute Health Impacts, Including Neurological Disorders, Development Abnormalities, Carcinogenicity, and Organ Damage

Sumit Mishra and Nipun Tirkey*
Birla Institute of Technology, Mesra, Ranchi, India

Soumitra Mishra
T.S. Misra Medical College and Hospital, Lucknow, India

9.1 INTRODUCTION

The term "heavy metals" is allocated to those elements which have high atomic weight (in the range 63.5–200.6 g mol^{-1}) or density greater than 5 g cm^{-3} [1]. Sometimes, even metalloids (such as selenium and arsenic) and lighter elements (like aluminium) are categorized as heavy metals as they are hazardous to the environment and human beings [2]. Elements which fall in the category of essential heavy metals are vital for the body and have a range of physiological and metabolic processes. Such metals are iron (Fe), copper (Cu), zinc (Zn), magnesium (Mg), manganese (Mn), chromium (Cr), cobalt (Co), vanadium (V), molybdenum (Mo) and selenium (Se). Additionally, these essential metals are constituents of enzymes, structural proteins and hormones. Even non-essential metals are heavy metals that do not have any role in the body; however, they pose toxic effects on the essential elements present in the body. This class of metal includes cadmium (Cd), arsenic (As), mercury (Hg), lead (Pb), nickel (Ni) and aluminium (Al) [3].

Heavy metals are dispersed into air, water and soil by release through various natural and anthropogenic sources [4]. Any essential or non-essential heavy metal present in excess has a toxic health impact. Human exposure to these heavy metals occurs via ingestion, inhalation or dermal contact which leads to bioaccumulation causing severe damage to our organs. This bioaccumulation may cause metabolic and physiological complications in the respiratory system, central nervous system, cardiovascular system, liver, kidney, lungs, gastrointestinal system, bones, brain, mucous tissues, skin, muscles, nails and hair.

The objective of this book chapter is to highlight the specifications of heavy metals, routes of human exposure to these heavy metals, their biochemistry, mechanism of metal toxicity, the impact of dosage and exposure time on metabolic disorders and the health impacts related to the toxicological effects of metal exposure on humans such as neurological disorder, development abnormalities,

* Corresponding author: sumitmishra1@gmail.com, smishra@bitmesra.ac.in

DOI: 10.1201/9781032685793-9

carcinogenicity, immunotoxicity, genotoxicity and damage of organs including bones, skin, kidney, liver and heart.

9.2 METAL SPECIFICATION

Metal specification refers to the physical as well as the chemical characteristics of a metal. The properties include solubility, volatile nature and the different organic, inorganic or elemental forms of a metal. Knowledge of these properties will further help us to elucidate the reaction, toxicity and mechanism of metals with different biomolecules and organisms. For instance, we can consider a few examples, such as mercury (Hg) is less toxic whereas methylmercury is highly toxic. Similarly, sea fish contains arsenic in the form of arsenobetaine (non-toxic) and not, toxic inorganic arsenite [5].

Metal specifications of arsenic, lead, mercury, cadmium, chromium, aluminium, selenium, manganese, cobalt, nickel, copper and zinc are discussed below:

9.2.1 Arsenic

Arsenic is a metalloid, also known as the king of poisons [6] and ranks 20th [7] in elemental abundance on Earth. It is brittle in nature and bright-silver grey in colour. Naturally, arsenic exists as the salt of calcium, copper, iron or sodium, as oxides or as sulphides [8] and in inorganic form, as arsenite (As^{3+}) and arsenate (As^{5+}) compounds. Other forms of arsenic include organic arsenic (methylated arsenic such as monomethyl arsenic acid [MMA], dimethyl arsenic acid [DMA] and trimethylarsine oxide) and arsine (AsH_3). Among them, arsine (AsH_3) is the most toxic form followed by inorganic arsenic ($As^{3+} > As^{5+}$), metalloids (As^0) and lastly organic arsenic [5, 9].

9.2.2 Lead

Lead is a naturally occurring slightly bluish, bright silver metal. Lead is used in batteries and its salts are utilized due to their physiochemical characteristics. It can also interact with both acids as well as bases and forms oxides, carbonates and sulphates by reaction with air and water [9, 10].

9.2.3 Mercury

Metallic mercury, is a silvery-white, odourless liquid metal, it turns into a colourless, odourless gas when heated. The three major forms in which mercury exists are inorganic (Hg^+, Hg^{2+}), metallic or elemental mercury (Hg^0) and organic (ethyl or methyl mercury). Among them methyl mercury is the most toxic followed by inorganic mercury (Hg^+, Hg^{2+}) and lastly metallic mercury (Hg^0) [11–13].

9.2.4 Cadmium

Cadmium ranks 7th among the most toxic heavy metals. It is silvery-bluish, soft, malleable and ductile metal. Its most stable oxidation state is +2 [14].

9.2.5 Chromium

Chromium is a shiny, steel-grey-coloured metal. It is hard in nature and reacts with most of the acids. Chromium can be generated from coal, petroleum and oil combustion. The three main states in which chromium exists are: naturally occurring trivalent chromium (Cr^{+3}), industry-generated elemental chromium and hexavalent chromium (Cr^{+6}). The trivalent and hexavalent forms of chromium are the most stable among them. However, hexavalent chromium is the most hazardous type of chromium, while trivalent chromium is a necessary micronutrient for our body [13, 15, 16].

9.2.6 Aluminium

Aluminium ranks 3rd among the most abundant elements in the Earth's crust. It is silvery-white in colour, lightweight, low density, soft, strong, oxidation resilient, corrosion resistant, malleable, easily deformable, non-magnetic, thermally conductive, highly ductile and durable heavy metal [13, 17].

9.2.7 Selenium

Selenium is a metalloid, found in rare minerals such as antimonselite, tiemannite, clausthalite, berzelianite and crookesite. It is present as a silver metal or in the form of red powder derived from copper electrolytic refining and is known to possess photovoltaic and photoconductive properties. Different allotropes of selenium are present in grey, red and black in colour and are interconvertible depending upon temperature changes and its rate. Furthermore, selenium has five stable isotopes with 16 radioactive isotopes. All living organisms possess selenium in their body, either in the inorganic forms (such as selenide [Se^{2-}], selenite [SeO_3^{2-}] and selenate [SeO_4^{2-}]) or in their organic form (such as selenomethionine, selenocysteine, methylselenocysteine, selenocystathionine). All organisms require selenium in trace quantity as it is present in selenocysteine (an amino acid) in the enzyme's active site, however, its higher concentration is toxic [18].

9.2.8 Manganese

Manganese, the 12th most abundant metal is an essential trace metal and has six oxidation states. It possesses properties such as brittleness and corrosion resistance. Manganese exists in the form of a manganese dioxide mineral named pyrolusite and a manganese carbonate mineral named rhodochrosite. It is essential for various cellular functions of the body and acts as an enzyme co-factor, for example, enzyme-glutamine synthetase and manganese superoxide dismutase [19].

9.2.9 Cobalt

Cobalt is a lustrous silvery-blue coloured metal, that ranks 32nd among the most abundant metal on earth's crust. It shows magnetic behaviour and is naturally present in ores such as cobaltite, erythrite and skutterudite [20].

9.2.10 Nickel

Nickel ranks 22nd among the most abundant metals and constitutes 3% of the earth's crust. Moreover, nickel is present in the form of minerals such as pentlandite and garnierite. It is a silver metal and corrosion-resistant at high temperatures. This metal is responsible for various vital functions of the body but toxic at higher doses. One of the sources of huge amounts of nickel is meteorites [14, 16, 21].

9.2.11 Copper

Copper is an essential element present in minerals (chalcopyrite and bornite). It is reddish-gold in colour and is a good conductor of heat as well as electricity [22].

9.2.12 Zinc

Zinc is an essential trace element. The two most common ores of zinc are zinc blende and calamine. It easily becomes ductile and malleable at temperatures above 110°C, and it is moderately reactive with respect to its reactivity with oxygen and other metals. Moreover, it functions as a reducing agent and a Lewis acid in hydrolytic reactions and is connected to several metalloenzymes, DNA binding, regulatory proteins and transcription factors. [4, 23].

9.3 ROUTES AND SOURCES OF HUMAN EXPOSURE TO HEAVY METALS

Natural routes of human exposure to heavy metals include volcanic eruptions, sea salt, forest fires, wind-blown soil particles, weathering of metal-containing rocks and leaching into the water bodies [24]. Mercury is one of the heavy metals that is discharged naturally into the environment via degassing of the earth's crust, evaporation from natural water bodies and volcanic emission [25]. Moreover, erosion, surface run-off and acid rain are also responsible for heavy metal pollution [24].

In the earth's crust, heavy metals are present in ores and recovered as minerals. During the mining and smelting activities, heavy metals are released from the ores and are deposited in water bodies, scattered in the air and also deposited in the soil [24, 26–28]. Furthermore, heavy metal pollution also occurs by combustion and discharge of effluents by utilization of heavy metals in industries (more information about the industrial sources of heavy metal is discussed in Table 9.1). As heavy metals are not degradable, high concentrations of heavy metals are present in both the treated as well as untreated sewage from industries. These heavy metals can then accumulate in aquatic life. Furthermore, consumption of contaminated seafood exposes humans to heavy metals.

Ingestion of contaminated food accounts for the overall intake of 90% of heavy metals into our body whereas skin and dermal contact with heavy metal accounts for the remaining 10% [26].

TABLE 9.1
Dietary and Industrial Sources of Heavy Metals

Heavy Metal	Dietary Sources	Industrial Sources and Uses
Aluminium [4, 17]	Cheese, tea, coffee and soft drinks, food additives (such as anticaking and baking agents, food preservatives and colourants)	• Manufacture of currency such as aluminium coins, kitchen utensils, foils, mirrors and mirrors of the telescope, packaging material (such as cans), explosives and fireworks. • Production of antacids, anti-aspirants, cosmetics and food additives. • It is applied in industries such as aerospace, transportation, construction, electrical and household appliances.
Vanadium [4, 12]	Grains, parsley, black pepper, mushroom, shellfish, wine and beer	• It is used for the production of vibrant resistant springs as steel shock and rust-resistant steel tools. • Vanadyl sulphate and sodium metavanadate are used as diet supplements. • Pigment of ceramic glass. • Superconducting magnets.
Chromium [4, 30, 31]	Grains, fruits and vegetables, meats and shellfish	• Stainless steel and alloy production are the main applications of elemental chromium. • Hexavalent chromium is utilized for the manufacture of textile dyes and wood protection. • Other uses of chromium are the manufacture of dye, paints, metal ceramics, ferrochrome, chrome pigments, synthetic rubies, electroplating, leather tanning, oil-well drilling, welding of stainless steel, colouring glasses, in pulp and paper industries and corrosion protection.

(Continued)

TABLE 9.1 *(Continued)*
Dietary and Industrial Sources of Heavy Metals

Heavy Metal	Dietary Sources	Industrial Sources and Uses
Manganese [4, 16, 26, 30]	Grains, soya beans, nuts, rice, herbs, parsley, olive oil, green beans, tea, Spinach and Oysters	• Alloys, packaging cans, fertilizers, pesticides, cosmetics and fireworks. • Manganese (IV) oxide – used as a rubber additive and to remove glass tints.
Cobalt [4, 20, 30]	Chocolate, butter, cheese and meat	• Utilized as magnets, electroplating, in jet and gas turbines and painting porcelain, pottery, enamel and glass. • *Medicinal purpose*: radioactive ^{60}Co is used for treating cancer, other use of this radioactive cobalt is as a food preservative.
Nickel [4, 16, 32]	Whole wheat, grains, millet, oats, rye, baked beans, kidney beans, soybeans, chickpeas, peas, nuts, soy products, baking powder, tea, cocoa, and chocolates.	• Nickel has its applications in manufacturing coins, jewellery, stainless steel alloy, rocket engines, boat propeller shafts, welding, plating of metals and armour plating. • Nichrome alloy is used for the manufacturing of non-corrosive heating appliances.
Copper [4, 16, 30]	Nuts and seeds, dark chocolate, Leafy green vegetables, mushrooms, shiitake, oyster, lobster, and Spirulina	• Manufacture of coins, copper alloys and wires, in barrier creams, preservatives of wood and fabric and fertilizers. • Copper sulphate is used in agriculture applications for the cure of mildew.
Zinc [4]	Cheese, sunflower seeds, lamb, beef, and herring	• Galvanization • Uses of oxides of zinc – personal care products, paints, weapons, electrical equipment, ink, batteries, plastics, textiles and rubbers. • Sulphides of zinc are used in X-ray screens, fluorescent lights and luminous paints.
Molybdenum [4, 30]	Grains, nuts and legumes	• Hardening of steel, increasing steel strength. • Applied in electrodes and in nuclear energy operations. • Acts as a catalyst in petroleum refinement.
Arsenic [5, 33]	Cereals and dairy products	• Wood preservatives, pesticides and sheep dips. • Medicinal products • Veterinary medicines • Manufacture of bronze, specific types of glass, pyrotechnics, lead alloys (such as car batteries and ammunition), doping agents in semiconductors and metalloid substances.
Cadmium [4, 33]	Mushrooms, shellfish, mussels, dried seaweed, and shrimps	• Electroplating industry, alloy production • Phosphate fertilizers, pesticides • PVC related products • Nuclear reactors • Rechargeable batteries (such as nickel-cadmium batteries), electrode components in alkaline batteries. • Anticorrosion agent • Glassware pigments, coating • Plastic stabilizer • It is also present in tobacco smoke.
Selenium [4, 18]	Grains, cereals, brazilian nuts, chocolate, yeast, broccoli, meats, fish, and eggs	• As pigments for glass, ceramics, plastics and paints. • Electronics • Animal feeds • Anti-dandruff shampoos

(Continued)

TABLE 9.1 *(Continued)*
Dietary and Industrial Sources of Heavy Metals

Heavy Metal	Dietary Sources	Industrial Sources and Uses
Mercury [4, 12, 13, 26]	Seafood and mushrooms	• Production of thermometers, rectifiers, electrical switches, as a component of batteries and as a catalyst. • Fluorescent light bulbs • Pulp and paper industries • Mercury arc lamps • Dental preparations such as amalgams. • Mining industries for the extraction of gold, chlorine manufacture, caustic soda production, in nuclear reactors, solvent for reactive and precious metals and wood preservatives as an antifungal agent. • The methyl and ethyl forms of mercury are utilized for the production of fungicides. • Skin lightning cream and soap contain mercury chloride ($HgCl_2$) as an active ingredient.
Lead [4, 21, 33]	Fruit, vegetables and grains.	• Production of lead-acid batteries, ceramic glazes, sheets for computer screens, cable sheets, sports equipment, stained glass windows, canisters for corrosive liquids, oxide synthesis for paints, ammunition, metal solders and pipes and X-ray shielding devices. • Lead acetates are used for the production of hair dyes, cotton dyes, chrome pigments, paint and pesticides. • Lead nitrates are utilized for the production of matches, textile printing, oxidizers, sensitizers and rodenticides.

Anthropogenic sources like wastewater irrigation, usage of pesticides and animal manure, dumping waste, residues of combustion from coal, oil and petroleum spillage, untreated sewage and wastewater from industries are responsible for heavy metal pollution. Mineral fertilizers and organic manure also have a high concentration of heavy metals present in them [24, 26–28]. Subsequently, the plants absorb heavy metals from contaminated soil resulting in contaminated fruits and vegetables. This leads to heavy metal bioaccumulation in our food chain. Furthermore, ingestion of heavy metals also occurs through dietary sources [4] (discussed in Table 9.1).

Thus, we can conclude that humans are exposed to heavy metals in the following four ways [29]:

1. Ingestion (through food contamination and dietary sources).
2. Inhalation or dermal contact from the atmosphere (it includes smoke, dust particles and chemical fumes from anthropogenic as well as natural sources).
3. Drinking contaminated water.
4. Skin contact (some of its sources are agricultural, pharmaceutical, manufacturing, residential, industrial and occupational) (Figure 9.1 and Table 9.1).

9.4 BIOCHEMISTRY OF HEAVY METALS

Toxicity mechanisms of heavy metals include inhibition of enzyme activity, reactive oxygen species (ROS) production, antioxidant damage, alteration in cellular metabolism, transcription and cellular

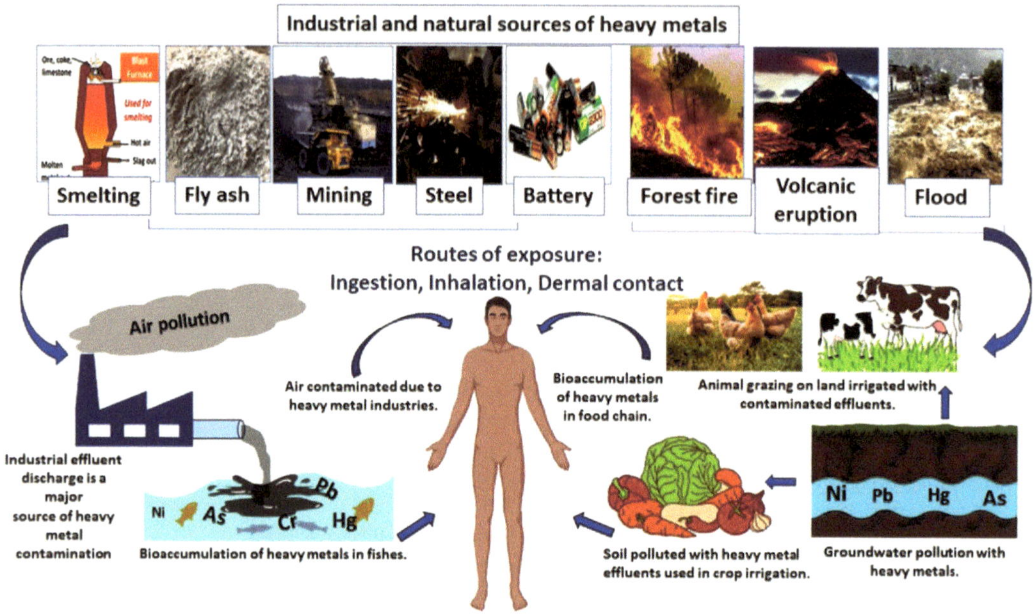

FIGURE 9.1 Sources of heavy metal and routes of exposure.

function and changes in the cell membrane permeability. Moreover, lipid peroxidation, synthesis of proteins and alterations in the function of nucleic acids are also involved [4]. The mechanistic action of heavy metal [16, 29] on humans is illustrated in Figure 9.2. Mechanisms of heavy metal toxicity such as enzyme inhibition, oxidative stress and lipid peroxidation are discussed as follows:

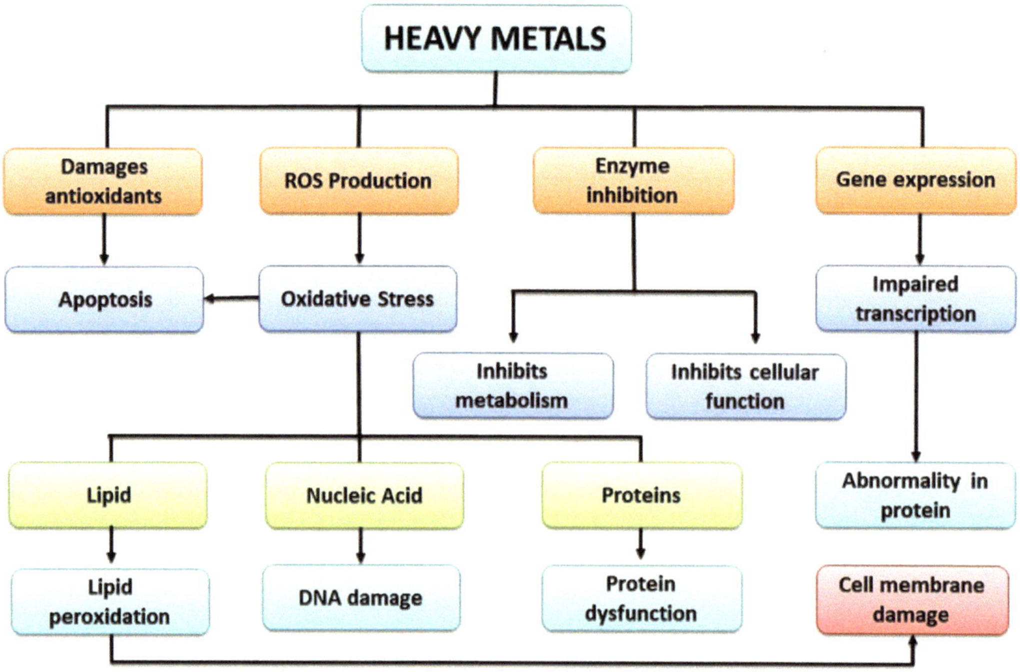

FIGURE 9.2 Mechanistic action of heavy metals.

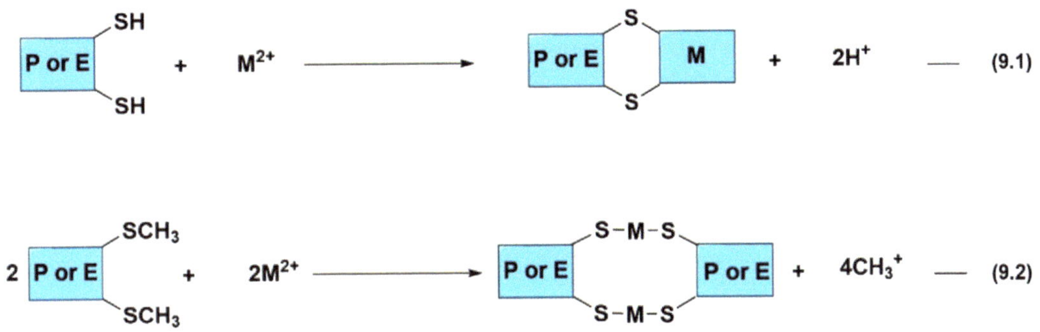

FIGURE 9.3 Reaction of enzymes or proteins with heavy metals. (9.1) Intermolecular bonding. (9.2) Intramolecular bonding. *Abbreviations*: E, enzymes; M, metal; P, proteins.

9.4.1 Inhibition of Enzyme Activity

On ingestion of heavy metals, the acidic medium of the stomach converts them to their most stable oxidation states (such as Ag^{+}, Cd^{2+}, As^{2+}, Hg^{2+}, Zn^{2+}, Pb^{2+} and As^{3+}) which is also their most toxic state. Furthermore, strong chemical bonds are formed due to the interaction of toxic metals with proteins and enzymes of the body. These toxic heavy metals have an affinity to replace hydrogen atoms and bond with sulphur atoms of sulfhydryl groups (–SH) and ($-SCH_3$) of cysteine and methionine, respectively (Figure 9.3), thus inducing inhibition of enzyme functioning.

Whereas, in the case of proteins, the protein-metal compound formed further reacts to form enzyme-substrate complex (E-S*) by the reaction of the substrate (S) and enzyme (E).

$$E + S \rightarrow E - S \rightarrow E - S^* \rightarrow E - P \rightarrow E + P \tag{9.3}$$

(E = Enzyme, S = Substrate, P = Product, E-S* = Activated enzyme substrate complex)

An enzyme that has already formed a complex E-S or E-S* or E-P cannot bond further with another substrate to form a complex until it is free. This is because they either fit like a lock and key or follow the induced-fit mechanism. Such a specific method for complex formation is followed due to the enzyme's chirality. In this case, the enzyme remains blocked and remains embedded in the tissues. This leads to bio-dysfunction and abnormalities in the body.

Furthermore, metalloenzyme-containing metal has the capability to replace another metal ion of similar size, for example, cadmium toxicity can be induced, where Cd^{2+} replaces Zn^{2+} in a dehydrogenating enzyme. This process completely inhibits enzyme activity, and the protein becomes bio-inactive. Furthermore, as the heavy metals form a very stable bio-toxic compound with the body's bio-molecules in their most stable oxidation states, it is difficult to extract them from the body [34].

An example of the biochemistry of metal on enzyme activity inhibition is demonstrated by cadmium. It inhibits human thiol transferases, such as glutathione, thioredoxin and thioredoxin reductase, by binding to their cysteine residues in an in vitro manner [35].

9.4.2 Oxidative Stress

Heavy metal toxicity generates free radicals, which may lead to oxidative stress, lipid peroxidation, modification in sulfhydryl homeostasis and DNA damage [36].

Heavy metals responsible for free radical generation are iron, copper, chromium, cadmium and vanadium. These metals follow the Fenton reaction (Eq. 9.6) for the generation of free radicals (such as superoxide radicals ($O^{2\bullet -}$) and hydroxide radicals ($^{\bullet}OH$)). Fenton's reaction is associated with the damage of mitochondria, peroxisomes and microsomes [37].

$$M^{(n+1)} + O^{2\bullet -} \rightarrow Mn + O^2 \quad (9.4)$$

$$2O^{2\bullet -} + 2H^+ \rightarrow H_2O_2 + O_2 \quad (9.5)$$

$$\text{Fenton reaction: } M^{n+} + H_2O_2 \rightarrow M^{(n+1)+} + OH^\bullet + OH^- \quad (9.6)$$

Iron (Fe^{2+}), copper (Cu^+), chromium (Cr^{4+}) and vanadium (V^{4+}) generates hydroxyl radical ($^{\bullet}OH$) whereas cobalt (Co^{2+}) and vanadium (V^{5+}) generates superoxide ($O^{2\bullet -}$) [37]. Both types of radicals are generated via hydrogen peroxide (H_2O_2) decomposition.

Arsenic is also responsible for free radical generation, such as superoxide ($O^{2\bullet -}$), nitric oxide (NO•), the peroxyl radical ($ROO^{\bullet}$), dimethylarsinic radical ($(CH_3)_2As^{\bullet}$) and dimethylarsinic peroxyl radicals ($(CH_3)_2AsOO^{\bullet}$). It also generates singlet oxygen ($^1O_2$) and hydrogen peroxide (H_2O_2) [38]. However, the mechanism of radical generation is not known.

The Hydroxyl radical ($OH^{\bullet}$) generated by the oxidation of metals is capable of damaging proteins, lipids and DNA by reacting with them. Furthermore, hydroxyl radical ($OH^{\bullet}$) generates 8-oxo-7, 8-dihydro-20-deoxyguanosine (8-oxo-dG) on reaction with guanine. This product signifies oxidative damage [39].

Moreover, the free radicals generated via metals cause DNA mutagenicity. This reveals a link between oxidative damage and carcinogenicity. DNA repair mechanism is inhibited by metals such as arsenic, cadmium and nickel. Furthermore, chromium and nickel are responsible for base modification. Nickel and oxidant along with copper or iron act as crosslinkers, and strand scission is carried out by cadmium, nickel and chromium along with oxidant. Lastly, metals responsible for depurination are copper, chromium and nickel. All these metals cause oxidative effects in DNA [37]. The oxidative stress pathway is illustrated in (Figure 9.4).

9.4.3 Lipid Peroxidation

ROS generated via metal can further attack polyunsaturated fatty acid-like phospholipids and result in lipid peroxidation [16].

$$\text{Initiation: } \text{Lipid} + R^\bullet/OH^\bullet \rightarrow \text{Lipid}^\bullet \quad (9.7)$$

$$\text{Propogation: } \text{Lipid}^\bullet + O_2 \rightarrow \text{Lipid-OO}^\bullet \quad (9.8)$$

$$\text{Lipid-OO} + \text{Lipid}^\bullet \rightarrow \text{Lipid-OOH} + \text{Lipid}^\bullet \quad (9.9)$$

$$\text{Termination: } \text{Lipid}^\bullet + \text{Lipid}^\bullet \rightarrow \text{Lipid-Lipid} \quad (9.10)$$

$$\text{Lipid-OO}^\bullet + \text{Lipid}^\bullet \rightarrow \text{Lipid-OO-Lipid} \quad (9.11)$$

In the initiation step (Eq. 9.7), a lipid radical is formed as a result of a free radical attack on the lipid membrane. Furthermore, the reaction of lipid radicals with dioxygen molecule (O_2) in the propagation step (Eqs. 9.8, 9.9) forms peroxyl lipid radicals. This initiates lipid molecule damage. In the termination step (Eqs. 9.10, 9.11), either a stable lipid molecule is formed via two lipid

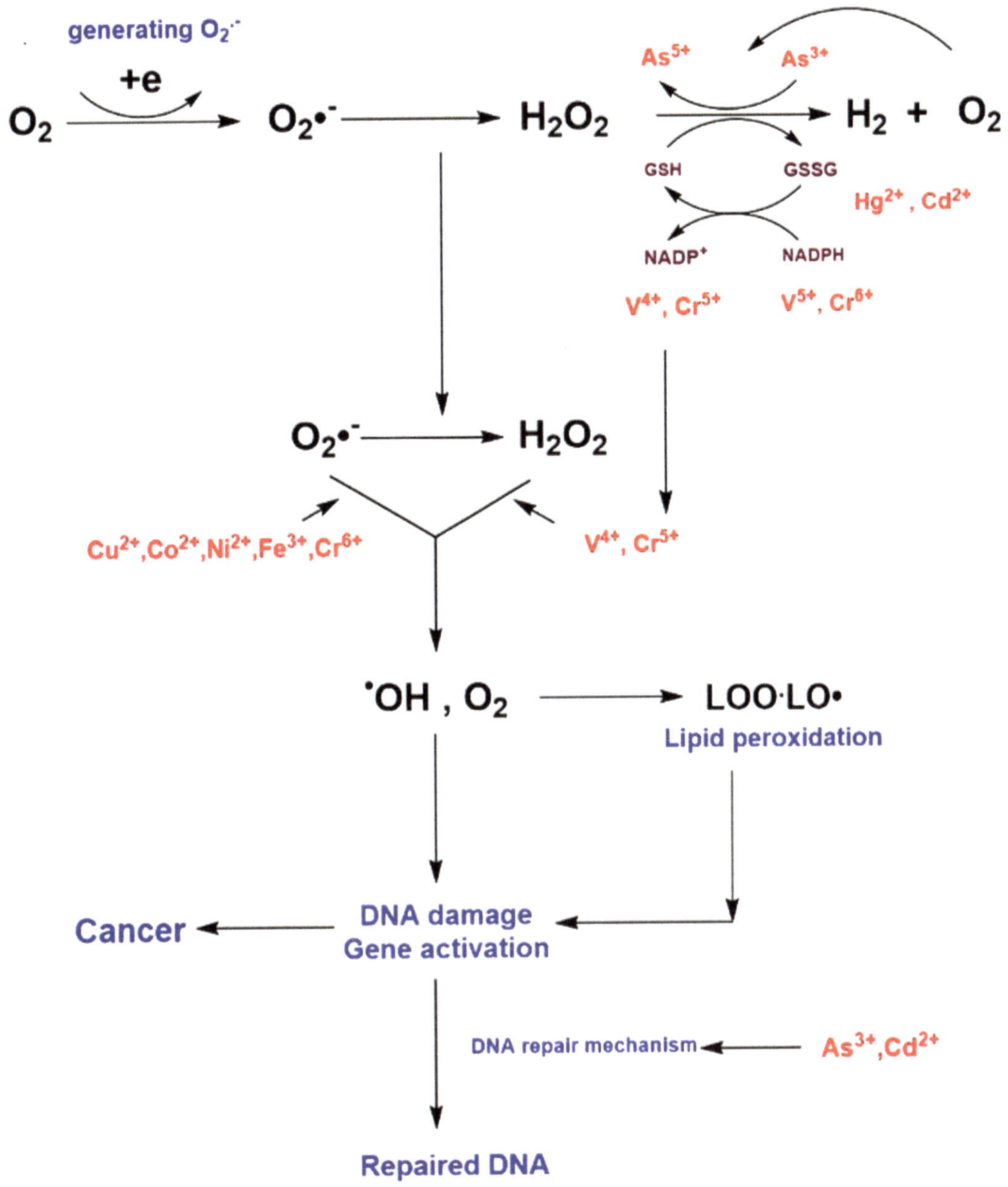

FIGURE 9.4 Oxidative stress pathway.

radical reactions or a lipid radical reacts along with lipid peroxyl radical to form malondialdehyde. Formation of this product indicates lipid peroxidation.

9.5 MECHANISM OF TOXIC ACTION OF HEAVY METALS

Heavy metals have a toxic effect on the body by interfering with the body's biological system and processes and by adapting several mechanisms. Herein, we have discussed the toxicity mechanism of several metals including mercury, iron, lead, arsenic, aluminium, cadmium, chromium, selenium, manganese, cobalt, copper and vanadium.

9.5.1 Mechanism of Mercury Toxicity

Mercury binds with different protein groups and deactivates it. These groups include amine, amide, carboxyl and phosphoryl as these groups have a high susceptibility to react with mercury. Mercury also has a high affinity towards sulfhydryl groups of the amino acids and T cell surfaces influencing its functioning [40].

One of the neurotoxicants of mercury is methylmercury. It is responsible for microtubule destruction, mitochondrial damage, lipid peroxidation and accumulation of neurotoxic molecules (such as serotonin, aspartate and glutamate). Other targets of mercury toxicity are nerves, kidneys and muscles. Mercury disturbs intracellular calcium homeostasis and membrane potential. Furthermore, mercury binds to thiols and damages the structure of tertiary and quaternary proteins. Bronchitis, asthma and temporary respiratory problems are induced by mercury vapours. Moreover, mercury damages the cellular function and structure by adhering to the sulfhydryl groups and selenohydryl [13]. It also intervenes with the transcription and translation process resulting in the disappearance of ribosomes and eradicating endoplasmic reticulum and the activity of natural killer cells. Also, free radical formation affects cellular integrity.

Effects: Mercury crosses the placental barrier during pregnancy which may lead to developmental anomalies in the foetus. Fat-soluble mercury may cross blood brain barrier.

9.5.2 Mechanism of Iron Toxicity

The concentration of iron in cells and biological fluids is affected due to over-absorption and saturation of iron in the body, this results in the inability of iron to bind with proteins and the generation of free radicals. These free radicals accumulate in the gastrointestinal tract, heart, kidney and liver and in turn lead to iron toxicity in these organs. Furthermore, free radicals result in lipid peroxidation causing damage to microsomes, mitochondria and cell organs. Moreover, attack on DNA by hydrogen free radicals causes cell damage, transformation of malignant and mutarotation further inducing several diseases [13].

Effects: It leads to hemosiderosis which has no visible effects of iron toxicity, whereas haemochromatosis is excess iron accumulation which is hepatotoxic and may lead to liver failure.

9.5.3 Mechanism of Lead Toxicity

The ionic mechanism of lead toxicity is responsible for the disturbance of the biological metabolism of the cell which includes apoptosis, regulation of enzymes, neurotransmitter release, adhesion of cells, protein folding and both intra and inter-cell signalling. The main cause of this mechanism is the ability of lead to replace monovalent and divalent cations (such as Na^+, Ca^{2+}, Mg^{2+} and Fe^{2+}) [41]. The heme synthesis pathway is the prime target of lead toxicity. Even at low concentrations, lead alters the morphology of RBC and inhibits haemoglobin and heme-biosynthesis. The action of lead on heme synthesis [42] is illustrated in Figure 9.5. Lead is responsible for the inhibition of the enzyme, δ-aminolevulinic acid dehydratase (ALAD). Additionally, it also inhibits ferrochelatase from binding iron into the protoporphyrin ring, in the last step of heme synthesis. Moreover, heme synthesis is reduced due to the inhibition of ALAD to convert two ALA molecules into porphobilinogen. This stimulates the accumulation of ALA synthetase and induces ROS generation, finally amplifying oxidative stress. Additionally, lead toxicity also targets the nervous system and organs such as bones, kidneys and liver.

Effects: The visible effects are porphyria on the skin (red-coloured patches) leading to protoporphyria, ALAdehydratase deficiency porphyria and related disorders.

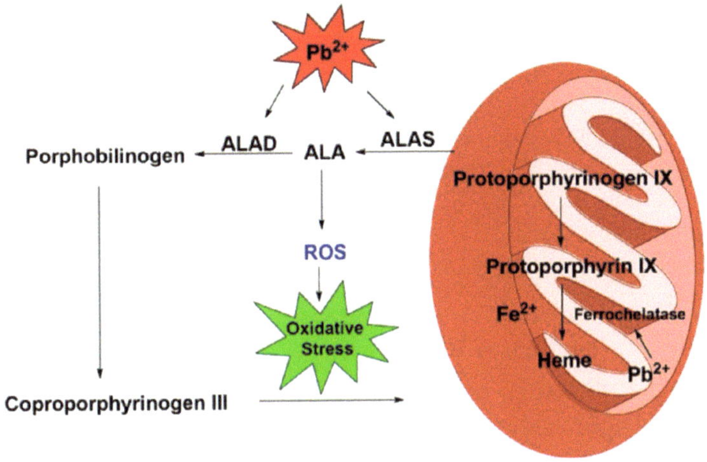

FIGURE 9.5 Action of lead toxicity on heme synthesis.

9.5.4 Mechanism of Arsenic Toxicity

The mechanistic action of arsenic toxicity can be explained via the biotransformation reaction (Eq 9.12) of inorganic arsenic (IAs). Inorganic arsenic is harmful in nature and is methylated enzymatically resulting in monomethylarsenic (MMA(III) and MMA(V)) and dimethylarsenic (DMA(V)).

$$\text{IAs}(\text{V}) \rightarrow \text{IAs}(\text{III}) \rightarrow \text{MMA}(\text{V}) \rightarrow \text{MMA}(\text{III}) \rightarrow \text{DMA}(\text{V}) \quad (9.12)$$

Effects: On prolonged exposure to arsenic, an intermediate product, MMA(III) remains in human cells (responsible for causing cancer), whereas the methylated products DMA(V) and MMA(V) are excreted [8].

9.5.5 Mechanism of Aluminium Toxicity

Consequences of human exposure to aluminium involve enzyme inhibition, alterations in the functioning of nucleic acid, permeability of the cell membrane, protein synthesis and the metabolism of triglycerides. Aluminium toxicity also arises due to the interaction of aluminium with apoplastic, symplastic and plasma membranes [43]. Furthermore, alterations in intercellular communication, the function of the secretory system and cell growth are induced due to the replacement of Mg^{2+} and Fe^{3+} by Al^{3+} [44, 45].

Effects: Alzheimer's disease due to aluminium deposition in the brain since it crosses blood brain barrier by choroid plexus from the nasal cavity or through the systemic circulation.

9.5.6 Mechanism of Cadmium Toxicity

Cadmium accumulates in the human body via inhalation, which is further absorbed by the intestine, lungs, liver and kidney. Transporter protein facilitates cadmium accumulation in the intestine whereas the faecal route facilitates its excretion. Furthermore, cadmium absorbed by the lungs is size-dependent. Small cadmium particles are deposited in the alveoli depending upon their solubility whereas large cadmium particles are absorbed by the nasopharyngeal along with tracheobronchial airways. Cadmium toxicity is associated with an enhancement in oxidative stress by ROS production, increased lipid peroxidation and depletion of sulfhydryl groups and protein-bound glutathione. Further apoptosis, transcription factors, DNA methylation and DNA damage are also induced due to cadmium toxicity. Hepatoxicity due to cadmium occurs on the formation of the cysteine-metallothionein complex by binding cadmium and metallothionein (a cysteine-rich protein). Furthermore, its accumulation in the renal tissues causes kidney damage. Cadmium toxicity can also take place via aspartate, cysteine, histidine and glutamate ligand binding. Moreover, genotoxicity can also be induced by cadmium via complex formation, ROS generation, cell death and multifactorial mechanisms [16, 35].

Effects: Nephrotoxicity and hepatotoxicity

9.5.7 Mechanism of Chromium Toxicity

Cr(VI) and Cr(III) are responsible for the toxicity of chromium in the body [46]. However, Cr(VI) is a strong oxidizing agent and possesses mutagenic properties and also induces chromosomal damage, alteration in oxidative protein and formation of adducts. It is transformable to its pentavalent, tetravalent and trivalent forms. Toxification of chromium occurs due to intracellular reduction away from the target site, however intracellular reduction at the target site leads to its detoxification. ROS such as (superoxide ion, hydrogen peroxide and hydroxyl radical) are generated on the reaction of Cr(VI) with ascorbate and thiols causing oxidative stress, protein and DNA damage [47].

Effects: Apoptosis and necrosis

9.5.8 Mechanism of Selenium Toxicity

Selenium at higher concentrations replaces sulphur atoms present in enzymes and biomolecules. Replacement of sulphur with selenium results in selenium toxicity as sulphur atoms are important for mitochondrial and microsomal electron transport. Selenium plays a major role in hepatic haem metabolism and is associated with lipid peroxidation. Moreover, selenium generates ROS by the redox catalysis of inorganic selenium with thiols present in the tissue. This results in the inhibition of thiol proteins.

Effects: Selenium compound, selenol is responsible for cellular apoptosis whereas the death of cancer cells is caused by selenium-methylselenocysteine [18].

9.5.9 Mechanism of Manganese Toxicity

Manganese enters the body and then gets absorbed by a mucosal layer of the gut either through simple diffusion or by the transport mechanism. Furthermore, Mn(II) binds to the plasma of the gastrointestinal tract, where the majority of the manganese is deposited in bile through the liver or gets oxidized to Mn^{3+} using ceruloplasmin.

Effects: Manganese also has a neurotoxic effect on the body and the mechanism involves inducing oxidative and endoplasmic reticulum stress, apoptosis, inflammation, autophagy as well as mitophagy. Moreover, mitochondrial dysfunction, excitotoxicity of glutamate and dysregulation of transcription also take place [4].

9.5.10 Mechanism of Cobalt Toxicity

Cobalt toxicity arises due to Co^{2+} both in the albumin-bound as well as unbound form. However, the unbound form is more toxic as it is bioavailable, and interacts with biomolecules, cellular receptors and ion channels. The mechanisms of Co^{2+} include ROS generation, lipid peroxidation, alteration of mitochondrial function and calcium and iron homeostasis. Furthermore, it triggers erythropoiesis, alters thyroid iodine uptake, affects the DNA repair process and induces a genotoxic effect. This mechanism is concentration-dependent and depends on cobalt content in blood and tissues [4].

Effects: Genotoxic, cell death and deranged thyroid function test

9.5.11 Mechanism of Copper Toxicity

Copper toxicity is induced due to Cu^{+}. Redox reaction carried out by cuproenzymes transforms Cu^{2+} to Cu^{+}. Furthermore, in the presence of glutathione and ascorbic acid, Cu^{2+} ROS can be reduced to Cu^{+} ROS. This state is more toxic due to ROS generation such as superoxide and hydroxyl radical formation by the Fenton reaction. These free radicals can further react with a variety of biomolecules [16].

Effects: Wilson disease, hepatotoxicity

9.5.12 Mechanism of Vanadium Toxicity

Vanadium is present in two oxidation states (V^{4+} and V^{5+}) in the body. It binds to transferrin, in the blood and is further taken up by erythrocytes. Since V^{4+} is oxidized to V^{5+}, the rate at which V^{5+} is taken up by erythrocytes is greater than V^{4+}. Furthermore, it follows a two-phase mechanism, firstly a fast phase reaction takes place followed by a slower phase reduction of V^{5+} to V^{4+}. Both enzymatic and non-enzymatic antioxidants are responsible for the reduction of vanadium. Vanadate is found to be more toxic than vanadyl as it inhibits $Na^{+}K^{+}$ ATPase (plasma membrane) and reacts with a variety of enzymes. Vanadium toxicity can be observed due to the reaction of vanadium with enzymes [36].

Effects: Lungs disorder, pulmonary oedema, upper respiratory congestion

9.6 HEAVY METAL CONCENTRATION AND METABOLIC DISORDER

Diseases may arise from an excess of any heavy metal. Humans need essential metals, but the body can be affected by both excess and lack of these elements. The most common cause of deficiencies is malabsorption. The consequences of metal excess and toxicity rely upon various factors, including the metal's intake or inhalation, dosage, duration of exposure, distribution and attained concentration, as well as biotransformation and excretion rate [4]. Table 9.2 discusses the health effects of heavy metals depending on dosage and time of exposure.

9.6.1 Minamata Bay Disaster

Minamata Bay disaster or Minamata disease was a case of methylmercury (MeHg) poisoning officially discovered in the year 1956 and was named after Minamata where the first medical case of this disease was reported. Methylmercury-containing waste water from the Chisso industrial plant was discharged and contaminated the Shiranui Sea located in the southwestern region of Japan's Kyushu Island.

The Chisso industrial plant initially started with fertilizer production, followed by expansion of the industry with the production of acetaldehyde, acetylene, acetic acid, vinyl chloride and octanol. In 1932, the Chisso industry started acetaldehyde production with 210 tons with an increment to 45,245 tons in the year 1960. The production of acetaldehyde began in 1951, it employed mercury sulphate as the catalyst and ferric sulphide as the co-catalyst, whereas before 1951, manganese dioxide was used as the co-catalyst. As a result, the side reaction produced methylmercury, an organic toxic mercury compound. This resulted in tons of effluent discharge in the sea as methylmercury.

The sea covered an area of 1200 km^2, and the coast of the sea and the nearby islands were habited by approximately 200,000 people, most of whom were engaged in the fishing industry.

TABLE 9.2
Health Effects of Heavy Metals Depend upon Dosage and Time of Exposure

			•	
			•	
			•	
			•	
			•	
			•	
			•	
			•	
			•	
			•	
			•	
			•	

Heavy Metal	Health Effects
Arsenic [8, 32, 48]	**Health Effects on Prolonged Exposure at Less Concentration:** Nausea, vomiting, decreased erythrocyte production, leukocytes, blood vessel damage, abnormal heart rate, reduced production of RBC and WBC, blood vessel damage, skin lesions, cancer, neurological disorder, pulmonary disease, hypertension, heart disease, diabetes mellitus, skin darkening, hyperpigmentation, hyperkeratosis, liver disease, organ cancer-kidney, lung, liver and urinary bladder, impairment of cognitive function, and Alzheimer's disease.
Cadmium [21, 49, 50]	**Health Effects on Prolonged Exposure at Less Concentration:** *Tubular impairment*: Fanconi's syndrome, abnormal urinary excretion, loss of zinc and copper bound to MT (metal binding protein), Kidney disease, Damage of lungs, Makes the bone fragile, Lung cancer, and Tubular dysfunction of the kidney. **Health Effects on Prolonged Exposure at High Concentration:** Itai-itai disease (renal impairment, osteomalacia, osteoporosis). **Health Effects on Short-Term Exposure at High Concentration:** Acute bronchitis, Pulmonary fibrosis, Emphysema, Lung cancer, Irritation of stomach, Causes vomiting and diarrhoea.
Chromium [30, 31, 51]	**Health Effects on Short-Term Exposure at High Concentration:** Impact on the nose-runny nose, nose ulcers, irritation in inner lining, Asthma, Cough, and Breathing difficulties. **Health Effects on Prolonged Exposure at Low Concentration:** Skin irritation, Liver damage, Kidney damage, and Damage of nerve tissues.
Silver [52]	**Health Effects on Prolonged Exposure at High Concentration:** Argyria, Discoloration of the skin and body tissues, Allergies such as rash, swelling, and inflammation, Breathing issues, Irritation in throat and lungs, and Stomach pain.
Mercury [34, 53]	**Health Effects on Short-Term Exposure at High Concentration:** Brain damage, Kidney damage, Damages foetus, Shyness, Tremors, Vision and hearing issues, and Memory problem. **Health Effects on Short-Term Exposure at High Concentration:** Lung damage, Blood pressure increase, Blood pressure increase, Rise in heart rate, Allergies, and Gastrointestinal dysfunctions.

(Continued)

TABLE 9.2 *(Continued)*

Health Effects of Heavy Metals Depend upon Dosage and Time of Exposure

Heavy Metal	Health Effects
Lead [21, 26, 31]	**Health Effects on Short-Term Exposure at High Concentration:** Brain damage, Kidney damage, Miscarriage, Damages male reproductive system, Impact on children-impaired development, learning disabilities, problems in coordination, cardiovascular disease risk, reduced intelligence, and short-term memory. **Health Effects on Prolonged Exposure at High Concentration:** Anaemia, Damages central nervous system, Inhibits motor function, Inhibits cognitive function, Causes coma, Kidney diseases, Lethargy, Inhibits heme-synthesis, and Rise in blood level.
Selenium [54, 55]	**Health Effects on Short-Term Exposure at High Concentration:** Gastrointestinal dysfunction. **Health Effects on Prolonged Exposure at High Concentration:** Selenosis: symptoms include hair loss, brittle nails, and abnormalities in neurological system. **Health Effects on Prolonged Exposure at Low Concentration:** Type 2 diabetes mellitus, Non-alcoholic fatty liver, Hyperlipidaemia and hyperglycaemia, and Cardiovascular disease.
Copper [16, 30]	**Health Effects on Short-Term Exposure at High Concentration:** Necrosis, Haemolysis, and Hepatocellular degeneration. **Health Effects on Short-Term Exposure at Low Concentration:** Acute gastroenteritis.
Zinc [4, 23]	**Health Effects on Short-Term Exposure at Low Concentration:** Nausea, Fatigue, Muscle soreness, Chest pain, Cough, Dyspnoea, and Bronchiolar leukocytes. **Health Effects on Prolonged Exposure at High Concentration:** Prostate cancer. **Health Effects on Short-Term Exposure at High Concentration:** Affects lymphocyte function.
Manganese [16, 26, 30]	**Health Effects on Prolonged Exposure at Low Concentration:** Manganism (neurological disorder), Parkinson's disease like symptoms-motor deficits, depression, anxiety, Impact on children-cognitive and memory deficit, Dementia, and Alzheimer's disease.
Iron [12, 13]	**Health Effects on Short-Term Exposure at High Concentration:** Haemochromatosis, and Thalassemia.
Nickel [12, 16]	**Health Effects on Short-Term Exposure at High Concentration:** Inhibits bone mineralization
Cobalt [20, 56]	**Health Effects on Prolonged Exposure at Low Concentration:** Respiratory problems, Gastrointestinal effects, Conjunctiva, Liver problems, Asthma, Reduced lung function, Cardiomyopathy and Dermatitis. **Health Effects on Short-Term Exposure at High Concentration:** Oedema, Congestion, Reduced ventilator function, and Lung haemorrhage.

The ingestion of contaminated fish led to the death of more than 100 people and thousands were permanently paralysed. Moreover, rapid advancement in the disease was observed when 16 out of the 34 early-diagnosed patients died within three months of mercury contamination. Furthermore, a pregnant woman who consumed contaminated fish gave birth to babies with genetic defects (such as congenital cerebral palsy, mental retardation, limb deformation, dysarthria, hyperkinesia, hypersalivation, strabismus, primitive reflex, cerebellar ataxia, paroxysmal and pyramidal symptoms, physical and nutrition growth disturbance). Even cats, fish and birds were affected by mercury poisoning.

Common symptoms observed in patients diagnosed with this disease included disturbance in audio, visual and senses, dysarthria, ataxia and tremor. Furthermore, characteristics of this disease included damage to the cerebral and cerebellar cortex of the central nervous system, damage in the calcarine region of the occipital lobe, the precentral and postcentral and temporal cortex, cerebellar atrophy, demyelination of dorsal root and its destruction.

In 1960, the mercury level in the sludge near the Chisso industrial plant was observed to be 2010 ppm. Moreover, mercury was found to be extensively present beyond Minamata Bay. A test conducted in 1959, on humans who were diagnosed and died with mercury poisoning had mercury concentration in the range 22.0–70.5 ppm, 2.6–24.8 ppm, 21.2–140.0 ppm in liver, brain and kidneys. Additionally, mercury concentration in marine life was found to be 35.7 ppm and 5.61 ppm in crabs and oysters, respectively [57].

9.7 SOME IMPORTANT TERMS

Toxicant: A toxicant is a harmful substance that is injurious to human health, as it has toxic effects on biological processes, tissues and organs. Whether a substance is poisonous or not, is dependent on several factors such as dosage, duration, rate, frequency of exposure and route of exposure such as inhalation (of gases, vapours, dust, fumes, mists), ingestion (through food or, drinking water) and exposure site (occupational or environmental).

Acute local exposure: Acute specifies a short time period of exposure, it may range from a few seconds to a few hours. Also, the term local refers to the specific site of exposure such as the eyes, skin or mucous membrane.

Chronic local exposure: In contrast to acute local exposure chronic specifies a longer time period of exposure, it may range from a few months to a few years. It occurs at the specific site of exposure such as the eyes, skin or mucous membrane.

Acute systemic exposure: Acute systemic exposure refers to a low dosage or short-term dosage of a toxicant in humans. This exposure usually occurs through inhalation and ingestion. Moreover, it affects organs not in direct contact/remote from the entry site of the toxicant, such as the liver.

Chronic systemic exposure: In contrast to acute systemic exposure, chronic systemic exposure refers to a high dosage or long-term dosage of a toxicant in humans. This exposure also occurs through inhalation and ingestion and affects organs not in direct contact/remote from the entry site of the toxicant, such as the liver.

9.8 HEALTH IMPACTS OF HEAVY METALS TOXICITY

This part of the chapter comprises the heavy metal-induced health impacts including neurological disorders, reproductive and developmental abnormalities, carcinogenicity and several organ damages such as bone, skin, immunological system, kidney, liver, heart and genes. Mechanism of heavy metal exposure on humans is illustrated Figure 9.6.

9.8.1 Neurological Disorder

Neurotoxicity occurs when exposure to neurotoxicants alters the normal activity of the nervous system. This eventually disrupts or kills neurons (nerve cells) which are essential for transmitting and processing signals in the brain and other parts of the nervous system [58]. Neurotoxicants involve heavy metals such as arsenic, aluminium, mercury, manganese, cadmium and vanadium, whereas the toxicity of iron, zinc and copper leads to neurodevelopment.

Aluminium toxicity in the nervous system arises due to free radical generation leading to ROS generation and oxidative stress. Furthermore, it alters the function of mitochondria and reduces its membrane potential. Thus, it inhibits enzymes, disturbing cellular signalling and transmission of neurons. Finally, it results in damage and degeneration of the nervous system and neural cell death [59].

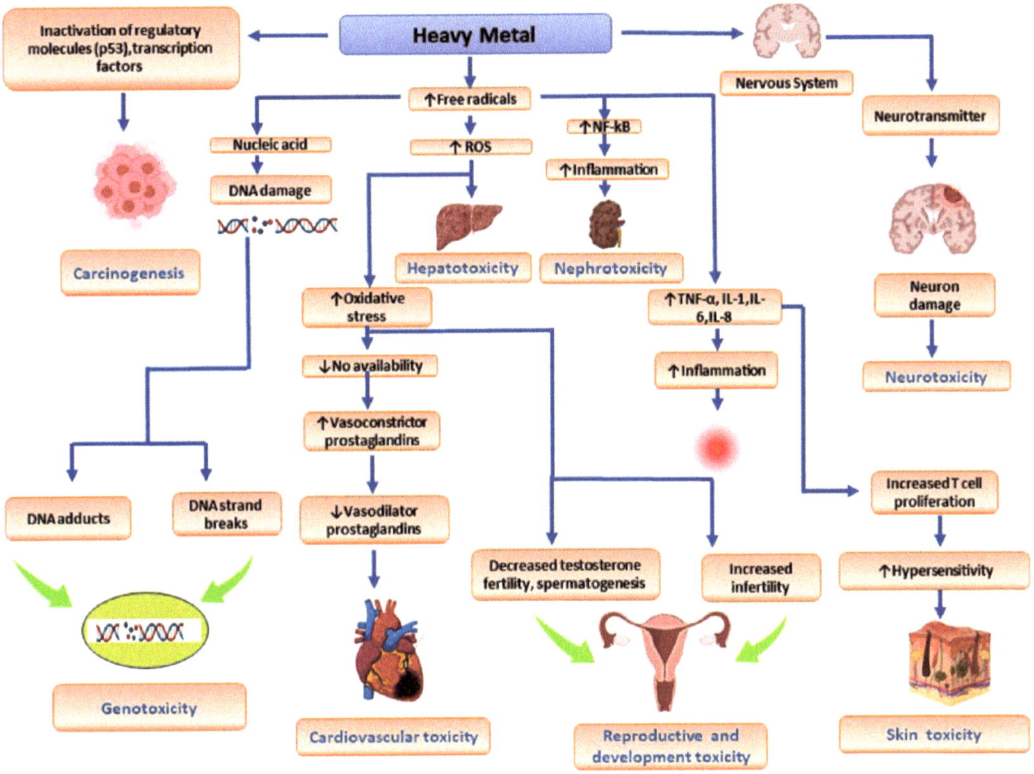

FIGURE 9.6 Mechanism of health impacts on heavy metal exposure.

Mechanism of cadmium neurotoxicity involves neuron gene expression inhibition, disruption of endocrine and epigenetic effects, finally causing neural cell death or apoptosis [60].

The organic form of mercury (Methylmercury) is a neurotoxicant, it damages the mitochondria and microtubule, causes lipid peroxidation and is responsible for the accumulation of neurotoxic molecules including aspartate, glutamate and serotonin in the nervous system. Metallic mercury induces toxicity as it enters the central nervous system by clinging onto the red blood cells and passing through the blood-brain barrier and placenta [61].

High dosage of manganese causes Alzheimer's and Parkinson's disease, it results in apoptosis (cell death) and homeostasis alterations. Apoptosis occurs via an increase in ROS formation leading to dysfunction of mitochondria, cytochrome c discharge, cleavage and binding of caspase-3 to pro-apoptotic protein, which further leads to DNA fragmentation causing apoptosis [62]. Impact of heavy metals on the neurological system is discussed in Table 9.3.

TABLE 9.3
Heavy Metal and Its Impact on the Neurological System

Heavy Metal	Neurological Impact	References
Aluminium	• Neuronal atrophy affects: locus coeruleus, substantia nigra and striatum • Neural damage • Neurodegeneration • Death of neural cells	[13, 63]

(Continued)

TABLE 9.3 *(Continued)*
Heavy Metal and Its Impact on the Neurological System

Heavy Metal	Neurological Impact	References
Arsenic	• Depression • Cognitive impairments • Alzheimer's disease • Nervous dysfunction – processing speed, language learning, mental retardation, development disability, executive functioning, neuropsychological, neural and development disabilities. • Parkinson's disease	[63, 64]
Iron	• Alzheimer's and Parkinson's disease • Aceruloplasminaemia • Huntington's disease • Amyotrophic lateral sclerosis • Neuroinflammation • Ageing • Epilepsy	[13, 63]
Mercury	• Alzheimer's disease • Cognitive impairment • Impact of impairment in neurological development in children – impaired nervous system, impaired cognitive development, low verbal intelligence quotient (IQ), lack of social development, lack of fine motor skills, mental retardation • Multiple sclerosis and Amyotrophic lateral sclerosis • Hearing loss • Cerebellar ataxia • Dysarthria	[12, 13, 31, 63]
Vanadium	• Neuroinflammation • Blood-brain barrier disruption • Loss of dendritic spine • Behavioural impairments • Cognitive and motor impairment • Memory alteration • Alzheimer's disease	[12]
Cadmium	• Alzheimer's disease • Etiopathogenesis • Neurodegenerative disease • Parkinson's disease • Huntington's disease • Amyotrophic lateral sclerosis and Multiple sclerosis • Cognitive function disorder • Impact on children – neurological disorders, learning disorders, behavioural difficulties	[12, 13, 16, 26, 31]
Copper	• Wilson's disease	[16, 30]
Thallium	• Damages peripheral nerve • Damages brain • Impact on cerebral hemisphere – oedema, vascular engorgement • Necrosis • Alterations in a capillary of the brain • Impact of cerebellar oedema – pyknotic Purkinje cells	[4, 26]

9.8.2 Reproductive and Development Abnormalities

According to the study done by the World Health Organization (WHO), it was observed that greater than 10% of women with heavy metal exposure (such as lead, cadmium and mercury) are at a risk of infertility. The major cause of female infertility is disruption of endocrine giving rise to hormonal imbalance. Furthermore, heavy metal exposure often leads to ovulation disturbance (such as absence, delay or irregularity of menstrual cycle) and oocytes affected by chromosomal abnormality [65].

Accumulation of cadmium in women is higher compared to men. This can be accounted by iron deficiency in women. However, cadmium toxicity in men is induced by inhibiting testicular functions (such as spermatogenesis and testosterone synthesis) by disrupting the blood-testis barrier. It also inhibits male fertility by disrupting the functions of the prostate gland thus leading to hormonal changes [66].

According to a report, mercury toxicity in women with exposure at a concentration of 0.01 mg m^{-3} had consequences of subfecundity, ovulation disturbances and adverse pregnancy outcomes [67]. In men, it was observed that the heavy metal Hg binds to the male sperm at the acrosomal cap, midpiece as well and tail [68]. Other targets of mercury toxicity were testis-supporting cells, seminal vesicles and epididymis, further causing disruption of the membrane of sperm, DNA synthesis and functions of mitochondria [68]. Impact of heavy metals on the reproductive system is discussed in Table 9.4.

TABLE 9.4
Heavy Metal and Its Impact on the Reproductive System

Heavy Metal	Impact on Reproductive Systems	References
Arsenic	• Impairment of the male reproductive system • Impact on male reproduction system – decreases the weight of testes and reduces the count of sperm present in the epididymis, causes variation in testosterone levels and gonadotropin levels and disturbs the steroidogenesis process. • Impact on females – endometrial cancer by impairment of endometrial angiogenesis • Impact of inorganic arsenic – route of prostate cancer due to effect on human prostate epithelial cells • Infant mortality	[12, 16, 32, 64]
Lead	• Disrupts brain development in foetus • Decreases male fertility – by altering sperm morphology • Stillbirth	[13, 16, 26, 31, 32]
Cadmium	• Toxicity in testes – disrupts blood-testis barrier • Loss of pregnancy	[21, 26, 31]
Mercury	• Menstrual disorder • Primary subfecundity • Adverse pregnancy outcomes • Infertility in men • Impacts epididymis, seminal vesicles, supporting cells in the testis • *Impact on children*: growth disorder, epilepsy, excessive salvation, limb deformity, dysarthria, misaligned eyes and primitive reflexes, chorea, athetosis, impaired kidneys, lungs, digestive system, immune system, nervous system, memory impairment, insomnia and tremors	[16, 26, 31]

9.8.3 Carcinogenicity

The term cancer can be defined as a condition wherein the body's own cells grow and replicate in an uncontrolled manner and cause harm to the body. Heavy metals such as chromium, cadmium, iron, nickel, mercury and arsenic possess a carcinogenic effect on human beings. According to [26], prolonged exposure to heavy metals for up to ten days raises the risk of oral cancer [69]. In a biopsy it was found that accumulation of heavy metals causes breast cancer. Heavy metals responsible for breast cancer include iron, nickel, chromium, zinc, cadmium, mercury and lead.

The carcinogenicity mechanism of cadmium involves cytotoxic effects such as ROS generation, oxidative stress, apoptosis, inflammatory process activation, impaired DNA repair processes causing DNA damage, abnormality in methylation of DNA, induction of immediate early response genes (IEGs), overexpression and stress response genes (such as genes inducing oxidative stress and GSH synthesis). In comparison to other carcinogens, cadmium is considered to be a weak mutagen [70].

Nickel-induced cancer involves several mechanisms such as the regulation of genes, management of transcription factors and generation of free radicals. Moreover, it regulates the expression of RNAs such as mRNAs, micro-RNAs and long non-coding RNAs. Furthermore, in order to raise the modulation of hypoxia-inducible factor-1, nickel takes part in promoter methylation and gene 3 (MEG3) downregulation [71].

Lead induces carcinogenicity by generating ROS, damaging DNA repair mechanisms and altering the structure and sequence of chromosomes and genes regulating cellular tumours. Moreover, it also removes zinc from its regulatory proteins and disrupts transcription [72].

In contrast to other carcinogenic heavy metals such as cadmium, nickel, arsenic and cobalt. Zinc is not a carcinogenic metal. However, zinc being replaced by carcinogenic metals from its zinc-binding structure causes cancer [23]. The carcinogenic impact of heavy metals on internal organs is discussed Table 9.5.

TABLE 9.5
Heavy Metal and Its Carcinogenic Impact on Internal Organs

Heavy Metal	Internal Organ Cancer	References
Chromium	• Lung • Nasal sinus	[12, 13, 31]
Cadmium	• Lung, breast and kidney cancer • Pancreatic and prostate cancer • Urinary bladder cancer • Nasopharynx • Impact on prostate gland – adenocarcinomas • Impact of epigenetic changes – breast cancer • Impact of Cd in blood – nasopharyngeal carcinoma	[12, 13, 21, 26]
Iron	• Prostate cancer • Pancreatic • Colorectal • Lungs • Breast	[12, 13]
Nickel	• Respiratory tract • Lungs • Nose and nasal sinus	[16, 71, 73]
Arsenic	• Lung, bladder cancer • Kidney and liver cancer • Skin cancer • Uterus and prostate cancer	[64]

9.8.4 Organ Damage

Herein, we have discussed the toxic effects and mechanism of heavy metals in organs including skin, bones, liver (hepatotoxicity), kidney (nephrotoxicity), heart (cardiovascular toxicity), immune system (immunotoxicity) and DNA (genotoxicity). The impact of heavy metals on organs are discussed in Table 9.6.

9.8.5 Bone Impact

Accumulation of lead in the human body can result in a lowering of bone density and its cortical width also enlarging the risk of bone fracture. Moreover, it has been observed that most of the lead in the human body is accumulated in bones. As lead is chemically similar to calcium it follows the same route as calcium and gets deposited in the bone structure [30].

Aluminium toxicity in the bone marrow gives rise to osteomalacia. Furthermore, inhibition of protein synthesis restricted the synthesis of haemoglobin in the bone marrow leading to microcytic hypochromic anaemia. Also, with an increase in aluminium content in the body, a decrease in osteoblasts and osteoclast count and a reduction in bone marrow remodelling occurs [74].

Interaction of calcium (present in the bones) with cadmium (dumped in the osteoid tissue) leads to disarrangement of bone metabolism by bone-remodelling and both calcification and decalcification. Moreover, bone demineralization is one of the impacts of cadmium interference with cell metabolism of bones (composed of calcium, copper, iron, magnesium and zinc) [75]. Furthermore, it also affects the ion's regulatory functions. Itai-itai disease [76] was one such disease that spread in Toyama, Japan due to the ingestion of cadmium-contaminated rice toxified by irrigation water. It was predominantly found in women. Characteristics of this disease involve joint pain, severe bone pain and complex fractures in the joint. The mechanism behind this disease was the inhibition of renal tubular cell functioning, which reduced vitamin D efficiency and consequently calcium absorption. Moreover, renal tubular dysfunction, osteomalacia and osteoporosis were also induced [76].

9.8.6 Skin Toxicity

Heavy metals such as chromium, mercury and arsenic are the major skin toxicants. Some major heavy metal-induced skin diseases include hyperpigmentation, keratosis, contact and systemic contact dermatitis. Hyperpigmentation [77] also termed raindrop-shaped discoloration, is a skin disease in which dark spots appear at the limbs and trunk of the skin. Simple keratosis resembles palm and sole thickening whereas nodular keratosis resembles small protrusions on palms and soles [78]. Sensitivity of the skin due to direct dermal contact to an allergen causes systemic contact dermatitis whereas delayed hypersensitivity due to repeated dermal contact causes contact dermatitis. Dermatitis is characterized by symptoms such as swelling, skin peeling, irritation and vesicle formation.

According to a study, mercury is the most common dermatological problem causing heavy metals [79]. Mercury induces skin diseases such as acrodynia, contact dermatitis and inflammation [80]. Skin inflammation can be observed within six months in people tattooed with red pigments of mercury and calcium sulphide [79].

Arsenic-induced skin disease is more prevalent in men rather than women [81]. The reason behind this is less effective arsenic methylation in men than in women. Methylated arsenic in comparison to inorganic arsenic is more hydrophilic so it is rapidly excreted from the body in the form of urine. Furthermore, the mechanism of arsenic toxicity involves ROS generation, repair of DNA and alteration in cell proliferation [82, 83].

9.8.7 Immunological Toxicity

The most common type of immunological toxic reaction of nickel in humans is an allergic reaction. Nickel causes either one or both immunomodulatory and immune toxic effects on humans. Cell phones, jewellery and hair clamps are some of the sources of nickel, and skin contact with these nickel-containing substances induces immunotoxicity [52].

Lead is responsible for harmful effects on the immune system including allergies, infections and autoimmunity. The mechanism behind immunological toxicity involves the production of B as well as T cells and MHC activity. Altering the function of T cells can affect humoral and cellular response and the body may become more autoimmune and hypersensitive [84].

Zinc must be available in adequate amounts for the immune system to function properly. As a result, it is essential for several cellular and molecular processes. An excess of zinc can inhibit the function of T cells and alter the function of lymphocytes as it is responsible for its response to cytokines and mitogens. Moreover, zinc acts as a co-factor for thymulin (thymic hormone) and helps in signal transduction of leukocytes [23].

9.8.8 Nephrotoxicity

Nephrotoxicity is a chronic kidney disease. Characteristics of nephrotoxicity include a decrease in glomerular filtration rate and permanent loss of nephrons. Furthermore, the decrease in glomerular filtration rate results in the accumulation of extraneous toxicants in blood. Arsenic, lead, mercury and cadmium are some known nephrotoxicants. These nephrotoxicants lead to modification of the kidney, such as kidney enlargement, nuclear damage, damage of mitochondria, reduction in antioxidant capacity and increment of malondialdehyde and metal content [85].

Aluminium leads to in vivo and in vitro oxidation of biological molecules. One such example is observed in renal tissues due to the accumulation of aluminium, where enhancement of lipid peroxidation takes place by lowering the content of glutathione. This affects cellular metabolism, generates oxidative stress and degenerates cells of renal tubules. Furthermore, alteration in the transport of p-aminobutyric acid, reabsorption of phosphate in the tubules of the kidney and imbalance in sodium and water content causes kidney damage or nephrotoxicity [45].

Thallium sulphate is highly injurious as after consumption, it accumulates in the body for up to two months. Albuminuria and haematuria are some of the indicators of thallium-induced nephrotoxicity. However, no severe damage to the renal function occurs [86].

Direct exposure of kidneys to cadmium impacts proximal tubular epithelium, a higher concentration of cadmium in urine, glucosuria, aminoaciduria as well as 32-microglobulinuria and inhibition of phosphate reabsorption by renal tubes [87]. Other impacts of cadmium on the kidney are Fanconi syndrome, glucosuria, renal failure, hypercalciuria and renal tubular acidosis [88].

9.8.9 Hepatotoxicity/Liver Damage

The liver is responsible for the performance of several metabolic functions of the body [89]. It plays a vital role in the metabolism such as the removal of toxic chemicals from the body. Hence, the liver is exposed to several toxic heavy metals. Toxicity of the liver is termed as hepatotoxicity.

The mechanism involved in copper hepatoxicity is oxidative stress therefore copper deposition is both pathogenic as well as pathognomonic. Toxicity of copper in the liver causes Wilson's disease. Moreover, Reduced biliary excretion of copper leads to cholestatic liver [90].

The chromium mechanism for inducing hepatoxicity involves increased ROS levels, lipid peroxidation, DNA and RNA suppression and DNA damage. Moreover, dysfunction of mitochondria, apoptosis, inhibition of cell growth and lowering activity of antioxidant enzymes are also responsible for causing harm to the liver. Steatosis of hepatocytes is one of the alterations in histopathology due to chromium toxicity [91].

Selenium is considered to be one of the reasons for non-alcoholic fatty liver disease (NAFLD), as selenium participates in lipid metabolism and antioxidant defence mechanisms. Eventually, the occurrence of oxidative stress leads to liver damage [92].

9.8.10 Cardiovascular Toxicity/Heart Damage

Increased OS, decreased NO availability, elevated vasoconstrictor prostaglandins, modifications to the renin-angiotensin system and decreased vasodilator prostaglandins are the mechanisms causing cardiovascular toxicity. Cardiac disease is also caused by suppression of smooth Ca^{2+} signalling, elevation of inflammation and endothelium-dependent vasorelaxation and modification of the vascular response to vasoactive agonists. Chronic exposure to lead may cause several diseases such as arteriosclerosis and atherosclerosis. Moreover, prolonged exposure to lead increases arterial pressure [93].

Cobalt exposure is a cause of both reversible systolic cardiac depression as well as cardiomyopathy disorders. Though cardiomyopathy is slow and fatal, survivors' cardiac function recovers [94].

Human exposure to methylmercury [95] is one of the causes of its adverse cardiovascular effects. Mercury-induced vascular effects raise oxidative stress and inflammation, whereas it reduces oxidative defence. Eventually, it leads to mitochondrial dysfunction, and thrombosis and the inner membrane of mitochondria is auto-oxidized. Mercury toxicity is related to coronary heart disease, myocardial infarction, hypertension and atherosclerosis. Additionally, the primary cardiovascular effects of mercury toxicity are an increase in blood pressure and variable heart rate.

Cadmium toxicity [96] causes ROS generation, membrane damage, apoptosis, oxidative stress and DNA damage. Moreover, this leads to an increase in cardiac markers. Furthermore, it generates antioxidants (such as glutathione and glutathione peroxidase) and enzyme activity is reduced. Eventually, it results in hypertension and cardiotoxicity. Other cardiovascular diseases are myocardial infarction and atherosclerosis.

9.8.11 Genotoxicity

Immunological signals are disrupted, DNA is destroyed and genetic information is lost when aluminium is present in the brain. DNA and RNA formation is prevented as they bind to aluminium salts. DNA can break into single strands and double strands at various stages of the cell cycle to accomplish this. Additionally, exposure to chemicals containing aluminium can alter the integrity of DNA structure and inhibit DNA repair [97].

Genotoxic effects of arsenic include alterations in deoxyribonucleic acid, abnormalities of chromosomes, micronuclei production, mutation and exchange of sister chromatid [98]. Mechanisms behind genotoxic effects involve oxidative stress generation and DNA repair disruption. Despite arsenic possessing weak mutagenicity, arsenic shows greater mutagenicity in human cells on exposure to UV light [99].

The transient species of chromium and the resulting ROS disrupt the regular physiological processes, oxidize proteins, induce damage to nucleic acids, induce lipid peroxidation and change gene expression. Moreover, oxidative stress causes DNA damage, modification of DNA bases, crosslink between DNA-DNA, DNA-protein, abasic sites lesions and sugar lesions. These dysfunctions resemble the genotoxic effect of chromium [100] (Table 9.6).

TABLE 9.6
Health Impact of Heavy Metals on Organs

Heavy Metal	Health Impact on Organs	References
Cadmium	**Health impact on bone** • Itai-itai diseases • Bone demineralization • Osteomalacia • Osteoporosis **Health impact on kidneys** • Glucosuria and hypercalciuria • Fanconi-like syndrome • Phosphaturia and Aminoaciduria • Renal tubular acidosis • Renal failure **Health impact on liver** • Targets renal cortex • Hepatic dysfunction • Liver failure hence increased risk of cancer **Health impact on heart** • Hypertension • Diabetes • Stroke • Carotid atherosclerosis • Peripheral arterial disease • Heart failure • Myocardial infarction **Health impact on the respiratory system** **Chronic health impact**: • Chronic bronchitis • Impaired sense of smell • Ulceration of nasal mucous membrane • Lung function abnormalities • Chest radiographs • Emphysema • Decreased olfactory functions **Acute health impacts**: • Metallic fever • Pulmonary oedema	[12, 13, 16, 26]
Nickel	**Health impact on bone** • Inhibits bone mineralization **Health impact on the immunological system** • Skin allergies • Lung fibrosis • Respiratory tract cancer • Facial dermatitis **Other health impacts** • Lung embolisms • Asthma • Heart disorders	[16]

(Continued)

TABLE 9.6 *(Continued)*

Health Impact of Heavy Metals on Organs

Heavy Metal	Health Impact on Organs	References
Manganese	**Health impacts** • Parkinson's disease mimicry • Manganism • Hypotension	
Vanadium	**Health impacts** • Kidney haemorrhage • Liver haemorrhage • Teeth as well as tongue blackening • Irritation in eyes and nose • Paralysis • Headache and nausea • Rashes	[12]
Lead	**Health impact on bone** • Decreases cortical width of bone • Decreases bone density • Bone fracture **Health impact on the immunological system** • Allergies • Infectious diseases • Autoimmunity • Cancer (lung, stomach, bladder) **Health impact on kidneys** • Proximal tubular dysfunction • Fanconi-like syndrome • Glomerulonephritis **Health impact on liver** • Chronic cirrhosis **Health impact on heart** • Arteriosclerosis • Hypertension • Thrombosis • Atherosclerosis • Increase in arterial pressure	[13, 16, 21, 26, 31]
Aluminium	**Health impact on bone** • Renal osteodystrophy • Osteomalacia • Osteoporosis • Inhibits bone mineralization **Other health impacts** • Coloitis • Pulmonary fibrosis • Hypo-parathyroid • Leukocytes	[13, 45]

(Continued)

TABLE 9.6 *(Continued)*
Health Impact of Heavy Metals on Organs

Heavy Metal	Health Impact on Organs	References
Mercury	**Health impact on bone** • Bone demineralization **Health impact on skin** • Type I and IV hypersensitivity • Dental amalgam • Acrodynia • Acute contact dermatitis • Oral lichen planus • Kawasaki disease • Allergy from tattoos **Health impact on the immunological system** • Amalgam (mercury alloy) – affects immune system **Health impact on kidneys** • Acute tubular necrosis • Altered mental status • Abdominal pain • Tremors • Hypotension • Epithelium and necrosis injury • Tubular failure • Acute dyspnoea **Health impact on heart** • Coronary failure • Atherosclerosis • Carotid artery stenosis • Myocardial infarction	[12, 13, 16, 26, 31]
Chromium	**Health impact on skin** • Type I and IV hypersensitivity • Skin cancer • Contact and systemic dermatitis • Eczema • Nasal septum perforation • Bronchial asthma • Allergic rhinitis **Health impact on the immunological system** • Obstruction of lymphocyte proliferation **Health impact on liver** • Causes histopathological changes **Impact of Cr(VI)** • Decrease in the phagocytic action of alveolar macrophages • Decrease in humoral immune response • Coma • Vertigo • Toxic nephritis • Ulcer	[16, 26, 30]

(Continued)

TABLE 9.6 (Continued)
Health Impact of Heavy Metals on Organs

Heavy Metal	Health Impact on Organs	References
	Other health impacts • Chrome hole • Pneumonia • Bronchitis • Gingivitis • Sinusitis	
Arsenic	**Health impact on skin** • Hyperkeratosis • Impact of prolonged exposure – hyperpigmentation • Skin cancer • Bowen's disease • Impact of chronic arsenic toxicity: pigmentation, keratosis **Health impact on the immunological system** • Diarrhoea • Respiratory tract infection – Influenza A and pulmonary tuberculosis • Lung cancer	[12, 32, 64]
Copper	**Health impact on liver** • Wilsons disease • Cholestatic liver disease **Other health impacts** • Metal fever • Hepatic disease such as hepatic cirrhosis • Poisoning • Wilson's disease • Kayser-Fleisher rings • Fatigue and Appetite loss • Insomnia • Demyelination	[22, 95]
Cobalt	**Health impact on heart** • Systolic cardiac depression • Cardiomyopathy **Other health impacts** • Oedema • Lung haemorrhage • Congestion • Lowering of ventilator function • Allergic dermatitis • Conjunctiva • Pneumonia and fibrosis	[20, 56, 101]
Selenium	**Health impact due to selenium deficiency** • Covid-19 • Keshan disease • Kashin-Beck disease • Infertility in men • Muscle necrosis • Hypothyroidism • Cancer • AIDS • Cardio-cerebrovascular disease	[102, 103]

(Continued)

TABLE 9.6 *(Continued)*
Health Impact of Heavy Metals on Organs

Heavy Metal	Health Impact on Organs	References
	Health impact due to excess intake • Selenosis – Symptoms include garlic like breath, hair loss, fingernail loss, myocardial infarction, renal failure, dermatitis • Hair loss, nail brittleness • Hyperlipidaemia • Non-alcoholic fatty liver • Cardiovascular disease • Type 2 diabetes mellitus (T2DM) • Hyperglycaemia	
Zinc	**Health impacts** • Impact on exposure to zinc-chloride constituting bombs-adult respiratory distress syndrome (ARDS) leading to death • Respiratory disorder • Inhaling zinc smoke causes metal fume fever • Acute impacts – fever, nausea, fatigue, muscle soreness, chest pain, cough, dyspnoea and bronchiolar leukocytes • Haemorrhagic pancreatitis • Renal failure • Prostate cancer • Neutropenia • Copper deficiency • Affects lymphocyte function	[23]

9.9 CONCLUSION

The majority of the heavy metals accumulate in the environment due to urbanization and industrialization. Humans are exposed to these heavy metals in their day-to-day lives via ingestion, inhalation or even dermal contact and through occupational hazards. Dosage and time of exposure play an important role in determining the adversity of the health impact. It is important to understand metal chemistry and the interaction of heavy metals with the biological system of our body. Over-exposure of metals to human beings leads to several kinds of metal toxicity including Alzheimer's, Wilson's, haemochromatosis, porphyrias, cholecystitis, chronic liver, kidney and lung disease, cancer through suppression of tumour suppressor genes and other developmental anomalies which may be fatal. Suitable prevention and remediation techniques must be employed at the source, during the process and medication must be available in case of emergency. Treatment strategies may include a combination of antidotes and chelation therapy along with symptomatic treatment and proper monitoring of the airway, breathing and circulation till recovery. Haemodialysis may be used in an emergency, if required.

ACKNOWLEDGEMENT

One of the authors, Nipun Tirkey, acknowledges the Council of Scientific & Industrial Research (CSIR), India, for the Junior Research Fellowship (JRF) provided by them during the tenure of this work. The authors are also grateful to Birla Institute of Technology, Mesra, for the necessary support for this work.

REFERENCES

1. N.K. Srivastava, C.B. Majumder, Novel biofiltration methods for the treatment of heavy metals from industrial wastewater. J. Hazard. Mater. 151 (2008) 1–8.
2. J.H. Duffus, Metales pesados: Un plazo significativo? Pure Appl. Chem. 74 (2002) 793–807.
3. M.A. Zoroddu, J. Aaseth, G. Crisponi, S. Medici, M. Peana, V.M. Nurchi, The essential metals for humans: A brief overview. J. Inorg. Biochem. 195 (2019) 120–129.
4. J. Briffa, E. Sinagra, R. Blundell, Heavy metal pollution in the environment and their toxicological effects on humans. Heliyon 6 (2020) e04691.
5. A.Q. Shah, T.G. Kazi, J.A. Baig, M.B. Arain, H.I. Afridi, G.A. Kandhro, S.K. Wadhwa, N.F. Kolachi, Determination of inorganic arsenic species (As3+ and As5+) in muscle tissues of fish species by electrothermal atomic absorption spectrometry (ETAAS). Food Chem. 119 (2010) 840–844.
6. D.K. Gupta, S. Chatterjee, Arsenic Contamination in the Environment: The Issues and Solutions, Springer, 2017.
7. D.E. Baker, L. Chesnin, Chemical monitoring of soils for environmental quality and animal and human health. Adv. Agron. 27 (1975) 305–374.
8. N. Singh, D. Kumar, A.P. Sahu, Arsenic in the environment: Effects on human health and possible prevention. J. Environ. Biol. 28 (2007) 359–365.
9. A.A. El-Kady, M.A. Abdel-Wahhab, Occurrence of trace metals in foodstuffs and their health impact. Trends Food Sci. Technol. 75 (2018) 36–45.
10. M.S. Collin, S.K. Venkatraman, N. Vijayakumar, V. Kanimozhi, S.M. Arbaaz, R.G.S. Stacey, J. Anusha, R. Choudhary, V. Lvov, G.I. Tovar, F. Senatov, S. Koppala, S. Swamiappan, Bioaccumulation of lead (Pb) and its effects on human: A review. J. Hazard. Mater. Adv. 7 (2022) 100094.
11. G. Bjørklund, M. Dadar, J. Mutter, J. Aaseth, The toxicology of mercury: Current research and emerging trends. Environ. Res. 159 (2017) 545–554.
12. Z. Haidar, K. Fatema, S.S. Shoily, A.A. Sajib, Disease-associated metabolic pathways affected by heavy metals and metalloid. Toxicol. Rep. 10 (2023) 554–570.
13. M. Jaishankar, T. Tseten, N. Anbalagan, B.B. Mathew, K.N. Beeregowda, Toxicity, mechanism and health effects of some heavy metals. Interdiscip. Toxicol. 7 (2014) 60–72.
14. G. Genchi, M.S. Sinicropi, G. Lauria, A. Carocci, A. Catalano, The effects of cadmium toxicity. Int. J. Environ. Res. Public Health 17 (2020) 1–24.
15. T. Pavesi, J.C. Moreira, Mechanisms and individuality in chromium toxicity in humans. J. Appl. Toxicol. 40 (2020) 1183–1197.
16. G.A. Engwa, P.U. Ferdinand, F.N. Nwalo, M.N. Unachukwu. Mechanism and Health Effects of Heavy Metal Toxicity in Humans. Poisoning Mod. World – New Tricks an Old Dog. 10 (2019) 70–90.
17. M.G. Soni, S.M. White, W.G. Flamm, G.A. Burdock, Safety evaluation of dietary aluminum. Regul. Toxicol. Pharmacol. 33 (2001) 66–79.
18. J. Alexander, Chapter 52-Selenium. Handbook on the Toxicology of Metals, 2015, 1175–208.
19. S.L. O'Neal, W. Zheng, Manganese toxicity upon overexposure: A decade in review. Curr. Environ. Health Rep. 2 (2015) 315–328.
20. L. Leyssens, B. Vinck, C. Van Der Straeten, F. Wuyts, L. Maes, Cobalt toxicity in humans—A review of the potential sources and systemic health effects. Toxicology 387 (2017) 43–56.
21. K. Rehman, F. Fatima, I. Waheed, M.S.H. Akash, Prevalence of exposure of heavy metals and their impact on health consequences. J. Cell. Biochem. 119 (2018) 157–184.
22. M. Bost, S. Houdart, M. Oberli, E. Kalonji, J.F. Huneau, I. Margaritis, Dietary copper and human health: Current evidence and unresolved issues. J. Trace Elem. Med. Biol. 35 (2016) 107–115.
23. L.M. Plum, L. Rink, H. Hajo, The essential toxin: Impact of zinc on human health. Int. J. Environ. Res. Public Health 7 (2010) 1342–1365.
24. S.S. Sonone, S. Jadhav, M.S. Sankhla, R. Kumar, Water contamination by heavy metals and their toxic effect on aquaculture and human health through food chain. Lett. Appl. NanoBioScience 10 (2021) 2148–2166.
25. M. S. Sankhla, M. Kumari, M. Nandan, R. Kumar, P. Agrawal, Heavy metals contamination in water and their hazardous effect on human health—A review. Int. J. Curr. Microbiol. App. Sci 5(10), (2016), 759–766.
26. S. Mitra, A.J. Chakraborty, A.M. Tareq, T. Bin Emran, F. Nainu, A. Khusro, A.M. Idris, M.U. Khandaker, H. Osman, F.A. Alhumaydhi, J. Simal-Gandara, Impact of heavy metals on the environment and human health: Novel therapeutic insights to counter the toxicity. J. King Saud Univ. – Sci. 34 (2022) 101865.
27. S. Kamran, A. Shafaqat, H. Samra, A. Sana, F. Samar, B.S. Muhammad, A.B. Saima, M.T. Hafiz, Heavy metals contamination and what are the impacts on living organisms. Greener J. Environ. Manag. Public Saf. 2 (2013) 172–179.

28. A. Selvi, A. Rajasekar, J. Theerthagiri, A. Ananthaselvam, K. Sathishkumar, J. Madhavan, P.K.S.M. Rahman, Integrated remediation processes toward heavy metal removal/recovery from various environments – A review, Front. Environ. Sci. 7 (2019) 66.
29. N. Singh, V.K. Gupta, A. Kumar, B. Sharma, Synergistic effects of heavy metals and pesticides in living systems. Front. Chem. 5 (2017) 70.
30. M.A.A. Wijayawardena, M. Megharaj, R. Naidu, Exposure, Toxicity, Health Impacts, and Bioavailability of Heavy Metal Mixtures, Elsevier Inc, Adv. Agron. 138 (2016) 175–234.
31. M. Balali-Mood, K. Naseri, Z. Tahergorabi, M.R. Khazdair, M. Sadeghi, Toxic mechanisms of five heavy metals: Mercury, lead, chromium, cadmium, and arsenic. Front. Pharmacol. 12 (2021) 1–19.
32. M.I. Khan, M.F. Ahmad, I. Ahmad, F. Ashfaq, S. Wahab, A.A. Alsayegh, S. Kumar, K.R. Hakeem, Arsenic exposure through dietary intake and associated health hazards in the Middle East. Nutrients 14 (2022) 1–22.
33. J.G. Paithankar, S. Saini, S. Dwivedi, A. Sharma, D.K. Chowdhuri, Heavy metal associated health hazards: An interplay of oxidative stress and signal transduction. Chemosphere 262 (2021) 128350.
34. J.O. Duruibe, M.O.C. Ogwuegbu, J.N. Egwurugwu, Heavy metal pollution and human biotoxic effects. Int. J. Phys. Sci. 2 (2007) 112–118.
35. C.A. Chrestensen, D.W. Starke, J.J. Mieyal, Acute cadmium exposure inactivates thioltransferase (Glutaredoxin), inhibits intracellular reduction of protein-glutathionyl-mixed disulfides, and initiates apoptosis. J. Biol. Chem. 275 (2000) 26556–26565.
36. M.E. Ingle, D. Watkins, Z. Rosario, C.M. VélezVega, A.M. Calafat, M. Ospina, K.K. Ferguson, J.F. Cordero, A. Alshawabkeh, J.D. Meeker, An exploratory analysis of urinary organophosphate ester metabolites and oxidative stress among pregnant women in Puerto Rico. Sci. Total Environ. 703 (2020) 1161–1208.
37. P.M. Hanna, M.B. Kadiiska, R.P. Mason, Oxygen-derived free radical and active oxygen complex formation from cobalt(II) chelates in vitro. Chem. Res. Toxicol. 5 (1992) 109–115.
38. J. Pi, S. Horiguchi, Y. Sun, M. Nikaido, N. Shimojo, T. Hayashi, H. Yamauchi, K. Itoh, M. Yamamoto, G. Sun, M.P. Waalkes, Y. Kumagai, A potential mechanism for the impairment of nitric oxide formation caused by prolonged oral exposure to arsenate in rabbits. Free Radic. Biol. Med. 35 (2003) 102–113.
39. M. Valko, M. Izakovic, M. Mazur, C.J. Rhodes, J. Telser, Role of oxygen radicals in DNA damage and cancer incidence. Mol. Cell. Biochem. 266 (2004) 37–56.
40. R.A. Bernhoft, Mercury toxicity and treatment: A review of the literature. J. Environ. Public Health (2012) 460508.
41. G. Flora, D. Gupta, A. Tiwari, Toxicity of lead: A review with recent updates. Interdiscip. Toxicol. 5 (2012) 47–58.
42. S.J.S. Flora, M. Mittal, A. Mehta, Heavy metal induced oxidative stress & its possible reversal by chelation therapy. Indian J. Med. Res. 128 (2008) 501–523.
43. M.R. Rahimzadeh, M.R. Rahimzadeh, S. Kazemi, R.J. Amiri, M. Pirzadeh, A.A. Moghadamnia, Aluminum poisoning with emphasis on its mechanism and treatment of intoxication. Emerg. Med. Int. 2022 (2022) 1–13.
44. W.M. El-Sayed, M.A. Al-Kahtani, A.M. Abdel-Moneim, Prophylactic and therapeutic effects of taurine against aluminum-induced acute hepatotoxicity in mice. J. Hazard. Mater. 192 (2011) 880–886.
45. S. Mahieu, M. del C. Contini, M. González, N. Millen, Melatonin reduces oxidative damage induced by aluminium in rat kidney. Toxicol. Lett. 190 (2009) 9–15.
46. Q. Zhang, Y. Song, K. Amor, W.E. Huang, D. Porcelli, I. Thompson, Monitoring Cr toxicity and remediation processes – Combining a whole-cell bioreporter and Cr isotope techniques. Water Res. 153 (2019) 295–303.
47. A.J. Paine, Mechanisms of chromium toxicity, carcinogenicity and allergenicity: Review of the literature from 1985 to 2000. Hum. Exp. Toxicol. 20 (2001) 439–451.
48. M.F. Naujokas, B. Anderson, H. Ahsan, H. Vasken Aposhian, J.H. Graziano, C. Thompson, W.A. Suk, The broad scope of health effects from chronic arsenic exposure: Update on a worldwide public health problem. Environ. Health Perspect. 121 (2013) 295–302.
49. A.G. Davison, A.J. Newman Taylor, J. Darbyshire, D.R. Chettle, C.J.G. Guthrie, D. O'Malley, H.T. Mason, P.M. Fayers, K.M. Venables, C.A.C. Pickering, D. Franklin, M.C. Scott, H. Holden, A.L. Wright, D. Gompertz, Cadmium fume inhalation and emphysema. Lancet 331 (1988) 663–667.
50. G.C. Lalor, Review of cadmium transfers from soil to humans and its health effects and Jamaican environment. Sci. Total Environ. 400 (2008) 162–172.
51. J. Fang, M. Lang, X. Ye, W. Zhang, K. Zhu, Non-isothermal crystallization behavior of polypropylene/zinc oxide composites. Sci. Eng. Compos. Mater. 23 (2016) 505–510.

52. H. Guo, H. Liu, Z. Jian, H. Cui, J. Fang, Z. Zuo, J. Deng, Y. Li, X. Wang, L. Zhao, R. He, H. Tang, Immunotoxicity of nickel: Pathological and toxicological effects. Ecotoxicol. Environ. Saf. 203 (2020) 111006.
53. K. Gulati, B. Banerjee, S.B. Lall, A. Ray, Effects of diesel exhaust, heavy metals and pesticides on various organ systems: Possible mechanisms and strategies for prevention and treatment. Indian J. Exp. Biol. 48 (2010) 710–721.
54. R. Dixit, D. Wasiullah, K. Malaviya, U.B. Pandiyan, A. Singh, R. Sahu, B.P. Shukla, J.P. Singh, P.K. Rai, H. Sharma, D. Lade, Paul, Bioremediation of heavy metals from soil and aquatic environment: An overview of principles and criteria of fundamental processes. Sustain 7 (2015) 2189–2212.
55. M. Vinceti, E.T. Wei, C. Malagoli, M. Bergoini, G. Vivoli, Adverse health effects of selenium in humans. Rev. Environ. Health 16 (2001) 233–251.
56. L.O. Simonsen, H. Harbak, P. Bennekou, Cobalt metabolism and toxicology – A brief update. Sci. Total Environ. 432 (2012) 210–215.
57. M. Harada, Minamata disease: Methylmercury poisoning in Japan caused by environmental pollution. Crit. Rev. Toxicol. 25 (1995) 1–24.
58. L.G. Costa, Biochemical and molecular neurotoxicology: Relevance to neurotoxicity testing and risk assessment. Toxicol. Lett. 95 (1998) 21.
59. Z. Wu, Y. Du, H. Xue, Y. Wu, B. Zhou, Aluminum induces neurodegeneration and its toxicity arises from increased iron accumulation and reactive oxygen species (ROS) production. Neurobiol. Aging 33 (2012) 199.e1–199.e12.
60. B. Wang, Y. Du, Cadmium and its neurotoxic effects. Oxid. Med. Cell. Longev. 2013 (2013) 898034.
61. R. da S. Raposo, D.V. Pinto, R. Moreira, R.P. Dias, C.A. Fontes Ribeiro, R.B. Oriá, J.O. Malva, Methylmercury impact on adult neurogenesis: Is the worst yet to come from recent Brazilian environmental disasters? Front. Aging Neurosci. 12 (2020) 1–7.
62. A.W. Dobson, K.M. Erikson, M. Aschner, Manganese neurotoxicity. Ann. N. Y. Acad. Sci 1012 (2004) 115–128.
63. M.S. Sankhla, K. Sharma, R.D. Kumar, Heavy metal causing neurotoxicity in human health. Int. J. Innov. Res. Sci. Eng. Technol. 6 (2017) 7721–7726.
64. M.F. Hughes, Arsenic toxicity and potential mechanisms of action. Toxicol. Lett. 133 (2002) 1–16.
65. P. Apostoli, S. Catalani, Metal ions affecting reproduction and development. Met. Ions Life Sci. 8 (2015) 263–303.
66. P. Massányi, M. Massányi, R. Madeddu, R. Stawarz, N. Lukáč, Effects of cadmium, lead, and mercury on the structure and function of reproductive organs. Toxics 8 (2020) 1–31.
67. C.M.Y. Choy, C.W.K. Lam, L.T.F. Cheung, C.M. Briton-Jones, L.P. Cheung, C.J. Haines, Infertility, blood mercury concentrations and dietary seafood consumption: A case-control study. BJOG. 109 (2002) 1121–1125.
68. E. Ernst, M. Christensen, J.G. Lauritsen, In vitro exposure of human spermatozoa to mercuric chloride – A histochemical study. Prog. Histochem. Cytochem. 23 (1991) 263–268.
69. C. Yang, S. Trent, V. Ionescu-Tiba, L. Lan, T. Shioda, D. Sgroi, E.V. Schmidt, Identification of cyclin D1- and estrogen-regulated genes contributing to breast carcinogenesis and progression. Cancer Res. 66 (2006) 11649–11658.
70. J. Huff, R.M. Lunn, M.P. Waalkes, L. Tomatis, P.F. Infante, Cadmium-induced cancers in animals and in humans. Int. J. Occup. Environ. Health 13 (2007) 202–212.
71. B. Zambelli, V.N. Uversky, S. Ciurli, Nickel impact on human health: An intrinsic disorder perspective. Biochim. Biophys. Acta 1864 (2016) 1714–1731.
72. E.K. Silbergeld, M. Waalkes, J.M. Rice, Lead as a carcinogen: Experimental evidence and mechanisms of action. Am. J. Ind. Med. 38 (2000) 316–323.
73. G. Genchi, A. Carocci, G. Lauria, M.S. Sinicropi, A.N. Catalano, Human health and environmental toxicology. Int. J. Environ. Res. Public Health 17 (2020) 679–700.
74. E.H. Jeffery, K. Abreo, E. Burgess, J. Cannata, J.L. Greger, Systemic aluminum toxicity: Effects on bone, hematopoietic tissue, and kidney. J. Toxicol. Environ. Heal. Part A 48 (1996) 649–665.
75. G. Kazantzis, Cadmium, osteoporosis and calcium metabolism. BioMetals. 17 (2004) 493–498.
76. M.R. Rahimzadeh, M.R. Rahimzadeh, S. Kazemi, A.A. Moghadamnia, Cadmium toxicity and treatment: An update. Casp. J. Intern. Med. 8 (2017) 135–145.
77. A. Nautiyal, S. Wairkar, Management of hyperpigmentation: Current treatments and emerging therapies. Pigment Cell Melanoma Res. 34 (2021) 1000–1014.
78. D.N. Guha Mazumder, R. Haque, N. Ghosh, K. De Binay, A. Santra, D. Chakraborty, A.H. Smith, Arsenic levels in drinking water and the prevalence of skin lesions in West Bengal, India. Int. J. Epidemiol. 27 (1998) 871–877.

79. A.S. Boyd, D. Seger, S. Vannucci, M. Langley, J.L. Abraham, L.E. King, Mercury exposure and cutaneous disease. J. Am. Acad. Dermatol. 43 (2000) 81–90.
80. Y.F. Mohamed, K. Manivannan, R.C. Fernandez, Bordetella pertussis. Trends Microbiol. 31 (2023) 1192–1193.
81. A.L. Lindberg, M. Rahman, L.Å. Persson, M. Vahter, The risk of arsenic induced skin lesions in Bangladeshi men and women is affected by arsenic metabolism and the age at first exposure. Toxicol. Appl. Pharmacol. 230 (2008) 9–16.
82. R.C. Yu, K. Hsu, C. Chen, Arsenic methylation capacity and skin cancer. Cancer Epidemiol. Biomarkers Prev. 9 (2000) 1259–1262.
83. L.M. Del Razo, E.A. García-Montalvo, O.L. Valenzuela, Arsenic exposure alters purine metabolism in rats, mice, and humans. In Proceedings of the Fifth International Conference on Arsenic Exposure and Health Effects Arsenic Exposure and Health Effects V, July 14–18, 2002, San Diego, California. Springer (2003) 135–145.
84. R.R. Dietert, J.E. Lee, I. Hussain, M. Piepenbrink, Developmental immunotoxicology of lead. Toxicol. Appl. Pharmacol. 198 (2004) 86–94.
85. S.E. Orr, C.C. Bridges, Chronic kidney disease and exposure to nephrotoxic metals. Int. J. Mol. Sci. 18 (2017).
86. T. Yumoto, K. Tsukahara, H. Naito, A. Iida, A. Nakao, A successfully treated case of criminal thallium poisoning. J. Clin. Diagnostic Res. 11 (2017) OD01–OD02.
87. J.L. Reyes, E. Molina-Jijón, R. Rodríguez-Muñoz, P. Bautista-García, Y. Debray-García, M.D.C. Namorado, Tight junction proteins and oxidative stress in heavy metals-induced nephrotoxicity. Biomed Res. Int. (2013) 730789.
88. R.A. Goyer, Mechanisms of lead and cadmium nephrotoxicity. Toxicol. Lett. 46 (1989) 153–162.
89. I. Campbell, Liver: Metabolic functions. Anaesth. Intensive Care Med. 7 (2006) 51–54.
90. L. Yu, I.W. Liou, S.W. Biggins, M. Yeh, F. Jalikis, L.N. Chan, J. Burkhead, Copper deficiency in liver diseases: A case series and pathophysiological considerations. Hepatol. Commun. 3 (2019) 1159–1165.
91. K. Renu, R. Chakraborty, H. Myakala, R. Koti, A.C. Famurewa, H. Madhyastha, B. Vellingiri, A. George, A. Valsala Gopalakrishnan, Molecular mechanism of heavy metals (lead, chromium, arsenic, mercury, nickel and cadmium) – induced hepatotoxicity – A review. Chemosphere 271 (2021) 129735.
92. X. Wang, Y.A. Seo, S.K. Park, Serum selenium and non-alcoholic fatty liver disease (NAFLD) in U.S. adults: National Health and Nutrition Examination Survey (NHANES) 2011–2016. Environ. Res. 197 (2021) 111190.
93. N.D. Vaziri, Mechanisms of lead-induced hypertension and cardiovascular disease. Am. J. Physiol. – Hear. Circ. Physiol. 295 (2008) 454–465.
94. A. Linna, J. Uitti, P. Oksa, P. Toivio, V. Virtanen, H. Lindholm, M. Halkosaari, R. Sauni, Effects of occupational cobalt exposure on the heart in the production of cobalt and cobalt compounds: A 6-year follow-up. Int. Arch. Occup. Environ. Health. 93 (2020) 365–374.
95. Ç. Sevim, E. Doğan, S. Comakli, Cardiovascular disease and toxic metals. Curr. Opin. Toxicol. 19 (2020) 88–92.
96. N.V. Solenkova, J.D. Newman, J.S. Berger, G. Thurston, J.S. Hochman, G.A. Lamas, Metal pollutants and cardiovascular disease: Mechanisms and consequences of exposure. Am. Heart J. 168 (2014) 812–822.
97. A. Lankoff, A. Banasik, A. Duma, E. Ochniak, H. Lisowska, T. Kuszewski, S. Góźdź, A. Wojcik, A comet assay study reveals that aluminium induces DNA damage and inhibits the repair of radiation-induced lesions in human peripheral blood lymphocytes. Toxicol. Lett. 161 (2006) 27–36.
98. J.S. Roy, D. Chatterjee, N. Das, A.K. Giri, Substantial evidences indicate that inorganic arsenic is a genotoxic carcinogen: A review. Toxicol. Res. 34 (2018) 311–324.
99. Y. Yin, F. Meng, C. Sui, Y. Jiang, L. Zhang, Arsenic enhances cell death and DNA damage induced by ultraviolet B exposure in mouse epidermal cells through the production of reactive oxygen species. Clin. Exp. Dermatol. 44 (2019) 512–519.
100. K.P. Nickens, S.R. Patierno, S. Ceryak, Chromium genotoxicity: A double-edged sword. Chem. Biol. Interact. 188 (2010) 276–288.
101. G. Nordberg, Assessment of risks in occupational cobalt exposures. Sci. Total Environ. 150 (1994) 201–207.
102. M.P. Rayman, Selenium and human health. Lancet 379 (2012) 1256–1268.
103. Y. Sun, Z. Wang, P. Gong, W. Yao, Q. Ba, H. Wang, Review on the health-promoting effect of adequate selenium status. Front. Nutr. 10 (2023) 1–15.

10 Overview of Risk Assessment Methodologies for Heavy Metals in the Environment

*Yash Raghav, Anjali Singh, Bhanu Pratap Singh, and Vishal Mishra**
School of Biochemical Engineering, IIT (BHU), Varanasi, Uttar Pradesh, India

10.1 INTRODUCTION

The fast expansion of industrialization over the past few decades is associated with a corresponding rise in soil, water, and air pollution (Hu et al., 2013). Heavy metal contamination occurrences have increased dramatically, leading to rapid heavy metal entry into the ecosystem at high concentrations, with devastating effects (Kamran et al., 2013). The term "heavy metal elements" is used to describe a class of metallic elements with a density greater than 4.5 g/cm^3 (Chen et al., 2020). The elements cadmium, chromium, mercury, lead, copper, arsenic, etc. make up a large portion of the heavy metal that causes pollution problems (Rahman & Singh, 2019). Heavy metals are one of the most dangerous contaminants since they are widespread (found in the air, water, and soil) and difficult to break down through biological processes (due to their high toxicity) (Oves et al., 2016). The primary sources of heavy metal contamination are derived from the extraction and processing of metallic elements, as well as the release of industrial and vehicle effluents (Timothy & Tagui Williams, 2019). These pollutants are typically deposited in rivers or leach into the soil through precipitation (Ciszewski & Grygar, 2016). Pollutants in the environment, such as heavy metals, can be breathed in, ingested, or absorbed through the skin, all of which cause a serious risk to human health (Fernández-Luqueño et al., 2013). The primary risks associated with specific heavy metals are as follows: Mercury (Hg) exhibits a direct tendency to accumulate in the liver after ingestion, resulting in notable impairments to the brain, nerves, and vision (Carocci et al., 2014). Cadmium (Cd), on the other hand, can induce elevated blood pressure, cardiovascular and cerebrovascular ailments, and harm various bodily structures, such as bones, liver, and kidneys (Thévenod & Lee, 2013). Exposure to lead has been found to have detrimental effects on the developing neurological system of the foetus, perhaps leading to congenital mental disabilities (Kinuthia et al., 2020). Levels of mercury (Hg), lead (Pb), cadmium (Cd), chromium (Cr), nickel (Ni), and thallium (Tl) were established in wastewater and soil samples obtained from eight sites in open drainage channels at Nairobi industrial area, Kenya. Ultra-trace inductively coupled plasma mass spectroscopy (ICP-MS) was used for metal analysis. Temperature, pH, and turbidity of wastewater ranged from 16.75 to 26.05°C, 7.28 to 8.78, and 160.33 to 544.69 ppm, respectively, and within World Health Organization (WHO) allowable limits. Wastewater conductivities in four sites ranged from 770 to 1074 μS/cm and above WHO limits at 25°C. The mean concentrations of the metals in wastewater ranged from 0.0001 to 0.015 ppm in an ascending order of Tl < Cd < Hg < Ni < Cr < Pb. Levels of Cd, Cr, and Ni in wastewater were within WHO, World Bank (WB), United States Environmental Protection Agency (US EPA), China, Kenya, and India's Central Pollution Control Board (CPCB)

* Corresponding Author: vishal.bce@itbhu.ac.in

DOI: 10.1201/9781032685793-10

limits, while Hg and Tl were above US EPA limits. The mean Pb levels in wastewater (five sites) were above WHO, US EPA, and Kenya allowable limits. Mean levels of the metals in soil samples ranged from 0.085 to 199.99 ppm, with those of Hg, Pb, Cr, Cd, and Ni being above WHO limits for agricultural soils. Positive correlations ($P < 0.05$) for Cd and Hg in soils and Tl (wastewater) and Cd (soil) were noted. In conclusion, wastewater in open waste channels at the Nairobi industrial area had elevated levels of Pb and Hg, while the soil from the same channels had high levels of Hg, Pb, Ni, Cr, and Cd. Good management of Nairobi industrial area effluents is inevitable since it borders densely populated informal settlements, which are likely to suffer exposure to toxic wastes. Effective wastewater treatment and reuse is highly recommended (Muchira, 2017). Levels of heavy metals in wastewater and soil samples from open drainage channels in Nairobi, Kenya (Kinuthia et al., 2020). Additionally, asbestos exposure can result in the development of systemic disorders, primarily causing harm to the skin. This chapter provides a brief overview of the various treatment options for removing metals from the environment. The primary goal is to disseminate useful data on heavy metals, their origins, their impacts on the environment, the extent to which they pollute soil, air, and water, and how to assess the risks linked with exposure to heavy metals in the environment (sources like the mining industry, agricultural wastewater, and municipal waste). It also covers the examination of regulatory frameworks and recommendations that are designed to regulate and mitigate exposure to heavy metals. It also addresses the existing issues and deficiencies in the current rules about heavy metal contamination, as well as possible possibilities for enhancement.

10.2 HEAVY METALS

The effluent produces pollutants that can be categorized into organic and inorganic pollutants, each exhibiting varying levels of toxicity (Akpor et al., 2014). The removal of organic pollutants commonly employs physical, chemical and biological methods (Ahmed et al., 2017). However, these methodologies are not effective for the treatment of inorganic pollutants such as heavy metals (Zamora-Ledezma et al., 2021). The heavy metal breakdown is a significant problem because of its properties such as oxidation-reduction characteristics, solubility, and complex formation (Xu et al., 2022). Elements with specific gravity above 5.0 and atomic weight between 63.5 and 200.6 are heavy metals (Kulhari et al., 2013). Heavy metals in wastewater have been a major environmental issue in recent years because they pose a substantial risk to human health and ecosystems even at low concentrations. Due to its aggregation, flexibility, non-biodegradability, and persistence, heavy metal pollution is a severe environmental issue. Industries, such as paper production, pesticide manufacturing, tanning, metal plating, and mining activities, release effluents containing heavy metals into the environment. These heavy metals have been identified as non-biodegradable and possess toxic properties that can be detrimental to human health and ecology (Mishra et al., 2019). However, it is possible to transform them into compounds that are less detrimental. The presence of hazardous metals in wastewater might provide challenges for their removal due to their ability to persist in either chemical or mixed forms (Ajiboye et al., 2021) (Figure 10.1 and Table 10.1).

10.2.1 Sources of Heavy Metals

10.2.1.1 Natural Source

Heavy metals are commonly found in nature through various sources, such as rock weathering, volcanic eruptions, biological activities, sea salt and forest fires (Bradl, 2005). Heavy metals are incorporated into ecosystems via various processes. Various factors, such as natural disasters, climatic change, the composition of soil, hyporheic exchange, and geological elements, influence the water environment. Natural catastrophes, including earthquakes, hurricanes, tornadoes, floods, and tsunamis, have been identified as significant contributors to heavy metal contamination in water

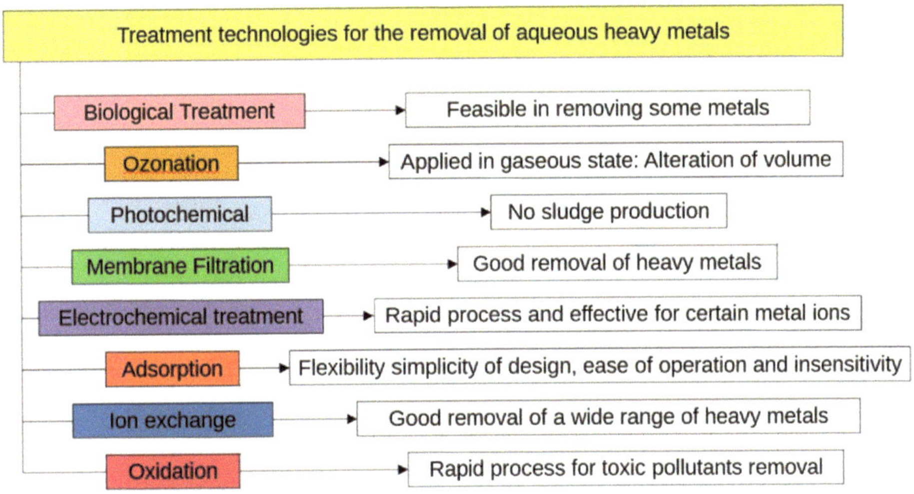

FIGURE 10.1 Methods for the removal of aqueous heavy metals.

TABLE 10.1
Various Heavy Metals' Toxicity Levels, Associated Risks, and Potential Origin

	Maximum Level, mg/L				
Ions	**US EPA by 2019**	**WHO by 2017**	**EU Standard**	**Associated Risks**	**Origin**
Hg	0.002	0.006	0.001	Abnormalities in nervous and digestive systems	Power plant, electronics, smelting, cement, etc.
Pb	0.0015	0.01	0.01	Impairment of the gastrointestinal, urinary, haemic, and neurological systems	Mining, glass manufacturing, battery, etc.
Zn	5.0	3.0	–	Damage organs like liver, kidneys	Mining, smelting, paper making, etc.
As	0.01	0.01	0.01	Adverse effects on the cardiovascular system and integumentary system	Mining, pigments, pesticides, etc.
Cd	0.005	0.003	0.005	Mutagenic and nephrotoxic	Chemical industry, electroplating, smelting, etc.
Cr	0.1	0.05	0.05	Skin diseases, such as eczema, psoriasis, and cancer	Mining, electroplating, smelting, painting, etc.
Cu	1.3	2.0	2.0	Symptoms like vomiting, stomach weakness, pain, dizziness, and trouble breathing	Metal processing, mining, machinery manufacturing, etc.

Source: Hoang et al. (2022).

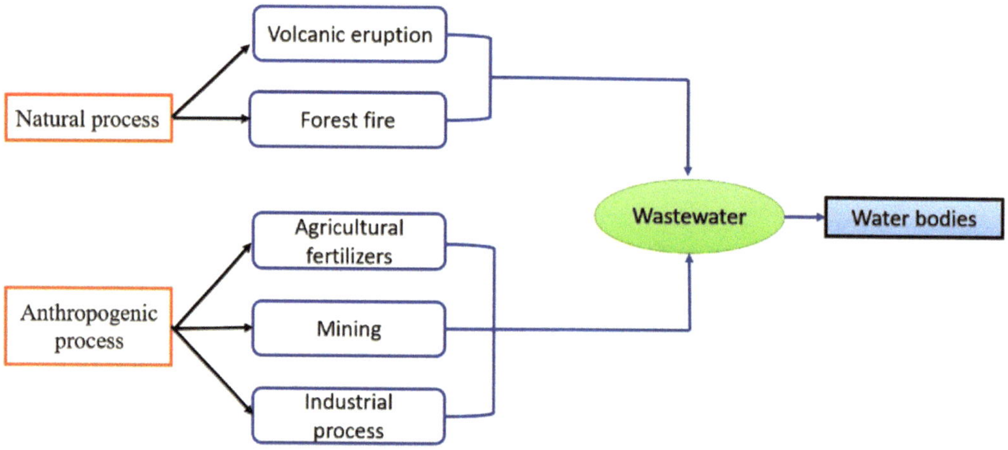

FIGURE 10.2 Sources of wastewater contamination.

environments (Akhtar et al., 2021). Among these disasters, floods have been identified as the primary cause of such pollution. Naturally occurring heavy metals are less hazardous than anthropogenic sources and can be absorbed by the environment.

10.2.1.2 Anthropogenic Sources

In ecological environments, heavy metal contamination is mostly caused by humans (Saleh & Aglan, 2018). The deposition of atmospheric particles, such as those resulting from fuel combustion, as well as the disposal of sewage sludge and effluents from many sources, such as industries, mining emissions, and automobiles, can lead to the eventual discharge of heavy metals from air and soil into water bodies through the process of rainfall (Singh & Steinnes, 1994). Table 10.1 compiles a variety of heavy metals' anthropogenic origins. It is clear that mining, smelting, batteries, chemical industry, agricultural, and other sectors are major contributors to the world's heavy metals problem. Mining is a major contributor to the anthropogenic load since it produces and releases such large quantities of heavy metals (Han et al., 2002). About 40%–73% of all man-made heavy metal emissions come from smelters, making it the industry's most significant contributor. Anthropogenic sources pose a serious threat, although they are not completely out of anyone's control. Because of this, human intervention behaviours like process optimization can be utilized to limit the probability of pollutant formation in the process of preserving the environment, in addition to efficiently managing pollution by existing technologies (Hui et al., 2023) (Figure 10.2).

10.3 TECHNIQUES FOR REMOVAL OF HEAVY METALS

Effective and reasonable management of heavy metal pollution remains a formidable task due to the complexity and variety of the damage it causes (Naja & Volesky, 2017). As a result, several heavy metal removal methods have been created and put into use in various contexts with the hopes of improving or resolving this issue. Firstly, we will discuss the conventional methods for remediating heavy metals along with their principles (Figure 10.3). In addition, Table 10.2 lists the benefits and drawbacks of each method.

10.3.1 Chemical Precipitation

Chemical precipitation has emerged as a highly efficient methodology to eliminate heavy metals, predominantly from the effluents generated by the paper manufacturing and electroplating sectors

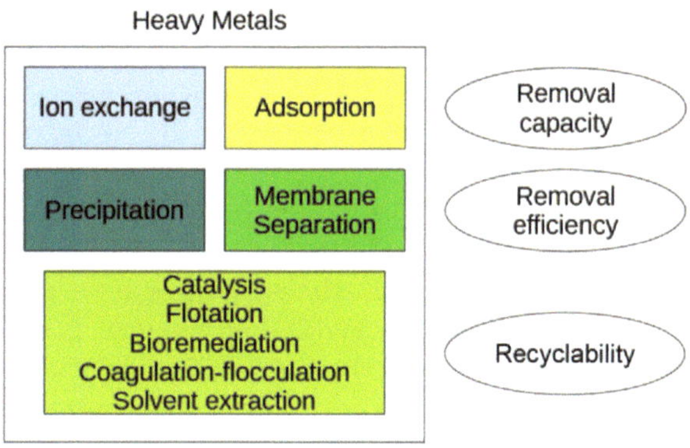

FIGURE 10.3 Techniques for removal of heavy metals.

TABLE 10.2
Conventional Methods for the Elimination of Heavy Metals from Wastewater: Their Benefits and Drawbacks

Conventional Treatment	Advantages	Disadvantages	References
Chemical precipitation	The treatment approach offers advantages, such as little capital investment, straightforward operation, and high potential for automation.	Low metal precipitation rates and long-term environmental implications are only some of the problems caused by producing vast volumes of sludge containing harmful compounds that need further treatment.	Tou et al. (2017), Delibacak et al. (2020), Moodley et al. (2018)
Coagulation/ Flotation	Easy to use and not too costly.	High operating costs owing to chemical use, sludge generation, and insufficient heavy metal removal are all problems.	Sun et al. (2020), Iwuozor (2019), Abood et al. (2014)
Membrane technologies	Reduced energy consumption, no secondary pollution, no chemical additives required, no phase shift, simple manufacturing, and low environmental impact.	Increased difficulty, more expense, fouling of membranes, and reduced permeate flow.	Esfahani et al. (2019), Urošević and Trivunac (2020)
Ion exchange	The removal efficiency is excellent, with fast kinetics, making it highly effective for treating inorganic effluent. Additionally, there is no need for sludge disposal, making it more convenient for fieldwork. It is also a time-saving process, using low-cost materials. Furthermore, the resin can be easily regenerated, adding to its overall efficiency.	The method exhibits nonselectivity, displaying a high sensitivity to changes in the pH of the solution. It is prone to fouling by metal ions and is primarily applicable to low concentrations of metals. Additionally, both the operating and capital costs associated with this method are significant.	Barakat (2011), Bolisetty et al. (2019), Zamora-Ledezma et al. (2021)
Electrochemical technologies	Simple, green, labour-efficient, and energy-saving.	Significant financial inputs, costly provision of energy, and limited efficacy in cases when the quantities of metal ions in aqueous solutions are minimal.	Amitesh et al. (2022), Chaplin (2019), Garcia-Segura et al. (2017), Shahedi et al. (2020), Tegladza et al. (2021), Tran et al. (n.d.)

(Azimi et al., 2017). The present study investigates the utilization of chemical precipitants, including alum, lime, iron salts, and certain polymers, to treat wastewater contaminated with heavy metals (Pohl, 2020). Through a series of reactions, these precipitants interact with the heavy metal ions in the wastewater, forming insoluble precipitates (Benalia et al., 2022). This process aims to effectively remove heavy metals from the wastewater, mitigating their potential environmental impact. Heavy metals in wastewater are often removed using chemical precipitation due to their low cost and easy application. The precipitating agent will be introduced when the wastewater pH adjusts to basic conditions. It forms insoluble precipitates with wastewater heavy metal ions. The precipitates may be sedimented or filtered. Conventional chemical precipitation comprises hydroxide and sulphide. Due to its cheap cost, simplicity, and pH control, hydroxide precipitation is the most extensively used chemical precipitation method for heavy metal removal from industrial effluent (Pohl, 2020). The pH range of 8.0–11.0 limits metal hydroxide solubility. Flocculation and sedimentation separate metal hydroxides. Due to their ease of use and cheap cost, several hydroxides precipitated heavy metals from wastewater. For most industrial wastewater treatment, lime-precipitated hydroxide is used. Hydroxide precipitators like NaOH and $Ca(OH)_2$ removed copper and chromium from wastewater. Adding chemical coagulants, such as ferrous salts, alum, and organic polymers, to hydroxide precipitation may improve the removal of heavy metals (Tang et al., 2016). Hydroxide precipitation was extensively employed, but huge volumes of low-density sludges caused dewatering and disposal challenges, limiting its industrial usage (Qrenawi & Rabah, 2021). A limited number of metal hydroxides are also amphoteric, and the presence of mixed metals complicates the application of hydroxide precipitation because the optimal pH for one metal may result in the reintroduction of another metal into the solution. Furthermore, the metal hydroxide precipitation will be impeded while complexing agents are present in the effluent.

Sulphide precipitation is effective for heavy metal remediation (Kumar et al., 2021). Metal sulphide precipitates are less soluble than hydroxide precipitates and are not amphoteric, making them important for use. Sulphide precipitation removes more metal than hydroxide precipitation across a broad pH range. Additionally, metal sulphide sludges exhibit superior thickening and dewatering properties in comparison to metal hydroxide sludges (Cao et al., 2021). Utilizing the sulphide precipitation process does, nevertheless, comprise certain potential hazards.

10.3.2 Chemical Coagulation/Flocculation

Heavy metal-contaminated effluent can be treated through coagulation/flocculation (Sun et al., 2020). Basically, the process of coagulation therapy involves the introduction of a chemical coagulant to induce the destabilization of colloidal particles, leading to subsequent sedimentation (Iwuozor, 2019). In general, the process of coagulation is often followed by the occurrence of flocculation, whereby the unstable aggregation of particles into larger floccules, hence augmenting the overall particle size (Abood et al., 2014). Finally, the sedimentation tank will settle the enormous floccules. Ferric chloride, ferrous sulphate, alum, and many other coagulants are generally used in conventional wastewater treatment procedures for pollutant removal (Aziz et al., 2007). Coagulation is a crucial process in wastewater treatment, primarily targeting the removal of suspended particles and hydrophobic colloids (Zhao et al., 2021). Flocculation refers to the process by which polymers facilitate the bonding of flocs, resulting in the formation of large agglomerates comprising interconnected particles (Abbasi Moud, 2022). The process of flocculation is used to aggregate suspended materials into larger particles, facilitating their subsequent removal using filtration, sedimentation, floatation, or straining techniques. Various forms of flocculants, such as polyacrylamide, poly-aluminium chloride, and polyferric sulphate, are often used in wastewater treatment processes (Abujazar et al., 2022). However, it should be noted that these flocculants are not directly suitable for the elimination of harmful metal ions from wastewater. The use of macromolecule flocculants has been shown to be a successful method for the removal of heavy metals from wastewater (El-Gaayda et al., 2021). In general, the entire removal of heavy metals cannot be achieved alone via the use of coagulation/flocculation treatment units. However, the integration of other treatment methods may lead to the complete removal of heavy metals in water. Despite its many benefits, coagulation/flocculation is subject to several limitations, such as the issue of high operating expenses resulting from the extensive use of chemicals (Wei et al., 2018).

10.3.3 Electrochemical Methods

Electrochemical technologies in industrial wastewater treatment are very successful in the removal of heavy metal ions (Tran et al., n.d.). This approach recovers heavy metals in elemental metallic form utilizing electrochemical cell anodic and cathodic processes. Due to its high capital and power requirements, this technology has limited industrial applicability (Chaplin, 2019). Scientists and researchers are drawn to electrochemical technologies due to strict wastewater disposal laws. For wastewater heavy metal ion removal, electrocoagulation, electrodeposition, and electro-flotation are used (Amitesh et al., 2022).

The electrocoagulation process is composed of electrodes that serve as the anode and cathode, respectively, where oxidation and reduction reactions occur (Garcia-Segura et al., 2017). The electrocoagulation procedure is governed by a multitude of physicochemical processes, including reduction, oxidation, coagulation, and adsorption. Electrocoagulation, akin to alternative treatment modalities, offers a cost-effective and operator-friendly approach to heavy metal removal at an industrial level. By electrically disintegrating iron or aluminium ions from electrodes composed of titanium or aluminium, this method generates coagulants in situ. The metal ion is generated at the anode, while the hydrogen gas is discharged from the cathode throughout the procedure. The hydrogen gas that resulted was utilized to lift the flocculated particulates above the water. Heavy metals, fluorides, dyes, nitrates, medicines, sulphates, and phenolic compounds have all been removed from wastewater successfully using this approach. The harmful heavy metals may be removed more effectively by an electrocoagulation technique, which is anticipated to be applicable to any treatment capacity (Tegladza et al., 2021). The electron alone serves as the necessary reagent in electrocoagulation; hence, no other chemicals are required. In comparison to the coagulation process, electrocoagulation results in less sludge that is more stable; hence, it is considered an environmentally benign technique (Shahedi et al., 2020). However, electrocoagulation has several drawbacks, the most notable of which are the greater operational and capital investment costs (Moussa et al., 2017). Electrochemical techniques for heavy metal removal from industrial effluent are fast, well-controlled, and use fewer chemicals, sludge, and metal ions (Fu & Wang, 2011). Electrochemical approaches were limited by significant initial capital investment and power supply costs.

10.3.4 Membrane Filtration

Membrane filtration techniques are categorized based on the specific membrane employed. Filtration has several advantages compared to other conventional methods, such as superior separation effectiveness, no alteration in phases, reduced energy consumption, straightforward expansion, and ecological compatibility (Esfahani et al., 2019).

10.3.5 Ultrafiltration

When it comes to removing dissolved and colloidal particles, ultrafiltration (UF) is a membrane technology that works at low transmembrane pressures (Urošević & Trivunac, 2020). UF membranes allow dissolved metal ions to pass through easily because their pore diameters are greater than the ions themselves or their low molecular weight complexes (Arthanareeswaran & Starov, 2011). It was proposed to use micellar-enhanced UF (MEUF) and polymer-enhanced UF (PEUF) to achieve high removal efficiency of metal ions (Yaqub & Lee, 2018).

Scamehorn et al. first created MEUF in the 1980s to purify water by removing organic molecules and multivalent metal ions (Şahin & Taşcıoğlu, 2016). This type of separation approach relies on the incorporation of surfactants into wastewater. Surfactant molecules in aqueous solutions will form micelles and bind metal ions to create sizable metal-surfactant complexes when the concentration of surfactants surpasses the critical micelle concentration (CMC). Micelles that contain metal ions can be effectively trapped by a UF membrane with hole diameters smaller than the sizes of the micelles (Yusaf et al., 2022). In contrast, the metal ions that are not trapped can easily travel through the UF membrane (Juang et al., 2003). To achieve maximum retentions, it is necessary to employ surfactants with an electric charge that is opposite to that of the ions that need to be eliminated (Kalam et al., 2021). Sodium dodecyl sulphate (SDS), a negatively charged surfactant, is commonly chosen to efficiently eliminate heavy metal ions in MEUF (Li et al., 2009). Landaburu-Aguirre et al. (2009) studied the elimination of zinc from artificial wastewater with the use of MEUF employing SDS. Rejection coefficients of up to 99% were attained when the molar ratio of surfactant to metal (S/M) exceeded 5. Samper et al. (2009) employed SDS and LAS in a lab-scale membrane system to remove Cu^{2+}, Cd^{2+}, Pb^{2+}, Zn^{2+}, and Ni^{2+} from synthetic water using MEUF. The surfactant-to-metal molar concentration ratio exceeds 5 in all studies. SDS concentrations lower than the CMC unexpectedly yielded metal retention above 90%, except for Ni^{2+}. Also, metal ions, save Ni^{2+}, may be completely removed at LAS concentrations below CMC.

Membrane-retained surfactants and heavy metals form the retentate. Since surfactants can make up a considerable part of operational costs, they must be recovered and reused cheaply. Surfactant and heavy metals produce secondary contamination if not disposed of (Sarubbo et al., 2015).

10.3.6 Nanofiltration

Intermediary between UF and reverse osmosis (RO) is nanofiltration (NF). Nickel, chromium, copper, and arsenic may be removed from wastewater using NF, a promising method. NF method has excellent pollution removal efficiency, reliability, easy operation, and low energy usage (Xiang et al., 2022). Figoli et al. (2010) tested two commercial NF membranes (NF90 and N30F) on synthetic water to remove pentavalent arsenic. Both membranes removed more As with higher pH, lower operating temperature, and lower arsenic feed concentration. Feed concentration affects As rejection and permeate stream production. Murthy and Chaudhari have focused on NF membrane heavy metal ion removal recently. Thin-film composite polyamide NF membranes were used to reject nickel ions from aqueous effluent (Siddique, 2023).

10.3.7 Reverse Osmosis

A semi-permeable membrane is used in the RO process, allowing the fluid to pass through while rejecting the impurities (Olusemire, 2022). In order to purify water, various methods exist, and RO is one of them. The percentage of global desalination capacity it provides is over 20%. Ni^{2+} and Cu^{2+} ions were successfully removed by the RO process, and the rejection efficiency of the two ions increased up to 99.5% by using Na_2EDTA, a small-scale membrane bioreactor system in conjunction with RO, and observed remarkably high rates of heavy metal removal (Bodzek, 2013). An inherent limitation of RO is its substantial energy consumption resulting from the need for high pumping pressures and membrane regeneration (Blandin et al., 2016).

10.3.8 Floatation

Currently, flotation is widely employed in the field of wastewater treatment (Kyzas & Matis, 2018). The technique of flotation, which utilizes bubble attachment, has been utilized to extract heavy metals from a liquid phase (Pooja et al., 2022). This method was initially developed in the field of mineral processing (Ata, 2012). The primary methods for eliminating metal ions from a solution are dissolved air flotation (DAF), precipitation flotation, and ion flotation (Amaral Filho et al., 2016). DAF is to let tiny air bubbles connect to the particles floating in the water. These bubbles will form clumps that are less dense than water, which will make the flocs rise to the surface and be easily removed as sludge. Heavy metal ion removal from wastewater is promising with ion flotation (Hoseinian et al., 2023). Ion flotation uses surfactants to make wastewater ionic metal species hydrophobic and air bubbles to remove them (Polat & Erdogan, 2007). Yuan et al. (2008) tested ion flotation with tea saponin to extract cadmium, copper and lead from a dilute aqueous solution. The highest removal of Pb^{2+}, Cd^{2+}, and Cu^{2+} at 3:1 collector-to-metal ratios is 89.95%, 81.13%, and 71.17%. The precipitate flotation process, which operates on the principle of precipitate attachment to air pockets for subsequent removal, is an alternative flotation method (Deliyanni et al., 2017). The mode of precipitation may vary depending on the metal solution concentration, progressing either through the formation of metal hydroxide or as a salt containing a particular anion (e.g., carbonate, sulphide) (Godelitsas & Astilleros, 2010). A laboratory-scale investigation was conducted into the removal of Cr(III) via precipitate flotation from diluted aqueous solutions, employing SDS as the anionic collector and ethanol as the frother (Medina et al., 2005). The findings indicated that a maximal removal of 96.2% was attained at a pH of approximately 8.0 (Zhong et al., 2020).

10.4 CONCLUSION

Heavy metals are chemicals that are released into wastewater. Heavy metals' contamination in wastewater has proved a problem for water treatment systems as they are found in limited quantities everywhere. They are present in high concentrations and have become a growing source of concern for public health as well as the environment. The physical, chemical, biological, and hybrid processes have been productive in handling a wide variety of heavy metals, but no complete elimination has been observed yet. Moreover, some heavy metals may remain in the effluents even after treatment and enter the environment after being converted into far more toxic forms due to various kinds of interactions. Even though there has been appreciable advancement in heavy metal treatment technologies since they were first discovered and their risks were identified, current heavy metal removal technologies are ridden with limitations, primarily in terms of efficacy. Nevertheless, biological treatments are associated with high retention times and issues with the final disposal of the sludge containing the heavy metals. However, most of the existing technologies are high energy and resource-intensive, and costly to maintain and operate, which may serve as a hindrance for developing nations to pursue initiatives to treat and reuse water and to protect human and environmental health. Therefore, the enhancement of WWTPs to achieve complete and efficient removal of heavy metals calls for immediate and thorough research and development, complemented by techno-economic feasibility studies of treatment systems.

ACKNOWLEDGEMENT

The financial and technical assistance provided by the Indian Institute of Technology (BHU) Varanasi, Varanasi, is acknowledged by the authors of this chapter.

REFERENCES

Abbasi Moud, A. (2022). Polymer based flocculants: Review of water purification applications. *Journal of Water Process Engineering*, *48*, 102938. https://doi.org/10.1016/j.jwpe.2022.102938

Abood, A. R., Bao, J., Du, J., Zheng, D., & Luo, Y. (2014). Non-biodegradable landfill leachate treatment by combined process of agitation, coagulation, SBR and filtration. *Waste Management*, *34*(2), 439–447. https://doi.org/10.1016/j.wasman.2013.10.025

Abujazar, M. S. S., Karaağaç, S. U., Abu Amr, S. S., Alazaiza, M. Y. D., & Bashir, M. JK. (2022). Recent advancement in the application of hybrid coagulants in coagulation-flocculation of wastewater: A review. *Journal of Cleaner Production*, *345*, 131133. https://doi.org/10.1016/j.jclepro.2022.131133

Ahmed, M. B., Zhou, J. L., Ngo, H. H., Guo, W., Thomaidis, N. S., & Xu, J. (2017). Progress in the biological and chemical treatment technologies for emerging contaminant removal from wastewater: A critical review. *Journal of Hazardous Materials*, *323*, 274–298. https://doi.org/10.1016/j.jhazmat.2016.04.045

Ajiboye, T. O., Oyewo, O. A., & Onwudiwe, D. C. (2021). Simultaneous removal of organics and heavy metals from industrial wastewater: A review. *Chemosphere*, *262*, 128379. https://doi.org/10.1016/j.chemosphere.2020.128379

Akhtar, N., Syakir Ishak, M. I., Bhawani, S. A., & Umar, K. (2021). Various natural and anthropogenic factors responsible for water quality degradation: A review. *Water, 13*(19), Article 19. https://doi.org/10.3390/w13192660

Akpor, O. B., Otohinoyi, D. A., Olaolu, D. T., & Aderiye, B. I. (2014). Pollutants in wastewater effluents: Impacts and Remediation Processes. *International Journal of Environmental Research and Earth Science*, *3*(3), Article 3.

Amaral Filho, J., Azevedo, A., Etchepare, R., & Rubio, J. (2016). Removal of sulfate ions by dissolved air flotation (DAF) following precipitation and flocculation. *International Journal of Mineral Processing*, *149*, 1–8. https://doi.org/10.1016/j.minpro.2016.01.012

Amitesh, A., Dohare, D., Rekhate, C., & Prajapati, A. (2022). Removal of heavy metal from electroplating wastewater using electrocoagulation: A review. *Desalination and Water Treatment*, *267*, 121–137. https://doi.org/10.5004/dwt.2022.28710

Arthanareeswaran, G., & Starov, V. M. (2011). Effect of solvents on performance of polyethersulfone ultrafiltration membranes: Investigation of metal ion separations. *Desalination*, *267*(1), 57–63. https://doi.org/10.1016/j.desal.2010.09.006

Ata, S. (2012). Phenomena in the froth phase of flotation—A review. *International Journal of Mineral Processing*, *102–103*, 1–12. https://doi.org/10.1016/j.minpro.2011.09.008

Azimi, A., Azari, A., Rezakazemi, M., & Ansarpour, M. (2017). Removal of heavy metals from industrial wastewaters: A review. *ChemBioEng Reviews*, *4*(1), 37–59. https://doi.org/10.1002/cben.201600010

Aziz, H. A., Alias, S., Assari, F., & Adlan, M. N. (2007). The use of alum, ferric chloride and ferrous sulphate as coagulants in removing suspended solids, colour and COD from semi-aerobic landfill leachate at controlled pH. *Waste Management & Research: The Journal of the International Solid Wastes and Public Cleansing Association, ISWA*, *25*(6), 556–565. https://doi.org/10.1177/0734242X07079876

Barakat, M. A. (2011). New trends in removing heavy metals from industrial wastewater. *Arabian Journal of Chemistry*, *4*(4), 361–377. https://doi.org/10.1016/j.arabjc.2010.07.019

Benalia, M. C., Youcef, L., Bouaziz, M. G., Achour, S., & Menasra, H. (2022). Removal of heavy metals from industrial wastewater by chemical precipitation: Mechanisms and sludge characterization. *Arabian Journal for Science and Engineering*, *47*(5), 5587–5599. https://doi.org/10.1007/s13369-021-05525-7

Blandin, G., Verliefde, A. R. D., Comas, J., Rodriguez-Roda, I., & Le-Clech, P. (2016). Efficiently combining water reuse and desalination through forward osmosis—Reverse osmosis (FO-RO) hybrids: A critical review. *Membranes*, *6*(3), 37. https://doi.org/10.3390/membranes6030037

Bodzek, M. (2013). Inorganic micropollutants removal by means of membrane processes – State of the art. *Ecological Chemistry and Engineering S*, *20*(4), 633–658.

Bolisetty, S., Peydayesh, M., & Mezzenga, R. (2019). Sustainable technologies for water purification from heavy metals: Review and analysis. *Chemical Society Reviews*, *48*(2), 463–487. https://doi.org/10.1039/C8CS00493E

Bradl, H. B. (2005). Chapter 1. Sources and origins of heavy metals. In H. B. Bradl (Ed.), *Interface Science and Technology* (Vol. 6, pp. 1–27). Elsevier. https://doi.org/10.1016/S1573-4285(05)80020-1

Cao, B., Zhang, T., Zhang, W., & Wang, D. (2021). Enhanced technology based for sewage sludge deep dewatering: A critical review. *Water Research*, *189*, 116650. https://doi.org/10.1016/j.watres.2020.116650

Carocci, A., Rovito, N., Sinicropi, M. S., & Genchi, G. (2014). Mercury toxicity and neurodegenerative effects. In D. M. Whitacre (Ed.), *Reviews of Environmental Contamination and Toxicology* (pp. 1–18). Springer International Publishing. https://doi.org/10.1007/978-3-319-03777-6_1

Chaplin, B. P. (2019). The prospect of electrochemical technologies advancing worldwide water treatment. *Accounts of Chemical Research*, *52*(3), 596–604. https://doi.org/10.1021/acs.accounts.8b00611

Ciszewski, D., & Grygar, T. M. (2016). A review of flood-related storage and remobilization of heavy metal pollutants in river systems. *Water, Air, & Soil Pollution*, *227*(7), 239. https://doi.org/10.1007/s11270-016-2934-8

Delībacak, S., Voronīna, L., & Morachevskaya, E. (2020). Use of sewage sludge in agricultural soils: Useful or harmful. *Eurasian Journal of Soil Science*, *9*(2), Article 2. https://doi.org/10.18393/ejss.687052

Deliyanni, E. A., Kyzas, G. Z., & Matis, K. A. (2017). Various flotation techniques for metal ions removal. *Journal of Molecular Liquids*, *225*, 260–264. https://doi.org/10.1016/j.molliq.2016.11.069

El-Gaayda, J., Titchou, F. E., Oukhrib, R., Yap, P.-S., Liu, T., Hamdani, M., & Ait Akbour, R. (2021). Natural flocculants for the treatment of wastewaters containing dyes or heavy metals: A state-of-the-art review. *Journal of Environmental Chemical Engineering*, *9*(5), 106060. https://doi.org/10.1016/j.jece.2021.106060

Esfahani, M. R., Aktij, S. A., Dabaghian, Z., Firouzjaei, M. D., Rahimpour, A., Eke, J., Escobar, I. C., Abolhassani, M., Greenlee, L. F., Esfahani, A. R., Sadmani, A., & Koutahzadeh, N. (2019). Nanocomposite membranes for water separation and purification: Fabrication, modification, and applications. *Separation and Purification Technology*, *213*, 465–499. https://doi.org/10.1016/j.seppur.2018.12.050

Fernández-Luqueño, F., López-Valdez, F., Gamero-Melo, P., Luna-Suárez, S., Aguilera-González, E. N., Martínez, A. I., García-Guillermo, M. D. S., Hernández-Martínez, G., Herrera-Mendoza, R., Álvarez-Garza, M. A., & Pérez-Velázquez, I. R. (2013). Heavy metal pollution in drinking water – A global risk for human health: A review. *African Journal of Environmental Science and Technology*, *7*(7), Article 7.

Figoli, A., Cassano, A., Criscuoli, A., Mozumder, M. S. I., Uddin, M. T., Islam, M. A., & Drioli, E. (2010). Influence of operating parameters on the arsenic removal by nanofiltration. *Water Research*, *44*(1), 97–104. https://doi.org/10.1016/j.watres.2009.09.007

Fu, F., & Wang, Q. (2011). Removal of heavy metal ions from wastewaters: A review. *Journal of Environmental Management*, *92*(3), 407–418. https://doi.org/10.1016/j.jenvman.2010.11.011

Garcia-Segura, S., Eiband, M. M. S. G., de Melo, J. V., & Martínez-Huitle, C. A. (2017). Electrocoagulation and advanced electrocoagulation processes: A general review about the fundamentals, emerging applications and its association with other technologies. *Journal of Electroanalytical Chemistry*, *801*, 267–299. https://doi.org/10.1016/j.jelechem.2017.07.047

Godelitsas, A., & Astilleros, J. M. (2010). Dissolution, sorption/(re)precipitation, formation of solid solutions and crystal growth phenomena on mineral surfaces: Implications for the removal of toxic metals from the environment. In M. Prieto & H. Stoll (Eds.), European Mineralogical Union Notes in Mineralogy (Vol. 10). GeoScience World. https://doi.org/10.1180/EMU-notes.10.8

Han, F. X., Banin, A., Su, Y., Monts, D. L., Plodinec, J. M., Kingery, W. L., & Triplett, G. E. (2002). Industrial age anthropogenic inputs of heavy metals into the pedosphere. *Naturwissenschaften*, *89*(11), 497–504. https://doi.org/10.1007/s00114-002-0373-4

Hoseinian, F. S., Ramshini, S., Rezai, B., Kowsari, E., & Safari, M. (2023). Toxic heavy metal ions removal from wastewater by ion flotation using a nano collector. *Minerals Engineering*, *204*, 108380. https://doi.org/10.1016/j.mineng.2023.108380

Hu, Y., Liu, X., Bai, J., Shih, K., Zeng, E. Y., & Cheng, H. (2013). Assessing heavy metal pollution in the surface soils of a region that had undergone three decades of intense industrialization and urbanization. *Environmental Science and Pollution Research*, *20*(9), 6150–6159. https://doi.org/10.1007/s11356-013-1668-z

Hui, C. X., Dan, G., Alamri, S., & Toghraie, D. (2023). Greening smart cities: An investigation of the integration of urban natural resources and smart city technologies for promoting environmental sustainability. *Sustainable Cities and Society*, *99*, 104985. https://doi.org/10.1016/j.scs.2023.104985

Iwuozor, K. (2019). Prospects and challenges of using coagulation-flocculation method in the treatment of effluents. *Advanced Journal of Chemistry-Section A*, *2*, 105–127. https://doi.org/10.29088/SAMI/AJCA.2019.2.105127

Juang, R.-S., Xu, Y.-Y., & Chen, C.-L. (2003). Separation and removal of metal ions from dilute solutions using micellar-enhanced ultrafiltration. *Journal of Membrane Science*, *218*(1), 257–267. https://doi.org/10.1016/S0376-7388(03)00183-2

Kalam, S., Abu-Khamsin, S. A., Kamal, M. S., & Patil, S. (2021). A review on surfactant retention on rocks: Mechanisms, measurements, and influencing factors. *Fuel*, *293*, 120459. https://doi.org/10.1016/j.fuel.2021.120459

Kamran, S., Shafaqat, A., Samra, H., Sana, A., Samar, F., Muhammad, B., Saima, A., & Tauqeer, H. M. (2013). Heavy metals contamination and what are the impacts on living organisms. *Greener Journal of Environmental Management and Public Safety*, *2*, 172–179. https://doi.org/10.15580/GJEMPS.2013.4.060413652

Kinuthia, G. K., Ngure, V., Beti, D., Lugalia, R., Wangila, A., & Kamau, L. (2020). Levels of heavy metals in wastewater and soil samples from open drainage channels in Nairobi, Kenya: Community health implication. *Scientific Reports*, *10*(1), Article 1. https://doi.org/10.1038/s41598-020-65359-5

Kulhari, A., Sheorayan, A., Bajar, S., Sarkar, S., Chaudhury, A., & Kalia, R. K. (2013). Investigation of heavy metals in frequently utilized medicinal plants collected from environmentally diverse locations of north western India. *SpringerPlus*, *2*(1), 676. https://doi.org/10.1186/2193-1801-2-676

Kumar, M., Nandi, M., & Pakshirajan, K. (2021). Recent advances in heavy metal recovery from wastewater by biogenic sulfide precipitation. *Journal of Environmental Management, 278*, 111555. https://doi.org/10.1016/j.jenvman.2020.111555

Kyzas, G. Z., & Matis, K. A. (2018). Flotation in water and wastewater treatment. *Processes, 6*(8), Article 8. https://doi.org/10.3390/pr6080116

Landaburu-Aguirre, J., García, V., Pongrácz, E., & Keiski, R. L. (2009). The removal of zinc from synthetic wastewaters by micellar-enhanced ultrafiltration: Statistical design of experiments. *Desalination, 240*(1), 262–269. https://doi.org/10.1016/j.desal.2007.11.077

Li, X., Zeng, G.-M., Huang, J.-H., Zhang, C., Fang, Y.-Y., Qu, Y.-H., Luo, F., Lin, D., & Liu, H.-L. (2009). Recovery and reuse of surfactant SDS from a MEUF retentate containing Cd^{2+} or Zn^{2+} by ultrafiltration. *Journal of Membrane Science, 337*(1), 92–97. https://doi.org/10.1016/j.memsci.2009.03.030

Medina, B. Y., Torem, M. L., & de Mesquita, L. M. S. (2005). On the kinetics of precipitate flotation of Cr III using sodium dodecylsulfate and ethanol. *Minerals Engineering, 18*(2), 225–231. https://doi.org/10.1016/j.mineng.2004.08.018

Mishra, S., Bharagava, R. N., More, N., Yadav, A., Zainith, S., Mani, S., & Chowdhary, P. (2019). Heavy metal contamination: An alarming threat to environment and human health. In R. C. Sobti, N. K. Arora, & R. Kothari (Eds.), *Environmental Biotechnology: For Sustainable Future* (pp. 103–125). Springer. https://doi.org/10.1007/978-981-10-7284-0_5

Moodley, I., Sheridan, C. M., Kappelmeyer, U., & Akcil, A. (2018). Environmentally sustainable acid mine drainage remediation: Research developments with a focus on waste/by-products. *Minerals Engineering, 126*, 207–220. https://doi.org/10.1016/j.mineng.2017.08.008

Moussa, D. T., El-Naas, M. H., Nasser, M., & Al-Marri, M. J. (2017). A comprehensive review of electrocoagulation for water treatment: Potentials and challenges. *Journal of Environmental Management, 186*, 24–41. https://doi.org/10.1016/j.jenvman.2016.10.032

Muchira, R. (2017). *Water and Wastewater Infrastructure Improvement Projects in Informal Settlements: A Case Study of Kisip Project in Kcc Village, Nairobi County* [Thesis, University of Nairobi]. http://erepository.uonbi.ac.ke/handle/11295/103345

Naja, G. M., & Volesky, B. (2017). Toxicity and sources of Pb, Cd, Hg, Cr, As, and radionuclides in the environment. In L. K. Wang, J. P. Chen, Y.-T. Hung, & N. K. Shammas, (Eds.), *Handbook of Advanced Industrial and Hazardous Wastes Management* (pp. 13–62). CRC Press.

Olusemire, J. O. (2022). Optimisation of Reverse Osmosis Reject in Water Treatment Operation: A Case Study of MTU Water Treatment Plant. http://localhost:8080/xmlui/handle/123456789/1063

Oves, M., Saghir Khan, M., Huda Qari, A., Nadeen Felemban, M., & Almeelbi, T. (2016). Heavy metals: Biological importance and detoxification strategies. *Journal of Bioremediation & Biodegradation, 07*, 334. https://doi.org/10.4172/2155-6199.1000334

Pohl, A. (2020). Removal of heavy metal ions from water and wastewaters by sulfur-containing precipitation agents. *Water, Air, & Soil Pollution, 231*(10), 503. https://doi.org/10.1007/s11270-020-04863-w

Polat, H., & Erdogan, D. (2007). Heavy metal removal from waste waters by ion flotation. *Journal of Hazardous Materials, 148*(1), 267–273. https://doi.org/10.1016/j.jhazmat.2007.02.013

Pooja, G., Kumar, P. S., & Indraganti, S. (2022). Recent advancements in the removal/recovery of toxic metals from aquatic system using flotation techniques. *Chemosphere, 287*, 132231. https://doi.org/10.1016/j.chemosphere.2021.132231

Qrenawi, L. I., & Rabah, F. K. J. (2021). Sludge management in water treatment plants: Literature review. *International Journal of Environment and Waste Management, 27*(1), 93–125. https://doi.org/10.1504/IJEWM.2021.111909

Rahman, Z., & Singh, V. P. (2019). The relative impact of toxic heavy metals (THMs) (arsenic (As), cadmium (Cd), chromium (Cr)(VI), mercury (Hg), and lead (Pb)) on the total environment: An overview. *Environmental Monitoring and Assessment, 191*(7), 419. https://doi.org/10.1007/s10661-019-7528-7

Şahin, D., & Taşcıoğlu, S. (2016). Removal of Cu(II) ions from single component and Cd(II) containing solutions by micellar enhanced ultrafiltration utilizing micellar effects on complex formation. *Desalination and Water Treatment, 57*(24), 11143–11153. https://doi.org/10.1080/19443994.2015.1042059

Saleh, H. E.-D. M., & Aglan, R. (2018). *Heavy Metals*. BoD – Books on Demand.

Samper, E., Rodríguez, M., De la Rubia, M. A., & Prats, D. (2009). Removal of metal ions at low concentration by micellar-enhanced ultrafiltration (MEUF) using sodium dodecyl sulfate (SDS) and linear alkylbenzene sulfonate (LAS). *Separation and Purification Technology, 65*(3), 337–342. https://doi.org/10.1016/j.seppur.2008.11.013

Sarubbo, L., Jr, R., Luna, J. M., Rufino, R., Santos, V. A., & Banat, I. (2015). Some aspects of heavy metals contamination remediation and role of biosurfactants. *Chemistry and Ecology*, *31*, 1–17. https://doi.org/10.1080/02757540.2015.1095293

Shahedi, A., Darban, A. K., Taghipour, F., & Jamshidi-Zanjani, A. (2020). A review on industrial wastewater treatment via electrocoagulation processes. *Current Opinion in Electrochemistry*, *22*, 154–169. https://doi.org/10.1016/j.coelec.2020.05.009

Siddique, J. A. (2023). Chapter 21—Nanofiltration membrane use for separation of heavy metals from wastewater. In A. Ahmad, R. Kumar, & M. Jawaid (Eds.), *Emerging Techniques for Treatment of Toxic Metals from Wastewater* (pp. 523–549). Elsevier. https://doi.org/10.1016/B978-0-12-822880-7.00010-8

Singh, B. R. & Steinnes, E. (1994). Soil and water contamination by heavy metals. In R. Lal, B. A. Stewart (Ed.), *Soil Processes and Water Quality* (pp. 1–271). CRC Press.

Sun, Y., Zhou, S., Pan, S.-Y., Zhu, S., Yu, Y., & Zheng, H. (2020). Performance evaluation and optimization of flocculation process for removing heavy metal. *Chemical Engineering Journal*, *385*, 123911. https://doi.org/10.1016/j.cej.2019.123911

Tang, X., Zheng, H., Teng, H., Sun, Y., Guo, J., Xie, W., Yang, Q., & Chen, W. (2016). Chemical coagulation process for the removal of heavy metals from water: A review. *Desalination and Water Treatment*, *57*(4), 1733–1748. https://doi.org/10.1080/19443994.2014.977959

Tegladza, I. D., Xu, Q., Xu, K., Lv, G., & Lu, J. (2021). Electrocoagulation processes: A general review about role of electro-generated flocs in pollutant removal. *Process Safety and Environmental Protection*, *146*, 169–189. https://doi.org/10.1016/j.psep.2020.08.048

Thévenod, F., & Lee, W.-K. (2013). Toxicology of cadmium and its damage to mammalian organs. In A. Sigel, H. Sigel, & R. K. Sigel (Eds.), *Cadmium: From Toxicity to Essentiality* (pp. 415–490). Springer Netherlands. https://doi.org/10.1007/978-94-007-5179-8_14

Timothy, N., & Tagui Williams, E. (2019). Environmental pollution by heavy metal: An overview. *International Journal of Environmental Chemistry*, *3*(2), 72. https://doi.org/10.11648/j.ijec.20190302.14

Tou, F., Yang, Y., Feng, J., Niu, Z., Pan, H., Qin, Y., Guo, X., Meng, X., Liu, M., & Hochella, M. F. (2017). Environmental risk implications of metals in sludges from waste water treatment plants: The discovery of vast stores of metal-containing nanoparticles. *Environmental Science & Technology*, *51*(9), 4831–4840. https://doi.org/10.1021/acs.est.6b05931

Tran, T.-K., Leu, H.-J., Chiu, K.-F., & Lin, C.-Y. (n.d.). Electrochemical treatment for wastewater contained heavy metal the removing of the COD and heavy metal Ions. *International Journal of Engineering Research*, *1*(9).

Urošević, T., & Trivunac, K. (2020). Chapter 3—Achievements in low-pressure membrane processes microfiltration (MF) and ultrafiltration (UF) for wastewater and water treatment. In A. Basile & K. Ghasemzadeh (Eds.), *Current Trends and Future Developments on (Bio-) Membranes* (pp. 67–107). Elsevier. https://doi.org/10.1016/B978-0-12-817378-7.00003-3

Wei, H., Gao, B., Ren, J., Li, A., & Yang, H. (2018). Coagulation/flocculation in dewatering of sludge: A review. *Water Research*, *143*, 608–631. https://doi.org/10.1016/j.watres.2018.07.029

Xiang, H., Min, X., Tang, C.-J., Sillanpää, M., & Zhao, F. (2022). Recent advances in membrane filtration for heavy metal removal from wastewater: A mini review. *Journal of Water Process Engineering*, *49*, 103023. https://doi.org/10.1016/j.jwpe.2022.103023

Xu, Z., Zhang, Q., Li, X., & Huang, X. (2022). A critical review on chemical analysis of heavy metal complexes in water/wastewater and the mechanism of treatment methods. *Chemical Engineering Journal*, *429*, 131688. https://doi.org/10.1016/j.cej.2021.131688

Yaqub, M., & Lee, S. H. (2018). Heavy metals removal from aqueous solution through micellar enhanced ultrafiltration: A review. *Environmental Engineering Research*, *24*(3), 363–375. https://doi.org/10.4491/eer.2018.249

Yuan, X. Z., Meng, Y. T., Zeng, G. M., Fang, Y. Y., & Shi, J. G. (2008). Evaluation of tea-derived biosurfactant on removing heavy metal ions from dilute wastewater by ion flotation. *Colloids and Surfaces A: Physicochemical and Engineering Aspects*, *317*(1), 256–261. https://doi.org/10.1016/j.colsurfa.2007.10.024

Yusaf, A., Usman, M., Ahmad, M., Siddiq, M., Mansha, A., Al-Hussain, S. A., Zaki, M. E. A., & Rehman, H. F. (2022). Highly selective methodology for entrapment and subsequent removal of cobalt (II) ions under optimized conditions by micellar-enhanced ultrafiltration. *Molecules*, *27*(23), 8332. https://doi.org/10.3390/molecules27238332

Zamora-Ledezma, C., Negrete-Bolagay, D., Figueroa, F., Zamora-Ledezma, E., Ni, M., Alexis, F., & Guerrero, V. H. (2021). Heavy metal water pollution: A fresh look about hazards, novel and conventional remediation methods. *Environmental Technology & Innovation*, *22*, 101504. https://doi.org/10.1016/j.eti.2021.101504

Zhao, C., Zhou, J., Yan, Y., Yang, L., Xing, G., Li, H., Wu, P., Wang, M., & Zheng, H. (2021). Application of coagulation/flocculation in oily wastewater treatment: A review. *Science of the Total Environment*, *765*, 142795. https://doi.org/10.1016/j.scitotenv.2020.142795

Zhong, C., Sun, S., Zhang, D., Liu, L., Zhou, S., & Zhou, J. (2020). Production of a bioflocculant from ramie biodegumming wastewater using a biomass-degrading strain and its application in the treatment of pulping wastewater. *Chemosphere*, *253*, 126727. https://doi.org/10.1016/j.chemosphere.2020.126727

11 Regulatory Frameworks and Guidelines Aimed at Controlling and Minimizing Heavy Metals Exposure

*Anjali Singh, Bhanu Pratap Singh, Yash Raghav, and Vishal Mishra**
School of Biochemical Engineering, IIT (BHU), Varanasi, Uttar Pradesh, India

11.1 INTRODUCTION

The rapid growth of industrialization in recent decades has been linked to an increase in air, water, and soil pollution (Swain, 2024). There has been a sharp rise in the frequency of heavy metal contamination, which has allowed heavy metals to quickly and in large quantities enter the ecosystem and have disastrous impacts (Hembrom et al., 2020). Since environmental pollutants, such as heavy metals, can be ingested, absorbed via the skin, or inhaled, they constitute a serious risk to human health (Briffa et al., 2020). Eliminating and controlling heavy metals is therefore essential for the ecosystem (Budovich, 2021). To assure safety, compliance, and standardization across many industries and sectors, numerous government agencies and organizations around the world create rules and guidelines to control and minimize heavy metals exposure (Edelstein & Ben-Hur, 2018). Government organizations in charge of enforcing rules for public health, environmental protection, and sector-specific laws are primarily in charge of overseeing regulations and guidelines for heavy metal contamination. Food producers are required by law to put controls in place as necessary to considerably reduce or avoid exposure to chemical risks and heavy metals (Nutrition, 2024d). Although heavy metal levels in food cannot be eliminated or prevented and can be decreased (Lambert et al., n.d.). This chapter provides a brief introduction of all the important regulatory bodies worldwide and a summary of their regulatory framework and guidelines for controlling and minimizing heavy metal exposure to the ecosystem. It also covers the examination of regulatory frameworks and recommendations that are designed to regulate and mitigate exposure to heavy metals.

Here are some noteworthy institutions and regulating bodies that deal with heavy metal exposure guidelines and regulations:

1. *World Health Organization (WHO)*: The WHO is an international public health specialized agency of the United Nations (WHO, 2024). Establishing rules and regulations for the control of heavy metals in different items, like food, water, cosmetics, and toys, is one of its duties. High atomic weight elements are known as heavy metals, and at certain exposure levels, they can be hazardous to both humans and the environment (Ali et al., 2019). The most prevalent and controlled heavy metals include nickel, mercury, chromium, cadmium, and lead.

* Corresponding Author: vishal.bce@itbhu.ac.in

DOI: 10.1201/9781032685793-11

2. *US Food and Drug Administration (FDA)*: The United States FDA (US FDA) conducts food safety tests for arsenic (As), mercury (Hg), lead (Pb) and cadmium (Cd) (Nutrition, 2024c). It also enforces FDA laws, monitors the safety of the country's food supply, and provides industry guidance and consumer counsel (Commissioner, 2023). The US FDA keeps an eye on and controls the amounts of mercury in food, supplements, and cosmetics (Nutrition, 2022). Foods that are frequently consumed by children under the age of two are one example of a specific food category that may be the focus of testing, as is a particular food or food group (WHO, n.d.). In response to reports of elevated contaminant levels in certain foods, testing may also be done. Testing may take place in state laboratories as part of cooperative agreements with states, FDA laboratories, or laboratories we contract with (US Food & Drug Administration, 2021). The FDA's objective is to reduce consumer exposure to arsenic by creating regulations, action levels, and consumer advice, with a particular emphasis on protecting the very young. The FDA published "Closer to Zero" in 2021 to detail the steps the agency is taking to lower the levels of lead, cadmium, arsenic, and mercury in the food that infants and young children eat (Nutrition, 2024a; US Food & Drug Administration, n.d.).
3. *US Environmental Protection Agency (EPA)*: Established by laws passed by Congress, the US EPA is an autonomous agency of the US federal government tasked with safeguarding the environment and public health through the creation and implementation of regulations. President Richard Nixon founded the EPA in 1970 to unify environmental regulations under a single agency. The organization carries out studies, keeps an eye on the state of the environment, and establishes and upholds national environmental protection regulations.
4. *European Union (EU)*: The maximum levels of cadmium (Cd), lead (Pb), arsenic (As), and mercury (Hg) in a variety of foods are set (see https://eur-lex.europa.eu/EN/legal-content/summary/maximum-levels-for-certain-contaminants-in-food.html). Lead (Pb) and cadmium (Cd) levels have most recently been updated in 2021, mercury (Hg) levels in 2022, and arsenic (As) levels in 2023. Heavy metal content is measured in milligrams per kilogram (mg/kg) of the product's wet weight.
5. *European Food Safety Authority (EFSA)*: An organization of the EU, the EFSA disseminates information on both current and new hazards connected to the food chain and offers unbiased scientific advice on matters about food safety. Its headquarters are in Parma, Italy, and it was founded in February 2002. EFSA employs 542 people and has a budget of €118.6 million for 2021. The EFSA's main goal is to guarantee consumer safety for food in Europe. It offers scientific guidance on plant protection, animal health and welfare, and food safety and nutrition. The European Parliament, the European Commission, and the EU Member States utilize EFSA's risk evaluations as a basis for developing food safety laws and policies. EFSA has a portal called OpenEFSA that hosts comprehensive information on its risk assessments, from the receipt of a mandate or dossier to the adoption of an output. The status of assessments, browsing data and studies, meeting agendas and minutes, and details about experts can be checked. The maximum levels of cadmium (Cd), mercury (Hg), arsenic (As), and lead (Pd) in a variety of foods are set by Regulation (EU) 2023/915. The most recent updates to the values were for lead (Pb) and cadmium (Cd) in 2021, mercury (Hg) in 2022, and arsenic (As) in 2023. Heavy metal content is measured in milligrams per kilogram of the product's wet weight.
6. *Food Safety and Standards Authority of India (FSSAI)*: The FSSAI is the main regulatory body in India for heavy metal regulations and standards. The Food Safety and Standards (Contaminants, Toxins, and Residues) Regulations, 2011 have been implemented, outlining the acceptable thresholds for distinct metal contaminants in several food categories. For example, when used in foods, heavy metals listed in Indian Pharmacopoeia monographs

must not include metal beyond the levels mentioned in the relevant monographs for the period that they are currently in effect. The regulations provide a detailed table specifying the maximum permissible limits of various metal contaminants (like lead, cadmium, and mercury) in different food items.

7. *Bureau of Indian Standards (BIS)*: Established under the BIS Act of 2016, this is the National Standards Body of India, whose purpose is to promote the harmonious growth of the marking, quality certification, and standardization of commodities. It functions under the direction of the Ministry of Consumer Affairs, Nutrition, and Public Distribution of the Indian government. The Bureau of Information Services (BIS) creates and disseminates Indian Standards, establishes and manages Conformity Assessment Schemes, certifies and oversees Conformity Assessment laboratories, employs Hallmarking, promotes consumer empowerment, carries out quality assurance capacity-building initiatives, and represents the nation in International Organization for Standardization (ISO) and International Electrotechnical Commission (IEC).

11.2 REGULATIONS AND GUIDELINES FOR A FEW HEAVY METALS

11.2.1 Arsenic (As)

11.2.1.1 WHO Guidelines on Arsenic (As)

The WHO has established a provisional guideline value for arsenic (As) in drinking water of 10 μg/L. The WHO drinking-water quality guidelines are meant to serve as the foundation for international standards and regulations.

11.2.1.2 US-FDA Guidelines on Arsenic (As)

The FDA keeps an eye on the overall amounts of both organic and inorganic arsenic (As) in food (Nutrition, 2023a). By creating regulations, action levels, and consumer guidance, the FDA hopes to reduce the amount of arsenic (As) that consumers are exposed to, with a particular emphasis on safeguarding the very young (Affairs, 2023). The maximum contamination level (MCL) of arsenic (As) in drinking water for public consumption, as well as bottled water, has been set by the US FDA at 10 ppb. People who acquire their water from private wells are urged to test for arsenic (As) and get in touch with their state's drinking water well program if they suspect an issue, as government organizations do not regularly test the water from private wells (US Food & Drug Administration, n.d.; Shen, 2023; US EPA, 2015a).

11.2.1.3 US EPA Guidelines on Arsenic (As)

The US EPA has established a limit of 0.01 parts per million (ppm) for arsenic (As) in drinking water to lessen the health concerns connected with arsenic (As) exposure from drinking water (US EPA, 2015c; US EPA, n.d.). Inorganic arsenic's EPA reference dose (RfD) is 0.0003 mg/kg/bw/d due to impacts on the skin like hyperpigmentation and keratosis and potential vascular effects from epidemiologic studies of drinking water exposure (US EPA, n.d.) (https://www.epa.gov/sites/default/files/2014-03/documents/arsenic_toxfaqs_3v.pdf). The EPA is seeking feedback on three possible regulation choices for arsenic (As) levels: 3 (feasible level), 10, and 20 μg/L (ATSDR, n.d.-a). The agency is also proposing a new drinking water standard of 5 μg/L. The EPA is recommending a Maximum Contaminant Level Goal (MCLG) of zero for arsenic (As) (epa.gov/sites/default/files/2021-04/documents/arsenic_april_2021.pdf).

11.2.1.4 EU Guidelines on Arsenic (As)

The majority of products that need to be tested for arsenic (As) are made of rice; additional products that need to be tested include baby food, fruit juices, and salt. In regulated products, the maximum amount of arsenic (As) that is permitted varies from 0.01 to 0.5 mg/kg (Markkinen, n.d.).

11.2.1.5 EFSA Guidelines on Arsenic (As)

The EFSA Panel on Contaminants in the Food Chain (CONTAM Panel) has set a reference point between 0.3 and 8 lg/kg bw per day as benchmark dosage lower confidence limit (BMDL01) for a 1% increased risk of cancer of the lung, skin, and bladder, as well as skin lesions (EFSA CONTAM Panel, 2009). Following that, based on epidemiological studies, the Joint FAO/WHO Expert Committee on Food Additives (JECFA) determined a BMDL05 of 3.0 lg/kg bw per day for an elevated risk of lung cancer (range 2–7 lg/kg bw per day) (JECFA, 2011).

11.2.1.6 BIS Guidelines on Arsenic (As)

The BIS specification for drinking water (IS10500:2003) specifies the maximum desirable limit and permissible limit values as 0.05 mg/L for Arsenic (As).

11.2.2 Lead (Pb)

11.2.2.1 WHO Guidelines on Lead (Pb)

The WHO has established time-weighted average (TWA) for inorganic lead compounds is 0.15 milligrams per meter cube (mg/m^3) and 0.05 mg/m^3 (TWA) for organic lead (Pb) compounds.

11.2.2.2 US FDA Guidelines on Lead (Pb)

The FDA has established lead (Pb) tolerances in a variety of food items. For instance, there is a 5 ppb tolerance for bottled water, 50 ppb for drinks, and 100 ppb for candies. A draft advice on action levels for lead (Pb) in food meant for infants and children has been released by the FDA to the industry (US Food & Drug Administration, n.d.; Environmental Defence Fund, n.d.; Nutrition, 2023b, 2024b, 2024e; Shen, 2023).

11.2.2.3 US EPA Guidelines on Lead (Pb)

The EPA has established new clearance levels of lead as 100 μg/ft^2 for window sill dust and 10 μg/ft^2 for floor dust. Lead (Pb) is controlled in the air under the Clean Air Act as a hazardous air pollutant and as one of the six common pollutants for which the EPA has established national ambient air quality standards (NAAQS). Under Title IV of the Toxic Substances Control Act (TSCA) and other provisions of the Residential Lead-Based Paint Hazard Reduction Act of 1992, the EPA additionally oversees lead-based paint dangers (US EPA, n.d., 2013c).

11.2.2.4 EU Guidelines on Lead (Pb)

In some baby foods, the maximum permitted lead (Pb) content is 0.01 mg/kg and in food supplements is 3 mg/kg. While the limits for fruit, vegetables, and fungus range from 0.1 mg/kg to 0.8 mg/kg, the majority of meat products, fats, and oils should not have more than 0.1 mg/kg of lead (Pb) (Markkinen, n.d.).

11.2.2.5 EFSA Guidelines on Lead (Pb)

The EFSA has established a provisional guideline value of 3.0 μg/kg/bw/d for lead (Pb) in food. Additionally, an opinion on potential health hazards associated with lead (Pb) contamination in food has been released by the EFSA. According to experts, the main dietary sources of lead (Pb) exposure for the majority of Europeans are vegetables, cereals, and tap water.

11.2.2.6 BIS Guidelines on Lead (Pb)

The BIS specification for drinking water (IS10500:2003) specifies the maximum desirable limit and permissible limit values as 0.01 mg/L for Lead.

11.2.3 Mercury (Hg)

11.2.3.1 WHO Guidelines on Mercury (Hg)

WHO set TWA for elemental mercury vapor is 0.05 mg/m^3; TWA for inorganic mercury compounds is 0.1 mg/m^3; TWA for organic mercury compounds is 0.01 mg/m^3.

11.2.3.2 US-FDA Guidelines on Mercury (Hg)

The most prevalent method that Americans are exposed to mercury (Hg) is through seafood consumption. The majority of seafood varieties don't contain amounts of mercury (Hg) that could be harmful to an adult's health. According to sections 502(a), 502(f)(1) and (2), and 502(j) of the Federal Food, Drug, and Cosmetic Act; any cosmetic product that contains mercury (Hg) as a skin-bleaching agent and is offered for sale as a skin-bleaching, beauty, or facial preparation is considered misbranded (US Food & Drug Administration, n.d.; Nutrition, 2023c; Shen, 2023).

11.2.3.3 US-EPA Guidelines on Mercury (Hg)

Under the Safe Drinking Water Act, the EPA has established the MCL for inorganic mercury (Hg) at 0.002 mg/L, or 2 ppb. The MCL for mercury (Hg) in drinking water must not be exceeded by public water systems. Guidelines on how to manage a mercury (Hg) spill at home are also available from the EPA (US EPA, 2015b).

11.2.3.4 EU Guidelines on Mercury (Hg)

Maximum mercury (Hg) levels are set for salt, food supplements, fish, and fishery products. Mercury (Hg) content in salt and supplements cannot exceed 0.1 mg/kg. The limit for fish and fisheries products varies depending on the species and might be anywhere between 0.3 and 1 mg/kg (Markkinen, n.d.).

11.2.3.5 EFSA Guidelines on Mercury (Hg)

The EFSA sets Tolerable Weekly Intakes (TWIs), sometimes known as "safe levels," to shield consumers from potential health risks linked with the presence of methylmercury and inorganic mercury (Hg), the two primary forms of mercury (Hg) found in food. A (TWIs) set by EFSA of 4 micrograms per kilogram body weight (µg/kg/bw) for inorganic mercury (Hg) and 1.3 µg/kg/bw for methylmercury. The purpose of TWIs is to safeguard consumers from the harmful health consequences that could arise from the potential presence of the primary forms of mercury (Hg) in food.

11.2.3.6 BIS Guidelines on Mercury (Hg)

The BIS specification for drinking water (IS10500:2003) specifies the maximum desirable limit and permissible limit values as 0.01 mg/L for mercury (Hg).

11.2.4 Cadmium (Cd)

11.2.4.1 WHO Guidelines on Cadmium (Cd)

The provisional guideline value for occupational exposure to cadmium (Cd) metal or its oxide has been established by the WHO at 0.01 mg/m^3 (TWA) and 0.005 mg/m^3 (TWA) for cadmium sulfide.

11.2.4.2 US FDA Regulations for Cadmium (Cd)

The FDA has established a 0.005 mg/L limitation for cadmium (Cd) in bottled water. The FDA has set a maximum limit of 15 ppm for cadmium (Cd) in color additives, such as zinc oxide, used in pharmaceuticals (ATSDR, n.d.-b; Nutrition, 2024d).

According to sections 502(a), 502(f)(1) and (2), and 502(j) of the Federal Food, Drug, and Cosmetic Act; any cosmetic product that contains cadmium (Cd) as a skin-bleaching agent and is offered for sale as a skin-bleaching, beauty, or facial preparation is considered misbranded (US Food & Drug Administration, n.d.; Shen, 2023).

11.2.4.3 US EPA Guidelines for Cadmium (Cd)

The EPA has designated cadmium (Cd) as a Group B substance, indicating its potential as a probable human carcinogen. The US EPA has established a RfD for cadmium (Cd) exposure in water at 5×10^{-4} mg/kg/d, and for food at 1×10^{-3} mg/kg/d. Under the Safe Drinking Water Act, the EPA has set a maximum contaminant level (MCL) for inorganic cadmium, which is an enforceable regulation. This (MCL is set at 0.002 mg/L or 2 ppb. Public water systems must guarantee that the concentration of cadmium (Cd) in drinking water is not above the (MCL. Cadmium (Cd) has been included in the EPA's National Emission Standards for Hazardous Air Pollutants (NESHAP) inventory, which comprises a comprehensive list of 189 substances known to pose a risk to air quality. According to the EPA biosolids rule, there exists a regulatory limit for the application of cadmium (Cd) to land, which is set at 85 mg/kg of fill material. The EPA also offers recommendations regarding the appropriate procedures for managing a cadmium (Cd) leak inside a residential setting (ATSDR, 2023).

11.2.4.4 EU Guidelines on Cadmium (Cd)

The permissible thresholds for cadmium (Cd) content range from 0.005 mg/kg in milk protein-based infant food products to 3 mg/kg in dietary supplements. The permissible cadmium (Cd) levels in fruit, vegetables, and fungi range from 0.02 to 0.5 mg/kg, while the acceptable limits for most meat and fish products fluctuate between 0.05 and 0.15 mg/kg (Markkinen, n.d.).

11.2.4.5 EFSA Guidelines on Cadmium (Cd)

The TWI for cadmium (Cd) has been established by the EFSA at a level of 2.5 μg/kg bw. The threshold of acceptable risk (TWI) denotes the point at which the occurrence of detrimental effects is not anticipated. According to the findings of the EFSA, the average exposure of adults throughout the EU is approximately equal to, or slightly surpasses, the acceptable weekly intake level. Additionally, the study determined that certain subpopulations, including vegetarians, children, smokers, and individuals residing in heavily polluted regions, may surpass the acceptable weekly intake by approximately twice the recommended amount.

11.2.4.6 BIS Guidelines on Cadmium (Cd)

The BIS specification for drinking water (IS10500:2003) specifies the maximum desirable limit and permissible limit values as 0.003 mg/L for cadmium (Cd).

FASSI Guidelines on Heavy Metals

FASSI guidelines on heavy metals are shown in Table 11.1.

Additional information regarding the acceptable thresholds of contaminants in food products, such as lead, arsenic, tin, and cadmium, can be accessed within the proposed revision of the Food Safety and Standards (Contaminants, Toxins, and Residues) Regulations.

Please note that these guidelines and regulations are subject to change, and it's always advisable to refer to the official websites of the respective agencies for the most up-to-date information.

11.3 CONCLUSION

Heavy metals are chemicals that are released into wastewater. Heavy metals contamination in wastewater has proved a problem for water treatment systems as they are found in limited quantities everywhere. They are present in high concentrations and have become a growing source of concern for public health as well as the environment. The physical, chemical, biological, and hybrid processes have been productive in handling a wide variety of heavy metals but no complete elimination has been observed yet. To assure safety, compliance, and standardization across many industries and sectors, numerous government agencies and organizations around the world create rules and

TABLE 11.1
FASSI Guidelines on Heavy Metals Permissible Limit

	Permissible Limit (ppm)			
Article of Food	Mercury	Cadmium	Lead	Arsenic
Infant food	0.01	0.005	0.01	–
Milk and milk products	0.01	0.01	0.02	–
Cereals and pulses	0.1	0.1	0.1	–
Fruits and vegetables	0.1	0.1	0.1	3
Meat and meat products	0.1	0.1	0.1	3
Fish and fish products	0.5	0.1	0.1	3
Edible oil and fat	0.1	0.1	0.1	–
Sugar and confectionery	0.1	0.1	0.2	–
Salt	2.5	2.5	2.5	–

guidelines to control and minimize heavy metals exposure. Government organizations in charge of enforcing rules for public health, environmental protection, and sector-specific laws are primarily in charge of overseeing regulations and guidelines for heavy metal contamination. Food producers are required by law to put controls in place as necessary to considerably reduce or avoid exposure to chemical risks and heavy metals. Although heavy metal levels in food cannot be eliminated or prevented and can be decreased. Hence, regulatory organizations like WHO, US FDA, US EPA, EU, EFSA, FASSI, and BIS provide guidelines to control heavy metals exposure. It is always advisable to check the official websites of the relevant agencies for the most recent information as these guidelines and regulations are subject to change.

ACKNOWLEDGMENT

The financial and technical assistance provided by the Indian Institute of Technology (BHU) Varanasi, Varanasi, is acknowledged by the authors of this manuscript.

REFERENCES

Ali, H., Khan, E., & Ilahi, I. (2019). Environmental chemistry and ecotoxicology of hazardous heavy metals: Environmental persistence, toxicity, and bioaccumulation. *Journal of Chemistry*, *2019*, e6730305. https://doi.org/10.1155/2019/6730305

ATSDR. (n.d.-a). Arsenic Toxicity: What Are the Standards and Regulation for Arsenic Exposure? | Environmental Medicine. Retrieved January 31, 2024, from https://www.atsdr.cdc.gov/csem/arsenic/standards.html

ATSDR. (n.d.-b). Cadmium Toxicity: What Are the U.S. Standards for Cadmium Exposure? Environmental Medicine. Retrieved January 31, 2024, from https://www.atsdr.cdc.gov/csem/cadmium/Safety-Standards.html

ATSDR. (2023, May 25). Cadmium Toxicity: What Are the U.S. Standards for Cadmium Exposure? | Environmental Medicine | https://www.atsdr.cdc.gov/csem/cadmium/Safety-Standards.html

Briffa, J., Sinagra, E., & Blundell, R. (2020). Heavy metal pollution in the environment and their toxicological effects on humans. *Heliyon*, *6*(9), e04691. https://doi.org/10.1016/j.heliyon.2020.e04691

Budovich, L. S. (2021). Effects of heavy metals in soil and plants on ecosystems and the economy. *Caspian Journal of Environmental Sciences*, *19*(5), 991–997. https://doi.org/10.22124/cjes.2021.5331

CFR - Code of Federal Regulations Title 21. (n.d.). Retrieved January 31, 2024, from https://www.accessdata.fda.gov/scripts/cdrh/cfdocs/cfcfr/CFRSearch.cfm?fr=73.1991

Edelstein, M., & Ben-Hur, M. (2018). Heavy metals and metalloids: Sources, risks and strategies to reduce their accumulation in horticultural crops. *Scientia Horticulturae*, *234*, 431–444. https://doi.org/10.1016/j.scienta.2017.12.039

Environmental Defence Fund. (n.d.). FDA is reevaluating its tolerances for lead in food, and food manufacturers should too. *EDF+Business*. Retrieved January 31, 2024, from https://business.edf.org/insights/fda-is-reevaluating-its-tolerances-for-lead-in-food-and-food-manufacturers-should-be-prepared/
FDA (2023, November 15). *What We Do*. FDA; FDA. https://www.fda.gov/about-fda/what-we-do
Hembrom, S., Singh, B., Gupta, S. K., & Nema, A. K. (2020). A comprehensive evaluation of heavy metal contamination in foodstuff and associated human health risk: A global perspective. In P. Singh, R. P. Singh, & V. Srivastava (Eds.), *Contemporary Environmental Issues and Challenges in Era of Climate Change* (pp. 33–63). Springer. https://doi.org/10.1007/978-981-32-9595-7_2
Lambert, M., Leven, B. A., & Green, R. M. (n.d.). *Environmental Science and Technology Briefs for Citizens*. https://doi.org/10.1080/15569543.2021.1891936
Markkinen, N. (n.d.). *Food Contaminant Analysis by EU Regulations | Measurlabs*. Retrieved January 31, 2024, from https://measurlabs.com/blog/eu-regulations-on-food-contaminants/#heavy-metals
Nutrition. (2022). FDA's Testing of Cosmetics for Arsenic, Cadmium, Chromium, Cobalt, Lead, Mercury, and Nickel Content. *FDA*. https://www.fda.gov/cosmetics/potential-contaminants-cosmetics/fdas-testing-cosmetics-arsenic-cadmium-chromium-cobalt-lead-mercury-and-nickel-content
Nutrition. (2023a). Arsenic in Food. *FDA*. https://www.fda.gov/food/environmental-contaminants-food/arsenic-food
Nutrition. (2023b). Lead in Food and Foodwares. *FDA*. https://www.fda.gov/food/environmental-contaminants-food/lead-food-and-foodwares
Nutrition. (2023c). Mercury in Food. *FDA*. https://www.fda.gov/food/environmental-contaminants-food/mercury-food
Nutrition. (2024a). Closer to Zero: Reducing Childhood Exposure to Contaminants from Foods. *FDA*. https://www.fda.gov/food/environmental-contaminants-food/closer-zero-reducing-childhood-exposure-contaminants-foods
Nutrition. (2024b). FDA Issues Guidance for Industry on Action Levels for Lead in Baby Foods. *FDA*. https://www.fda.gov/food/cfsan-constituent-updates/fda-issues-guidance-industry-action-levels-lead-baby-foods
Nutrition. (2024c). Testing Results for Arsenic, Lead, Cadmium and Mercury. *FDA*. https://www.fda.gov/food/environmental-contaminants-food/testing-results-arsenic-lead-cadmium-and-mercury
Nutrition. (2024d, January 17). *Environmental Contaminants in Food*. FDA; FDA. https://www.fda.gov/food/chemical-contaminants-pesticides/environmental-contaminants-food
Nutrition. (2024e, January 26). *Draft Guidance for Industry: Action Levels for Lead in Food Intended for Babies and Young Children*. FDA. https://www.fda.gov/regulatory-information/search-fda-guidance-documents/draft-guidance-industry-action-levels-lead-food-intended-babies-and-young-children
Shen, Y. (2023, April 11). *Heavy Metals Regulations in the United States: An Overview*. Compliance Gate. https://www.compliancegate.com/heavy-metal-regulations-united-states/
Swain, C. K. (2024). Environmental pollution indices: A review on concentration of heavy metals in air, water, and soil near industrialization and urbanisation. *Discover Environment*, *2*(1), 5. https://doi.org/10.1007/s44274-024-00030-8
US EPA. (n.d.). *Technical Fact Sheet*. Proposed Rule for Arsenic in Drinking Water and Clarifications to Compliance and New Source Contaminants Monitoring | Arsenic | Retrieved January 31, 2024, from https://archive.epa.gov/water/archive/web/html/regulations_pro-techfactsheet.html
US EPA. (n.d.). *Lead Regulations* [Overviews and Factsheets]. Retrieved January 31, 2024, from https://19january2017snapshot.epa.gov/lead/lead-regulations
US EPA. (2013a, January 29). *Our Mission and What We Do* [Overviews and Factsheets]. https://www.epa.gov/aboutepa/our-mission-and-what-we-do
US EPA. (2013b, January 29). *The Origins of EPA* [Collections and Lists]. https://www.epa.gov/history/origins-epa
US EPA. (2013c, February 12). *Lead Regulations* [Overviews and Factsheets]. https://www.epa.gov/lead/lead-regulations
US EPA. (2015a, May 7). *Private Drinking Water Well Programs in Your State* [Collections and Lists]. https://www.epa.gov/privatewells/private-drinking-water-well-programs-your-state
US EPA. (2015b, September 9). *What EPA Is Doing to Reduce Mercury Pollution, and Exposures to Mercury* [Overviews and Factsheets]. https://www.epa.gov/mercury/what-epa-doing-reduce-mercury-pollution-and-exposures-mercury
US EPA. (2015c, October 13). *Chemical Contaminant Rules* [Other Policies and Guidance]. https://www.epa.gov/dwreginfo/chemical-contaminant-rules

US Food & Drug Administration. (n.d.). CFR – Code of Federal Regulations Title 21. Retrieved January 31, 2024, from https://www.accessdata.fda.gov/scripts/cdrh/cfdocs/cfcfr/CFRSearch.cfm?fr=73.1991

US Food & Drug Administration. (2021, December 3). *Laboratory Accreditation for Analyses of Foods.* Federal Register. https://www.federalregister.gov/documents/2021/12/03/2021-25716/laboratory-accreditation-for-analyses-of-foods

WHO. (2024). *About WHO.* https://www.who.int/about

WHO. (n.d.). *Infant and Young Child Feeding.* Retrieved January 31, 2024, from https://www.who.int/news-room/fact-sheets/detail/infant-and-young-child-feeding

12 Analytical Techniques for Heavy Metal Analysis

*Sneh Lata and Nasreen Ghazi Ansari**
Analytical Chemistry Division, Regulatory Toxicology Group,
CSIR-Indian Institute of Toxicology Research (CSIR-IITR),
Lucknow, Uttar Pradesh, India
and
Academy of Scientific and Innovative Research (AcSIR), Ghaziabad, India

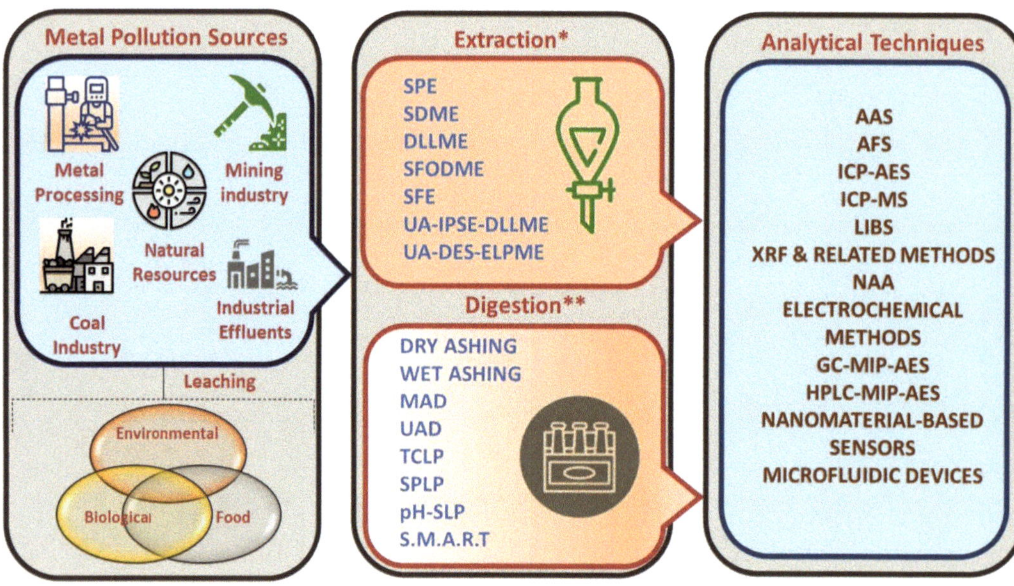

12.1 INTRODUCTION

Analytical chemistry is pivotal in analysing heavy metals (HMs), offering the tools and techniques to accurately assess and quantify these elements in diverse matrices, including environmental, food, and biological tissues. HMs, characterised by their high atomic weights and densities, can harm living organisms in higher concentrations. They can infiltrate the environment through natural processes and anthropogenic activities, threatening ecosystems and human well-being (Tarekegn et al. 2022). The scrutiny of trace metals in the environment has drawn substantial consideration in current years, and this trend is expected to persist due to the increasing presence of metals in our surroundings. Approximately 30 metals and metalloids, out of the 92 naturally occurring elements, have a harmful ability to mankind. This group encompasses D-block (Titanium, Cobalt, Nickel, Vanadium, Chromium, Copper, Molybdenum, Manganese, Palladium, Silver, Cadmium, Tungsten, Platinum, Gold, Mercury), P-block (Boron, Aluminium, Arsenic, Selenium, Tin, Antimony,

* Corresponding Authors: nasreen@iitr.res.in

DOI: 10.1201/9781032685793-12

TABLE 12.1
Maximum Contaminants Limit (MCL) for Heavy Metals in Different Matrices by Regulatory Agencies Worldwide

Matrices	Range of MCL	Regulatory Agencies
Food and beverages	0.05–0.1 mg/kg	Codex Alimentarius
	0.020–0.1 mg/kg	European Food Safety Authority (EFSA)
	1–10 mg/kg	Food and Drug Administration (FDA)
Drinking water	0.001–2 mg/L	World Health Organization (WHO)
	0.001ppb–15 mg/L	Environmental Protection Agency (EPA)
Soil and sediments	0.05–2 mg/kg	EPA Soil Screening Levels
	0.05–2 mg/kg	EU Soil Framework Directive
Air quality	100–1000 mg/m^3	National Ambient Air Quality Standards (NAAQS)
		OSHA
	10–5000 mg/m^3	NIOSH
		EU Ambient Air Quality Directive
Pharmaceuticals and medicinal products	1–10 mg/L	Pharmacopoeias
Cosmetics	1–10 mg/L	EU Cosmetic Regulation

Sources: Abah et al. (2016); Briffa et al. (2020).

Tellurium Lead, and Bismuth) and S-block (Beryllium, Strontium, Lithium, Caesium, Barium) (Pasinszki, Prasad, and Krebsz 2023).

Regulatory guidelines for HMs in various matrices can differ based on the country or region. Table 12.1 presents some commonly referenced regulatory guidelines for HMs in different matrices.

12.2 SAMPLING

The initial phase of any analytical procedure involves sampling, wherein the sample of the targeted matrix is procured for analysis. This process ensures that the collected sample is a true reflection of the original material. All potential sources of variations in elemental concentration must be considered. Sampling of various matrices can follow a simple, random, stratified random sampling or systematic sampling method from the surface or subsurface from the target area. Whenever, feasible it is advisable to gather samples by facing in the upstream direction and upwind to reduce the risk of contamination when working either from a boat or on land, employing a grab sampling method. The fundamental idea behind the grabbing technique is to swiftly immerse a sample bottle into the water and promptly seal it to minimise any potential exposure to airborne particulate matter. In the case of subsurface samples, they are obtained by either suctioning the sample into an immersed sample bottle or pumping it to the surface. Before sample collection, all sampling equipment and sample containers undergo thorough cleaning in a laboratory or specialised cleaning facility using a combination of detergents, mineral acids, and reagent water (Kinuthia et al. 2020). When collecting biological samples, such as blood, urine, hair, or tissues, it is essential to follow strict protocols to minimise contamination and preserve the sample's representative. Samples should be collected using clean, sterile equipment to prevent external contamination, and collection containers must be free from any trace of HMs. Sampling personnel should use clean, non-contaminated equipment and containers to avoid introducing external contaminants. Depending on the specific food product, samples may need to be collected from different locations and batches to account for potential variations. Care should be taken to minimise cross-contamination between samples during collection and transportation (El Hosry et al. 2023; López-Botella et al. 2023).

12.3 SAMPLE PRESERVATION

The subsequent crucial phase involves sample preservation, especially considering the time lapse between sample collection and analysis. Effective sample preservation must ensure that the sample's inherent physical and chemical attributes remain intact, providing the study accurately reflects the subject under investigation. Alterations in a sample's composition post-collection can arise from physical, chemical, and biological processes. The former encompasses volatilisation, diffusion, and surface adsorption. Potential chemical modifications contain photochemical reactions, oxidation, and precipitation, while biological processes include enzymatic reactions and biodegradation. Standard practices in sample preservation include temperature regulation, incorporating preservatives, suitable container selection, and adherence to recommended storage duration. These durations vary based on the target analyte and the specific sample matrix. Table 12.2 overviews some conventional preservation methods employed in metal analyses (Furness, Thompson, and Walsh 2018).

12.4 SAMPLE PREPARATION

Sample preparation represents the final and essential step before analysis, primarily because most samples aren't immediately suitable for directly introducing analytical instruments. Sometimes, it is necessary to liberate the components within the sample mixture through processes like extraction and digestion to achieve a clear solution. The sampling process involves several steps based on the type matrix, analyte concentration levels, and sample nature (Operacz et al. 2022).

12.4.1 Extraction Procedures

Given the intricate nature of matrices and the relatively low levels of HMs, effective techniques like extraction are often necessary for separating and concentrating the desired metal. The primary extraction methods include Solid-Phase Extraction (SPE) (El Ouardi et al. 2021) and Liquid-Phase Microextraction, which encompasses Soxhlet Extraction, Single-Drop Microextraction (SDME), Solvent Microextraction (SME), Supercritical Fluid Extraction (SFE), Dispersive Liquid-Liquid Microextraction (DLLME), Ultrasound Assisted-Deep Eutectic Solvent-Based Emulsification Liquid-Phase Microextraction (UA-DES-ELPME), Solidified Floating Organic Drop Microextraction (SFODME), Ultrasound-Assisted Ion Pair-Based Surfactant-Enhanced Dispersive Liquid-Liquid Microextraction method (UA-IPSE-DLLME) (Song and Huang 2022). These analytical approaches are characterised by their ease of operation, speed, sensitivity, and compatibility with high-throughput analysis using various advanced instruments (Golui et al. 2020).

TABLE 12.2
Conventional Sample Preservation Methods are Employed in Heavy Metal Analyses

Sample/Matrix	Preservation	Storage Container	Storage Time
Environmental	Filter onsite using 0.45/.22-μm filters and acidify to pH 2 with HNO_3 (water)	Glass bottle/FEP/PTFE	Up to 1 year
	Exclude moisture and air contact (soil and sediments)	Plastic	Years
Biological	Lithium heparin anticoagulant (blood) Urine (−20°C)	Heparinised vacutainer tubes (blood) Polyethylene containers (urine)	Years
Food	4°C	FEP, PTFE	Few days to a week

Source: USEPA (1669 Method).

12.4.2 Digestion Procedures

12.4.2.1 Standard Procedures

The standard procedure given by various agencies such as USEPA methods 200.7, 6020, 3051A, 3052, and 3050B (Inductively Coupled Plasma Mass Spectrometry [ICP-MS]) for metals in multiple matrices), APHA (American Public Health Association), SR ISO 8288/2001 standard for water and wastewater analysis, ISO methods:16968 and 14780 (Metal Analysis from various matrices), FDA-Elemental Analysis Manual, and so on for digestion of HM, involves a series of chemical processes designed to break down complex samples, such as environmental or biological samples, to release the HM ions for subsequent analysis.

12.4.2.2 R&D Procedures

12.4.2.2.1 Dry Ashing

Dry ashing is a widely employed technique for assessing HM content in a variety of samples, including biological tissues, food items, and environmental specimens. In this procedure, the sample undergoes heating within a muffle furnace, typically within 400–600°C. This elevated temperature leads to the complete combustion of organic constituents, leaving behind an inorganic residue or ash. Subsequently, this ash can be dissolved in a suitable acidic solution, enabling the subsequent analysis of HMs through such as atomic spectroscopic methods. Dry ashing is esteemed for its simplicity and proficiency in eliminating organic interferences, rendering it an invaluable approach for precisely determining HM concentrations within various sample types. However, taking precautions during the process is crucial to diminish the contamination risk and the potential loss of analytes (El Hosry et al. 2023).

12.4.2.2.2 Wet Ashing

This digestion procedure was performed using one of the following combinations depending on the type of matrix:

a. Mixture of Oxi-acidic (HNO_3:H_2SO_4:H_2O_2) in a 4:1:1 by volume ratio.
b. Concentrated HNO_3: HCl which were added in a 3:1 by volume ratio.
c. Concentrated HCl: HNO_3: $HClO_4$) were used in a 3:1 by volume ratio.

The combination was heated to 85–150°C for 3 hours to liquefy completely, then diluted to the required volume using acidify MQ. The same procedure was adapted for blank digestion (Asher, Samuel, and Mary 2020).

12.4.2.3 Microwave-assisted Digestion (MAD)

Microwave-assisted methods have emerged as highly efficient and rapid techniques for HM analysis. Microwave irradiation accelerates the digestion of samples using a mineral acid solution, allowing the release and subsequent quantification of HMs. The process significantly reduces analysis time, minimises chemical reagent consumption, and enhances sample throughput, making it ideal for environmental monitoring and industrial quality control. Microwave-assisted methods offer a reliable and cost-effective approach to HM analysis, contributing to improved analytical precision and resource conservation in various scientific and industrial applications (Oreste et al. 2020).

12.4.2.4 Ultrasound-assisted Digestion (UAD)

It is a process in which ultrasound waves enhance or facilitate a sample's digestion, typically in analytical chemistry. Applying ultrasound energy helps break down complex samples into simpler components, making them more accessible for analysis or further processing. The known amount of samples will be measured and placed in 5 ml of micro-centrifuge tubes. Overall 10% v/v HNO_3 was added to these tubes according to sample mass. These samples underwent sonication.

Subsequently, the phases were separated through centrifugation and dispensed to the autosampler cups for analysis (Mayotha, Chantiratikul, and Siriangkhawut 2023).

12.4.2.5 Toxicity Characteristic Leaching Procedure (TCLP)

The Toxicity Characteristic Leaching Procedure (TCLP) method is employed to assess the leaching of organic and inorganic substances in various types of waste, including liquid, solid, and multiphasic waste, by the USEPA SW-846 Method 1311, ASTM 1994, and CCPSA 1998 guidelines. When the sample pH exceeds 5, Reagent 1(100 µM of glacial acetic acid). The pH of this extracting solvent is adjusted between 2 and 3 by adding acid and base as necessary. In cases where the sample pH is below 5, Reagent 2 (Glacial acetic acid in NaOH maintaining a pH value of approximately 4–5). To begin the extraction process, a centrifuge tube containing a known quantity of sample and extractant solution was filled with the mixture and spun for 18–21 hours. Following which, each tube's liquid was vacuum-filtered to separate it from the solid phase. TCLP extract's pH was then determined, and each section was acidified with 1% nitric acid (v/v) before being analysed using the proper method (Golui et al. 2020).

12.4.2.6 Synthetic Precipitation Leaching Procedure (SPLP)

Regulatory authorities frequently use the SPLP (Synthetic Precipitation Leaching Procedure) test to assess how waste materials may leach into the environment. This experiment seeks to mimic any possible leaching resulting from genuine precipitation. The US EPA developed a standardised batch leaching process for trash categorisation called Method 1312 (USEPA, 1994). The extraction fluid for the SPLP test is a mixture of diluted nitric (three parts) and sulphuric acid (two parts). Keeping the extraction solution's pH around four to five takes great precipitation. The resultant leachate solution is filtered using a glass fibre filter and kept in containers for further analysis (Sharma et al. 2021).

12.4.2.7 pH-Static Leaching Procedure (pH-SLP)

The pH Static Leaching Procedure (pHh-SLP) is designed to investigate how the pH of the extracting solution affects the release of metals from the matrix into solutions with varying pH levels. Three distinct solutions were prepared with pH values of 3, 7, and 9 for acidic, neutral, and alkaline leaching conditions. These pH solutions were created using strong acids and bases in deionised water. The liquid-to-solid ratio for this test was kept at 20:1. Throughout 18 hours of rotation, the solution's pH was checked and adjusted at a speed of 30 ± 2 rpm every 30 minutes. The solutions were passed through a 0.22 µm filter and procured in 1-L plastic containers for instrumental analysis (Atique et al. 2020).

12.4.2.8 Small Mass, Affordable, Rapid, Transfer-Less Digesting Technique (S.M.A.R.T.)

A compact, cost-effective, and swiftly transfer-free digestion method has been developed to quantify the concentrations of HMs in small amounts of material. This technique involves a digestion process using a hot water bath, where sample digestion and dilution occur within the same sample tube. This methodology outpaces conventional approaches in terms of speed and minimises potential error sources. Remarkably, it can successfully digest and analyse sample masses as minuscule as 0.01 grams. Furthermore, the sample preparation and digestion process have been reduced from 10 to 4 hours (French et al. 2017).

12.5 ANALYTICAL TECHNIQUES

Atomic spectroscopic methods are crucial in qualitative and quantitative investigation of a broad spectrum of elements. Typically, these methods exhibit remarkable sensitivity, detecting minute concentrations from parts per million to parts per billion and sometimes at parts per trillion levels. Atomic Spectroscopy offers several advantages, including rapidity, ease of use, and high selectivity.

These techniques can be broadly grouped into two main categories: optical atomic spectrometry and atomic mass spectrometry (MS). To analyse atomic species using spectroscopy, the sample must exist in a gaseous state, where individual atoms or elementary ions are well-separated (Arjomandi and Shirkhanloo 2019). Therefore, the initial step in all atomic spectroscopic procedures involves atomisation, which transforms the sample into vapour and breaks it down to generate atoms or ions in a gaseous phase. The atomisation process's efficiency and consistency significantly impact the analytical method's sensitivity, precision, and accuracy. Atomisation is pivotal in atomic spectroscopy (Gong and Xie 2020).

Various techniques are available for sample atomisation in atomic spectroscopy. The most commonly employed methods include inductively coupled plasmas, flames, and electrothermal atomisers. Atomic Absorption Spectrometry (AAS) often uses flames and electrothermal atomisers, whereas nuclear MS and optical emission use inductively coupled plasmas.

12.5.1 Atomic Absorption Spectroscopy (AAS)

AAS, pioneered in the 1950s, serves as a rugged analytical technique for qualitative and quantitative assessment of HMs. It relies on distinctive spectral characteristics and varying degrees of attenuation of characteristic spectral lines for precise measurement (Gandhi et al. 2022).

12.5.2 Principle and Mechanism

AAS consists primarily of five integral components: a light source, an atomiser, a spectroscopic system, a detector, and a modulated synchronous system for power supply. The resonance line emitted from a sharp-line light source is particular to the analysed element. The atomisation system's primary goal is to supply the necessary energy to allow the sample solution to be atomised, evaporated, and dried to produce fundamental atomic vapour. The resonance line of the target element must be successfully isolated from nearby spectral lines by the spectroscopic equipment. By converting optical signals into electrical signals, the power synchronous modulation system enables the display of absorbance. Finally, the detecting system is utilised to reduce background emissions' interference, improve the S/N ratio, and lengthen the life of the light being used in the process (Welz and Vale 2019).

12.5.3 Sample Introduction and Atomisation

As samples commonly exist in liquid or solid forms, atomising and vapourising the analyte atoms or ions becomes necessary. Numerous AAS techniques are defined based on sample introduction and atomisation approach. Flame AAS (FAAS) (Arı, Bakırdere, and Ataman 2020). Electrothermal atomisers (ETAAS) (Bustos et al. 2022), Hydride generation AAS (HGAAS) (Juhászová et al. 2019), and Cold vapour AAS (CVAAS)(Yu 2021) have been extensively detailed. FAAS involves nebulising the liquid sample, mixing the aerosol with acetylene, and introducing it into a flame atomiser. ETAAS, which focuses on graphite furnace AAS (GFAAS), produces atom concentration in a smaller space than a flame, resulting in increased absorption and lower detection limits than FAAS. HGAAS reduces the analyte to a volatile hydride, atomised using various methods. CVAAS is unique to mercury because of its high atomic vapour pressure at ambient temperature. AAS is generally a single-element approach, albeit it has the potential for simultaneous multielement analysis. It is tedious for multielement analysis in voluminous samples. AAS offers various detection limits for different elements, as illustrated below (Figure 12.1):

- *FAAS:* 10–1000 μg/L
- *GFAAS:* 0.01–0.5 μg/L
- *HGAAS:* 0.1 μg/L
- *CVAAS:* Below 0.02 μg/L

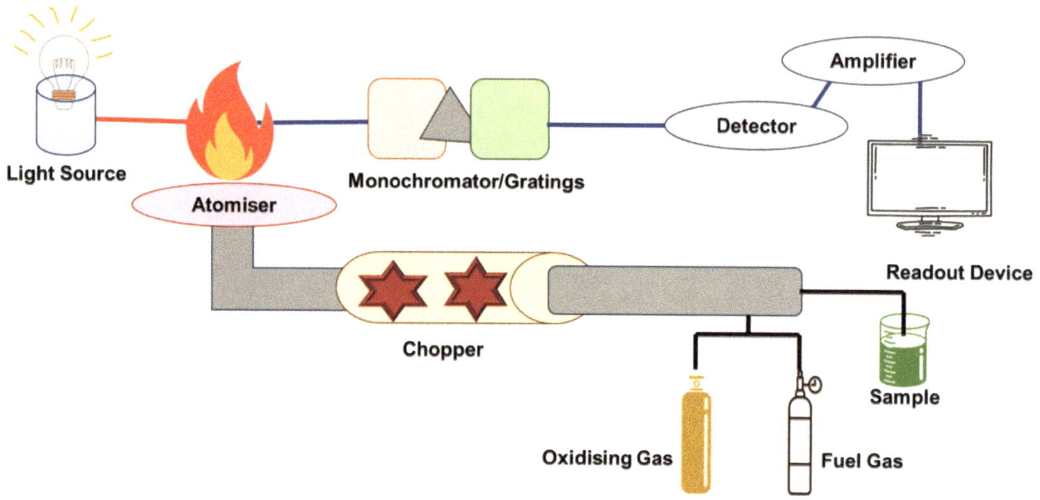

FIGURE 12.1 Diagram illustrating the operational concept of atomic absorption spectroscopy.

12.5.3.1 Applications

AAS finds utility across numerous elements with suitable light sources. FAAS is recommended for elements like Mn, Fe, Ni, Cu, Zn, Cd, Li, Na, Mg, Al, K, Ca, Ba, and Pb. ETAAS offers enhanced sensitivity and favours elements such as Be, Ag, In, Si, Mo, Sn, Sb, Au, Cr, Co Tl, and Bi. HGAAS excels in determining elements forming volatile hydrides. CVAAS is optimal for mercury analysis in sediments and tissues. Various U.S. EPA methods include 7000 series, 200.9, 245.1, 245.5, and 245.6, and outlined AAS protocols for diverse elements, serving applications in groundwater, surface water, wastewater, and more.

12.5.4 Inductively Coupled Plasma-Atomic Emission Spectrometry (ICP-AES)

Inductively Coupled Plasma-Atomic Emission Spectrometry (ICP-AES) is a method for analysing components in various samples in both a qualitative and quantitative manner. Because of its ability to analyse multiple elements and its high sensitivity is emerging as a viable alternative to FAAS.

12.5.4.1 Principle and Mechanism

Atomic Emission Spectroscopy, or ICP-AES, works by observing the optical emissions of excited atoms to ascertain the concentration of analytes. Light-emitting high-temperature atomisation sources raise the atoms to high-energy levels and then induce them to fall back to lower energy states. Desolvation, vapourisation, excitation, and ionisation of atoms are all successfully carried out using inductively coupled plasma, an extremely high excitation source (usually approximately 7000–8000 K). Each element has its characteristic wavelengths of released photons, and the emission strength is often proportional to the quantity of that element's atoms in the initial sample. The simultaneous or sequential quantification of 20 or more components is possible with ICP-AES and other atomic emission methods, attaining sensitivities similar to AAS's. ICP-AES has an impressive dynamic range since its calibration curves may retain linearity across various orders of magnitude. Additionally, ICP-AES can measure several nonmetals, such as phosphorus. One of the critical advantages of AES, including ICP-AES, over AAS is its capability to detect and measure all the atoms in a sample simultaneously, offering a substantial advantage in its ability to analyse multiple elements simultaneously (Figure 12.2).

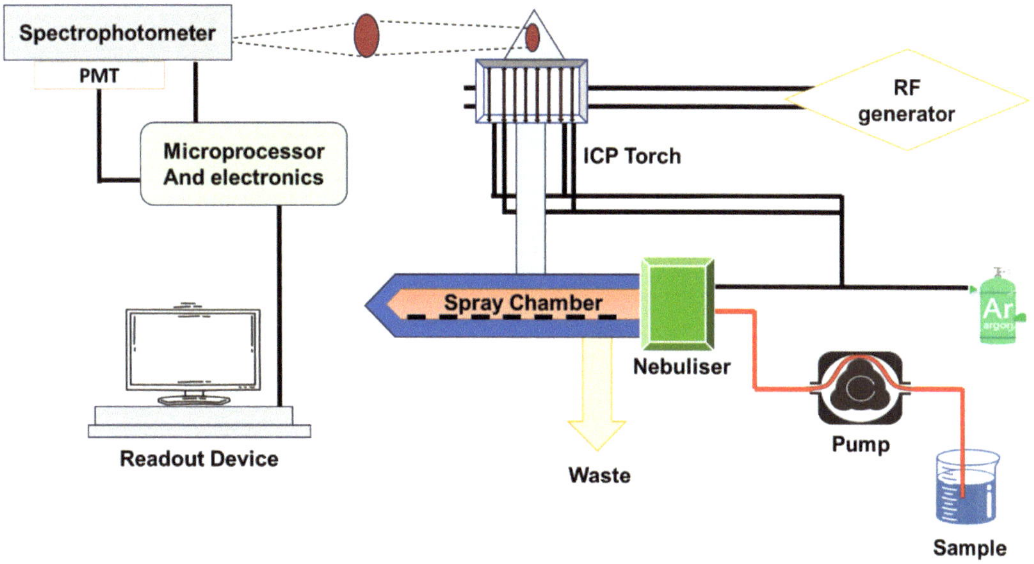

FIGURE 12.2 Schematic representation of inductively coupled plasma-atomic emission spectrometry (ICP-AES).

12.5.4.2 Application

Biological tissues, soils, sediments, natural waterways, seas, and airborne particulate matter are just a few of the materials used in environmental applications that use ICP-AES for metal identification. Following the sample preparation procedures provided in USEPA Method 200.2 is essential for analysing water, wastewater, and solid samples. U.S. EPA Method 200.7 offers thorough instructions for measuring metals and certain nonmetals using ICP-AES (Clarck, 1994). The analytical process for measuring metallic and metalloid particles in occupational environments is defined by OSHA Method ID-125G (OSHA 1994). The 13 elements found in welding smoke, including Beryllium, Antimony, Lead, Bismuth, Vanadium, Chromium, Manganese, Iron, Cobalt, Nickel, Copper, Zinc, and Cadmium, may be precisely determined using this approach (Ortega 2002).

12.5.5 Inductively Coupled Plasma-Mass Spectrometry (ICP-MS)

The fusion of ICP-MS results in the formation of ICP-MS. This method effectively disintegrates the material into its constituent atoms, employing an ICP argon plasma as the ion source. The ICP's ionisation conditions lead to highly effective ionisation, which is notable for creating primarily singly charged ions (Rittirong and Saenboonruang 2018).

12.5.5.1 Principle and Mechanism

Ions are separated by mass analysis according to their mass-to-charge ratio (m/z). Standard mass analysers in ICP-MS systems are the quadrupole and magnetic sectors. The four metal rods that make up a quadrupole have a hyperbolic cross-section. These rods have radiofrequency (RF) and direct-current (DC) voltages applied to them, generating an electric field inside their borders. This electric field permits ions in a small m/z range (usually 0.7–0.9 m/z) to flow through depending on the RF/DC ratio. As a result, the quadrupole may be successively scanned throughout a range of m/z values by accurately altering the RF/DC ratio, providing the impression of practically simultaneous mass analysis. The quadrupole mass analyser is widely used because of its affordability, stability,

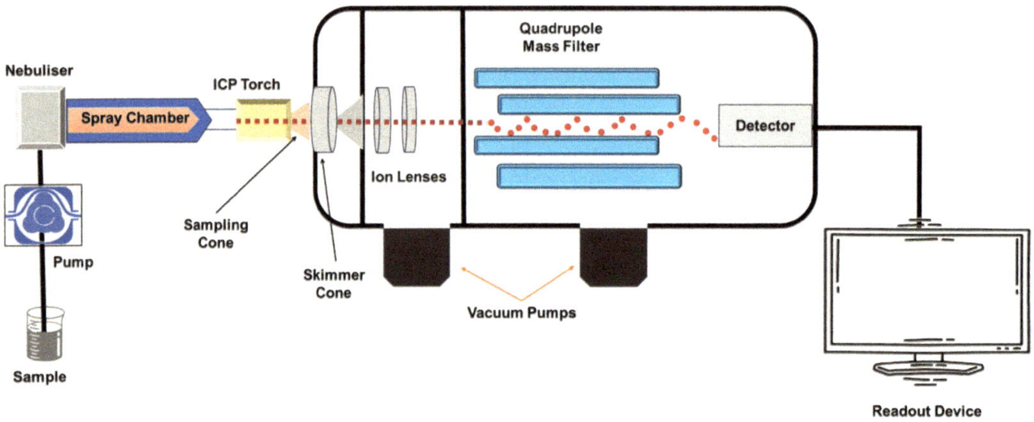

FIGURE 12.3 Diagrammatic illustration of inductively coupled plasma-mass spectrometry (ICP-MS).

compact size, and sufficient mass resolution for elemental analysis. Among mass analysers, it is the most often used.

A magnetic sector comes into play in situations demanding exceptionally high resolution or accurate concurrent mass analysis. The underlying principle of magnetic sector mass analysers involves the deflection of ions by an electric field. These ions, gathered from the plasma, undergo acceleration in standard commercial instruments before traversing by an electric sector, serving as an energy filter. Subsequently, they encounter a magnetic field, causing deflection in a single direction, with the extent of deflection escalating with higher m/z values. Scanning the magnetic field generates a mass spectrum. Accurately concurrent mass measurement may be achieved by maintaining steady magnetic and electric fields while arraying several detectors. Magnetic mass analysers are more costly, uncommon, and difficult to use than quadrupoles. They cannot be scanned as quickly as quadrupoles and have a restricted number of masses that can be evaluated simultaneously. Nonetheless, a magnetic sector's most significant benefit is its ability to obtain a high resolution. ICP-MS quadrupoles generally have resolutions ranging from 15 to 400, depending on the m/z, resulting in peak widths of 0.75–0.85. Magnetic sectors, on the other hand, may attain resolutions over 10,000. While quadrupoles provide sufficient resolution for most applications when spectroscopic interferences pose substantial problems, the more excellent resolution possible with a magnetic sector becomes desired (Meermann and Nischwitz 2018) (Figure 12.3).

ICP-MS enables the rapid assessment of multiple elements in a sample, comparable to Inductively Coupled Plasma Atomic Emission Spectroscopy (ICP-AES) and significantly faster than GF Atomic Absorption Spectroscopy (GFAAS). An advantageous feature distinguishing it from other spectroscopy methods is the capacity to acquire isotope information. Despite its speed, ICP-MS faces challenges, mainly when interferences arise due to a species sharing an identical nominal m/z (mass-to-charge ratio) as the target analyte. The detection threshold for Quadrupole ICP-MS ranges from 1 to 10 ng/L, while Magnetic Sector ICP-MS achieves a lower range of 0.01–0.1 ng/L.

12.5.5.2 Applications

Increased sensitivity has led to its usage in assessing ultra-low levels of trace elements; ICP-MS applications closely match those of ICP-AES. ICP-MS offers a wide range of uses in analysing various components in environmental samples. It is exceptionally well suited for the examination of HMs in groundwater samples. U.S. EPA method 200.8 outlines the utilisation of a quadrupole mass analyser in scanning mode to quantify the levels of dissolved elements in groundwater, surface waters, and drinking water. Additionally, it can assess the total recoverable concentrations of elements in these fluids and in wastewater, sludge, and soil samples.

12.5.6 Atomic Fluorescence Spectrometry

12.5.6.1 Principle and Mechanism

The most recent innovation in optical atomic spectroscopic methods is nuclear fluorescence spectrometry (AFS). Like AA, it ignites the element of interest using an external source. Instead of capturing radiation directly from the source, AFS measures the radiation often emitted at right angles. Atomic fluorescence methods utilise conventional hollow-cathode or electrodeless-discharge sources. Like AA, it ignites the element of interest using an external source. Instead of capturing radiation directly from the source, AFS measures the radiation often emitted at right angles. Atomic fluorescence methods utilise conventional hollow-cathode or electrodeless-discharge sources. Consequently, creating commercial Atomic Fluorescence equipment has been a protracted process. However, sensitivity improvements have been seen for some elements, including Hg, Sb, As, Se, and Te (Zou et al. 2018).

Laser-induced atomic fluorescence spectrometry, particularly when combined with electrothermal atomisation, demonstrates the capability to attain remarkably higher Sensitivity. Numerous elements have been proven to have detection limits as low as femtograms (10^{-15} g) to attograms (10^{-18} g). However, commercial equipment for laser-based AFS has not been widely adopted, most likely due to its high price and the nonstandard nature of high-powered lasers. It's important to remember that atomic fluorescence has a drawback in that it can only be used with one element (Zhang et al. 2019). While AFS primarily operates as a single-element method, it can attain multi-element analysis by employing a continuous source that sequentially adjusts with numerous wavelengths. The detection limit of AFS is 0.5 μg/L.

12.5.6.2 Application

The study of various environmentally essential elements, Including Mercury, Arsenic, and Selenium, is now possible using AFS equipment with hydride production capacity.

12.5.7 Laser-Induced Breakdown Spectroscopy (LIBS)

The laser-induced breakdown spectroscopy (LIBS) technique is a qualitative and quantitative way to analyse the elements of different materials. Applying a high-energy laser pulse to a sample involves producing a micro-plasma or spark on its outermost layer. The identification and quantification of elements in the sample become achievable by utilising the distinctive light wavelengths emitted during the return of ions and atoms within the sample to their ground state (Wang et al. 2021).

12.5.7.1 Principle and Mechanism

Laser-induced plasma and atomic emission spectroscopy serve as its foundation. In laser ablation, a concentrated pulsed laser beam emits particles, atoms, electrons, and ions into the gas phase at a small location on a solid sample. A laser with a wavelength of 1.064 nm called an Nd: YAG (neodymium-doped yttrium aluminium garnet) is most frequently utilised. A 10-ns pulse with an energy output of mJ and a spot width of 50 mm has a power output of 50 GW/cm^2. Typically, each pulse removes material up to 0.02–5 mm depth. The approach is almost non-destructive since each pulse only releases a few nanograms of material. Ar sweeps the ablation product created in a sealed chamber or He into the plasma through a Teflon-coated tube in atomic emission. By using consecutive pulses that probe more profoundly, it is possible to do depth profiling and quantify the elemental concentrations that change with depth. Each laser pulse creates a momentary, 10,000–20,000 K plasma where material dissociates into excited atoms and ions. The plasma's first optical emission is a continuum with little discernible information. Discrete atomic emission lines are seen after 1–10 seconds after the plasma expands supersonically and cools to those temperatures. After the laser pulse, the detector is gated to record atomic emission for a few microseconds. Through seven optical fibres connected to seven polychromators, each of which has a linear 2048-pixel charge-coupled

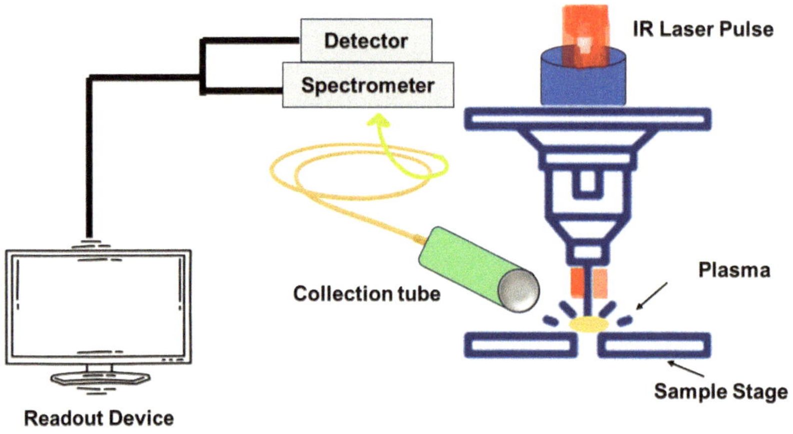

FIGURE 12.4 Schematic representation of laser-induced breakdown spectroscopy.

device to see a distinct spectral range, one commercial system monitors plasma emission. Over a range of 200–980 nm, the spectral resolution is 0.10 nm. Elements can be selectively ablated, selectively transferred to the plasma, or selectively atomised in the plasma, which presents a significant difficulty for quantitative analysis by laser ablation (Fabre 2020; Shah et al. 2020). (Figure 12.4).

12.5.7.2 Applications

X-ray fluorescence (XRF), ICP-MS, AES, and AAS, among other analytical technologies currently used for environmental monitoring, have limitations when it comes to measuring light elements, sample preparation, and on-site analysis. An innovative method known as LIBS allows for the quick, real-time, in-situ, remote, and simultaneous detection of numerous elements with minimal sample preparation. Environmental monitoring is one of the many domains where LIBS has seen extensive use. It is crucial for assessing soil, water, and air constituents. However, there haven't been any thorough summaries of the study on the LIBS technique for environmental analysis (Zhang et al. 2021) (Table 12.3).

12.5.8 X-Ray Methods

In 1914, the British physicist H. G. J. Moseley made a significant discovery. He found that analysing the secondary X-rays emitted when X-rays are applied to a material can reveal the components in the sample. This discovery was important since it gave the periodic table of elements its final

TABLE 12.3
Comparison of Different Atomic Spectroscopic Techniques

Parameters	FAAS	GFAAS	ICP-AES	ICP-MS	LIBS
Detection limits (ug/kg)	10–1000	0.01–1	0.1–10	0.00001–0.0001	1–100
Linear range	10^2	10^2	10^5	10^8	10^5
Sample screening/element	10–15 sec.	3–4 min.	6–60 min.	All elements in 2–5 min.	10–15 sec.
Volume needed	Large	Very small	Medium	Medium	Small
Isotopic study	Yes	No	No	Yes	No

Source: Seidel et al. (2021).

structure. These days, this process is also often employed as a non-destructive evaluation method. When atoms are exposed to radiation of a specific energy, such as that produced by the bombardment of ions, photons, or electrons, electrons from their inner orbital shells are evacuated. The outermost orbital electrons have X-ray radiation by refilling the gaps in these orbital shells. Measuring the energy and intensity of these emitted X-rays is the fundamental concept of XRF spectrometry (Allegretta et al. 2022).

12.5.8.1 X-Ray Fluorescence (XRF)

Utilising primary X-rays, typically produced by X-ray tubes or radioisotopes, XRF spectrometry causes the emission of secondary X-rays that are different in wavelength from the components in the sample. The wavelength/energy of the produced X-rays is used to identify the elements, and the X-ray intensity is used to calculate the elements' concentrations. Two types of detectors are employed to collect and examine the secondary radiation. Wavelength-dispersive XRF spectroscopy diffracts X-rays using a crystal while scanning distinct angle sites with a proportionate sensor. Energy-dispersive XRF spectrometry uses a solid-state detector to generate peaks reflecting the X-ray spectra's pulse-height distributions. Compared to conventional analytical techniques, sample preparation for XRF analysis is often minimal. However, it is typical to generate a homogenous sample for quantitative analysis by fusing solid materials using a borate flux for solid samples (Cherkashina and Pellinen 2021). Simultaneously, determine most elements, except those with atomic numbers below 8. The detection limits of XRF for Soil are 10–100 ng/kg and 0.5–10 ug/g for water matrix (Figure 12.5).

12.5.8.1.1 Applications

EDXRF effectively determines the major constituents of soils, while wavelength-dispersive XRF is preferred for water analysis. Additionally, XRF can analyse traces of metals in ambient airborne particles. The use of portable field XRF spectrometry for on-site screening and rapid elemental evaluation of dangerous materials is growing in popularity. The U.S. Environmental Protection Agency (EPA) has issued a standardised operating procedure for elemental analysis through the utilisation of field XRF analysers. Lead in paint analysis, in situ, metal analysis of soil and sediments, and monitoring of air filters are examples of applications. However, there are sensitivity and accuracy issues with portable XRF equipment (Ding et al. 2018).

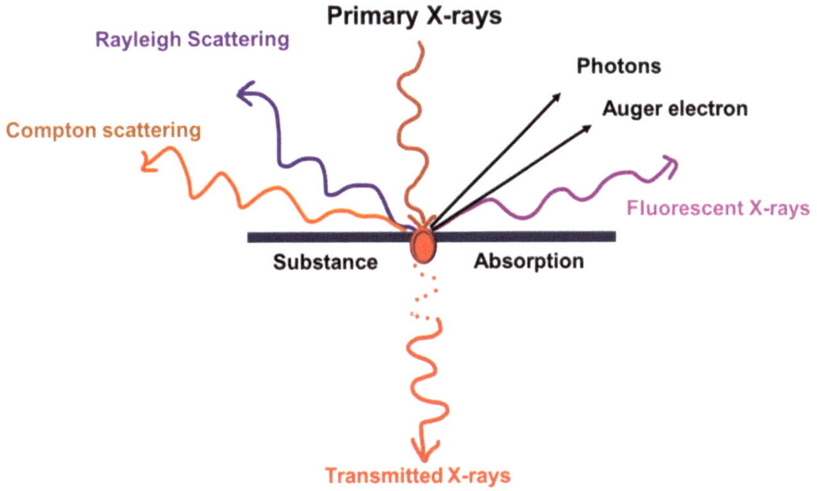

FIGURE 12.5 Diagrammatic illustration of X-ray emission phenomena.

12.5.8.2 Particle-Induced X-Ray Emission (PIXE)

In PIXE, the analyte in the sample is heated to produce heavy charged particles, generally 0.5–5 MeV protons, which emit X-rays. Energy-dispersive detectors are commonly used to find these released X-rays (Ene et al. 2019). PIXE can be measured with a detection limit of 0.1–10 μg/g.

12.5.8.2.1 Applications

Environmental research makes substantial use of PIXE, notably for detecting HMs in biological samples and aerosols. Compared to traditional energy-dispersive XRF, it is quicker and needs a smaller sample mass. It also offers lower detection limits. Although it needs a MeV particle accelerator and may not be widely available, microbeam PIXE (micro-PIXE) offers spatially resolved analysis with micrometre resolution (Ishii 2019).

12.5.8.3 Total-Reflection X-Ray Fluorescence (TXRF)

Utilising the whole excitation beam reflection from conventional radiation sources onto a flat platform is the fundamental principle of TXRF. To achieve TXRF, an extremely narrow beam with an angular deviation of less than 1 m rad is needed. This remarkable reduction in the signal-to-noise-ratio enables absolute detection limits up to three orders of magnitude lower than conventional XRF technologies. The standard sample preparation method for liquid samples is dropping a drop of solution onto a clean surface. The residue can be irradiated and then submitted for analysis following solvent evaporation (Beltrán et al. 2020) (Costa, Prais, and Nascentes 2022). TXRF can measure multielement simultaneously with a detection limit of 0.1 μg/g.

12.5.8.3.1 Applications

TXRF has seen an increase in usage in the study of several components in a variety of biological and environmental materials. These materials cover many specimens, including ice, airborne particles, seawater, river water, rainfall, seawater, seaweed, biopsy samples, food, and humic compounds.

12.5.8.4 Electron Probe X-Ray Microanalysis (EPXMA)

Finding out a specimen's elemental composition via an electron microscope is possible using the potent technology known as EPXMA. When a photon beam hits a specimen, a variety of interactions might happen, but one, in particular, causes the emission of X-rays due to the ejection of an inner-shell electron from the section; outer-shell electrons rearrange themselves. The elemental makeup of the exposed area of the specimen is revealed by the X-rays released from it (Xhoffer and Van Grieken 2018). Energy-dispersive X-ray spectrometers (EDX) or wavelength-dispersive X-ray spectrometers (WDX) may simultaneously detect all elements with atomic numbers (Z) more than ten or successively identify elements with Z larger than five utilising EPXMA's multielement capability. Detection limits of EPXMA typically range from 0.1 to 1 ng/g with EDS and 0.01–0.1 ng/g with WDS.

12.5.8.4.1 Applications

The examination of particulate materials frequently uses EPXMA. To enable both morphological and chemical characterisation, It comprises examining each particle in aerosols and sediments under a microscope and analysing them using energy-dispersive X-ray technology. Comparing samples with those found in particulate matter from suspected sources using this information is helpful. Particle associations with certain elements can provide information on where they came from; for instance, lead and bromine-containing particles have been related to automobile exhaust emissions, while lead-to-silicon and Pb/K ratios have been linked to soil and trash burning, respectively. For statistically significant findings, it is necessary to analyse a large number of particles, albeit this might take some time. Analysis times have greatly decreased because of computer-controlled EPXMA automation. Additionally, EPXMA has a wide range of uses in examining biological materials, especially when looking at the mineral makeup of lung particles and how it relates to lung illnesses. It has also been applied to research pollutant HMs' tissue localisation, absorption, and dispersion in plants and animals (Ingram and LeFurgey 2017).

12.5.9 Micro-PIXE Analysis

In Micro-PIXE analysis, ion beams focus on small cross-sectional areas, typically a few μm^2 in size. This makes it possible to examine small areas of a sample. Comparing Micro-PIXE to EPXMA, sensitivity levels are about 100 times higher due to using a proton beam compared to an electron beam. These nuclear microprobes, also referred to as ion microprobes, are capable of using a number of microspectrometries, including PIGE, PIXE, ERDA, RBS, and other nuclear reaction spectrometries. These methods work well for identifying lightweight components and complement micro-PIXE analysis for concurrently identifying all elements with Z >10. Detection limits for micro-PIXE analysis typically range from 1 to 10 μg/g. Micro-PIXE achieves spatial resolutions in the range of 0.3–3 μm (Nagao et al. 2023).

12.5.9.1 Applications

Applications for Micro-PIXE include the study of individual particles in samples of atmospheric aerosols and medium-sized fly ash. Additionally, it has been used to identify airborne particles inside the respiratory system. Furthermore, utilising micro-PIXE, researchers have thoroughly investigated the distribution of HMs, which act as indicators of exposure to these metals, in various organs or cells (Pongrac et al. 2020).

12.5.10 X-Ray Absorption Fine-Structure Spectroscopy (XAFS)

X-ray photons are absorbed via photoelectric contact, and their energy is subsequently transformed into the kinetic energy of orbit electrons within the absorber atoms. For this interaction to occur, the original photon's energy needs to be at least as high as the binding power of the released electron. This means that the photoelectric cross-section exhibits sharp shifts, or absorption edges, at photon energies that match the higher energy levels of the atom. The distinctive oscillations observed in condensed material around these absorption edges are called the X-ray absorption fine structure (XAFS). The region stretching from 0 to 40 eV beyond the actual edge is known as the X-ray absorption near-edge structure, or NEXAFS (near-edge XAFS). It is possible that tiny characteristics of the atom will also be seen here. The fine structure forms as a result of interference between waves scattered from neighbouring atoms and the returning photoelectric wavefront. Measuring coordination numbers and interatomic distances around the absorbing atom is possible by precisely analysing its microscopic structure. As a result, XAFS emerges as a crucial structural study tool that is especially effective when looking at non-crystalline materials. To facilitate extremely monoenergetic X-ray beams, synchrotron radiation sources are often used for XAFS investigations (Winterer 2024; Mastelaro and Zanotto 2018).

Finding the level of HM oxidation in solid samples may be done effectively using XANES. For instance, Chromium and Arsenic ratios in different oxidation stages during fly ash deposition have been quantitatively determined by XANES spectrometry. Similarly, it has been used to evaluate the Cr (VI)/Cr (III) ratios in soils with Chromium contamination. However, by exposing details, EXAFS spectra offer vital structural information such as the separation between the centre atom and its neighbours, the nature of these neighbours, the presence of local disorder, and the number of nearby atoms. Through the use of different kinds of matrices, EXAFS has proved helpful in identifying HM speciation. Recent applications include molecular-level speciation investigation of Lead, Arsenic, and Selenium in polluted soils and mining tailings. EXAFS has also been suggested as a valuable method for evaluating chemical remediation tactics in soils with chromium contamination. It has also made it possible to distinguish between lead contamination sources by doing lead speciation analyses in polluted soils. The structure of metal ion complexes with humic compounds is one

TABLE 12.4
Comparison of Different X-ray Spectroscopic Techniques

Techniques/Parameters	Detection Limits	Linear Range	Sample Screening	Sample Volume
XRF	10–100 μg/g	10^6	Z < 8	Small
PIXE	0.1–10 μg/g	10^9	Multielement	Small
TXRF	0.1 μg/g	10^6	Multielement	Small
EPXMA	10–1 ng/g	10^6	Z > 10	Small
Micro-PIXE	1–10 μg/g	10^9	Z > 10	Small

Sources: Igwebike-Ossi et al. (2017); Madariaga (2021).

example of how EXAFS is helpful in researching the complexation of metal cations with natural chemicals (Table 12.4) (Popmintchev et al. 2018; Xinxin et al. 2021).

12.5.11 Neutron Activation Analysis (NAA)

Neutron activation analysis (NAA) is a very sensitive method for measuring the quantity of chemical substances in various materials. NAA depends on converting stable atomic nuclei into radioactive ones and then measuring the distinctive nuclear radiation that these nuclei release with radioactivity. Activation is the result of a nuclear reaction that yields a radioactive nucleus. Generating rapid neutrons with energy within a few MeV ranges with a neutron producer or isotopic neutron source is one method of producing the incoming neutrons needed for activation. The resulting radionuclide emits distinctive radiation, frequently in the form of gamma radiation, as it decays into a stable atomic nucleus is possible to identify the kind of active element and the emitting radionuclide using decay techniques and gamma radiation energy calculations. According to the direct inverse correlation between radioactivity and the quantity of targeted nuclei in the irradiated sample, quantitative analysis depends on detecting the radiation intensity. The two primary drawbacks of NAA are its relatively high cost and restricted availability of required facilities. Radiochemical NAA is an option when the sensitivity of instrumental activation analyses is insufficient. This scenario involves chemical separation of the radionuclides corresponding to the target elements after irradiation. You can utilise various separation methods, such as ion exchange, chromatography, precipitation, electrolysis, and distillation. Both the matrix composition of the material and the constituents to be measured are taken into account when designing these separation strategies. These plans typically encompass a small number of components and organise the radionuclides in a way that makes it possible to determine them independently (Das, Sharma, and Chawla 2023; Krishnan et al. 2022). With some restrictions, most elements can be identified. The detection limit of NAA is 10–100 μg/g.

12.5.11.1 Applications

NAA-measured activated radioisotopes of several different HMs may have their gamma activity, including chromium (^{51}Cr but with potential interference from mercury), vanadium (^{52}V), manganese (^{56}Mn), cobalt (^{60}Co after separation from ^{59}Fe), nickel (^{65}Ni), zinc (^{65}Zn), selenium (^{75}Se), arsenic (^{76}As), technetium (^{99}mTc for molybdenum determination from activation of ^{99}Mo), silver (^{110}mAg with radiochemical separation required), cadmium (^{115}Cd), antimony (^{122}Sb with interference from ^{76}As or ^{124}Sb), barium (^{131}Ba with low sensitivity), gold (^{198}Au with high sensitivity), gold (^{199}Au for platinum determination from activation of ^{198}Pt), and mercury (^{203}Hg). For instance, radiochemical NAA can achieve a 100-fold improvement, while instrumental NAA only achieves a sensitivity of about 0.3 g/L for Chromium. The analysis of water and the study of trace elements in animal or human tissues, bodily fluids, foods, and other nutritionally significant materials typically use radiochemical NAA. Despite being less prevalent, radiochemical NAA offers useful information on lesser known elements like vanadium and molybdenum, among others (La and Joshi 2021).

12.5.12 Electrochemical Methods

Electroanalysis encompasses a wide range of techniques distinguished by the controlled variable, which can be either voltage or current. Usually, one of these variables is monitored while the other is applied to a mixture that contains the analyte species. It is possible to learn essential details about the solution's electrically charged species' type and concentration by graphing the measured variable versus the applied variable. Only a handful of the several electrochemical methods—voltammetry, amperometry, and potentiometry are often used in environmental studies (Baldo et al. 2020; Vasanthi Sridharan and Mandal 2022).

12.5.12.1 Voltammetry

Voltammetry is the name given to a group of methods where current is measured as a function of applied potential. Using a functioning electrode and polarised circumstances, the study is conducted inside an electrochemical cell. Both electrolysis and stripping are phases in the procedure. Prior to being stripped off and analysed, the target analyte is first reduced and collected at the working electrode. Electroanalytical methods are renowned for their efficiency in environmental analysis and provide both quantitative and qualitative data. Several voltammetric ways are used to measure the amounts of HMs (Honeychurch and Piano 2018). The primary use of voltammetry is to allow for the sequential study of individual metal ions. However, one benefit of ASV is that it can differentiate between multiple states of oxidation of the same metal. The voltammetric method's detection range for HMs is between nanomolar and micromolar.

12.5.12.1.1 Cyclic Voltammetry

By changing the potential within a given potential range, the current is monitored both forward and backwards. An oxidation or reduction peak will appear at a certain possibility during redox when a metal ion performs a Faradaic reaction, as shown in a cyclic voltammogram. This method also enables us to evaluate the Faradaic reaction's reversibility. The working electrode in a three-electrode system can be either carbon paste, gold, platinum, or glassy carbon electrode (GCE). A wire made of platinum acts as the auxiliary electrode, and a calomel or silver electrode acts as the reference electrode. Anchidin-Norocel et al. (2021) describe the examination of Ni in food on urease-alginate biosensor affixed to silver SPEs cyclic voltammetry. Similarly the absorption of Cr(VI)on glutaraldehyde-crosslinked chitosan beads with a gold electrode is also archieved by cyclic voltammetry (Watwe and Kulkarni, 2021).

12.5.12.1.2 Pulse Voltammetry

In pulse voltammetry, a potential sweep is induced by a sequence of superimposed voltage pulses. Metal ions will experience redox reactions as a result of the applied voltage, which will produce measurable Faradaic currents. Differential pulse voltammetry (DPV) is a popular choice among several pulse voltammetric techniques, as opposed to conventional and reverse voltammetry (Baluta et al. 2022).

12.5.12.1.3 Square Wave Voltammetry

Another widely used voltammetric method for measuring HM ions (HMIs) is SWV. Compared to DPV, It now features a quicker scanning rate, minimal adsorption on the functioning electrode, and faster measurement periods. It is feasible to concurrently detect many HMIs when using cyclic voltammetry (CV), DPV, and serial pulse voltammetry (SWV). Typically, electrochemical deposition followed by stripping is combined with voltammetric methods used for HMI measurement (Leon-Medina et al. 2022).

12.5.12.1.4 *Stripping Voltammetry*

A two-step process is involved in stripping voltammetry. HMIs are electrodeposited onto the working electrode surface as the first stage using the electrolyte solution as a medium. Using one of the voltammetric techniques previously discussed, the HMIs on the electrode are stripped into the electrolyte solution in the second stage. Depending on whether anodic or cathodic scan direction is used or which stage the stripping happens at, the technique is referred to as either anodic stripping voltammetry or cathodic stripping voltammetry (Nodehi, Baghayeri, and Veisi 2021).

12.5.12.1.5 *Applications*

Electrochemistry measurement of metal ions finds extensive application across diverse matrices, encompassing saltwater, fresh and potable water, wastewater, air, soil, sewage, sediment, dust, ash, and additional mediums. These methods effectively identify metals and metalloids, ranging from bismuth (Bi) to titanium (Ti). The high salt concentration and saltwater's corrosiveness make it challenging to analyse metal ions. Metal ion analysis in seawater often uses stripping techniques like ASV etc. U.S. EPA methods 7063 and 7472 describe how to use electrochemical processes to identify the presence of HMs in aqueous samples.

12.5.13 On-Line Coupled Methods for Metal Speciation Analysis

Environmental speciation research increasingly uses directly linked devices, combining the empathetic detection ability of atomic spectroscopy with the separation power of chromatography; these hybrid techniques combine an elemental detection process with an online elemental analysis. The main hyphenated methods developed in various laboratories include GC, LC, and very sensitive element-selective detectors; one instrument, the GC-Microwave-Induced Plasma [MIP]-Atomic Emission Spectroscopy [AES], is the only equipment available commercially for online speciation analysis of organometallic compounds. Thus, it's essential to keep that in mind. As a result, the analytical laboratories have experience developing these hyphenated procedures. Additionally, chemical alteration of the target metal might be used when the sensitivity attained by direct detection utilising GC, LC, or coupled techniques is insufficient for real-sample analysis. By derivatising or complexing the existing metal states, this alteration changes them into a different state and makes it easier to analyse them.

12.5.14 Gas Chromatography (GC)

Gas chromatography (GC) is a reliable separation method that works well with individual element techniques such as atomic emission spectroscopy (AES), AAS, and MS without causing any sample loss. Many environmental investigations have used hyphenated methods that combine GC with AAS, MIP-AES, and ICP-MS. Typical applications of GC-AAS include investigating volatile or thermally stable substances like lead, tin, and mercury alkyl compounds. Since MIP-AES can identify almost all metals and metalloids with absolute detection limits in the sub-picogram range for several elements (i.e., mercury, tin, and lead), on the other hand, excellent sensitivity and selectivity are possible using GC-MS. To explore volatile environmental components of selenium, tin, mercury, and lead, GC systems in conjunction with ICP-MS have been used, with detection limits as low as 0.8, 0.2, 0.05, and 0.08 pg/L in water samples. Additionally, a method called GC-atomic fluorescence spectroscopy (AFS) has been created to identify organomercury chemicals in a variety of environmental samples (Clases et al. 2021; Faßbender et al. 2021)

12.5.15 Liquid Chromatography (LC)

The separation step in the majority of hyphenated environmental metal speciation techniques is the use of liquid chromatography (LC). There is much documentation on using hyphenated LC methods to separate various metals. High-Performance Liquid Chromatography (HPLC) and Atomic Spectrometry procedures have been documented in various combinations, including HPLC-HGAAS, HPLC-MIP-AES, HPLC and ICP-AES, and HPLC-ICP-MS. LC effectively separates organometallics because it is suited for separating ionic, polar, and nonpolar chemicals. For instance, HPLC and ICP-MS may reach very low detection limits (0.03–0.05 g/L). With detection limits as low as 0.2 g/L, other techniques, such as HPLC-HG-AFS, have found species like As (III), As(V), and organo-arsenic compounds. There is a constant pursuit of improving spectroscopic and electrochemical methods' size, sensitivity, and selectivity. An emerging trend with advantages in portability and usability is the application of colourimetric and electrochemical technologies based on paper, which has made it possible to identify HM ions with the unaided eye (Hechavarría-Hernández et al., 2023).

12.6 OPTICAL SENSING OF HEAVY METAL IONS

Organic molecules, nanomaterials, hybrids, and quantum dots are only a few of the materials used in the process of detecting HM ions (HMIs) utilising colour or fluorescence signals. While optical approaches often result in exclusive signal creation owing to the target HMI, electrochemical sensing frequently permits simultaneous signal generation for many HMIs. According to the type of material utilised for HM sensing, the following parts are divided into categories:

12.6.1 Cyclodextrins

Cyclodextrin-based sensors are cutting-edge devices for locating HMs in various environmental, industrial, and analytical applications. Cyclodextrins, cyclic oligosaccharides comprising glucose units, are employed in industrial quality control, environmental monitoring, and food safety testing because of their distinctive structural ability to form inclusion complexes with guest molecules, such as HM ions. Modifying -cyclodextrin (CD) with multiwalled carbon nanotubes (MWCNT) provides a simple and affordable method of detecting lead (Pb). These techniques have benefits, including high specificity, sensitivity, and real-time monitoring capability. They aid in attempts to guarantee adherence to regulatory guidelines for the limit of HM pollution in various industries and support efforts to safeguard the environment and human health (Gao and He 2021).

12.6.2 Organic Molecules

Organic molecules used for HM sensing often contain functional groups or ligands such as EDTA (Ethylenediaminetetraacetic Acid) that can selectively bind to specific metal ions, leading to detectable changes in their properties, such as fluorescence or colour, for example, fluorescent probes based on fluorescein, rhodamine, dithizone, and calcein to detect various metal ions such as Ca^{2+}, Hg^{2+}, Pb^{2+} through fluorescence changes (Iftikhar et al. 2022).

12.6.3 Carbon Quantum Dots (CQDs)

Carbon quantum dots (CQDs) have become popular as opt sensors for HMIs because of the intrinsic functional groups on their surface. Charge carriers are produced due to the interaction between light and CQDs and are used by HMIs for sensing. Phosphorescence, photo-induced charge transfer, quenching and amplification of fluorescence, and inner filter effects are among the mechanisms involved. Differently sourced CQDs are capable of sensitively detecting Pb^{2+} ions (Shen et al. 2022).

12.6.4 Nanoparticles

Nanoparticles, notably Ag, Au, and Pt nanoparticles, have been extensively studied for HMI (Heavy Metal Ion) sensing based on surface plasmon resonance. These nanoparticles generate colour changes or aggregation-based signals upon HMI interaction. Nanoparticles of Ag, Au, and Pt have been used for optical sensing of several HMIs. In particular, Ag, Au, and Pt nanoparticles have been the subject of substantial research for surface plasmon resonance-based HMI sensing. Upon contact with the HMI, these nanoparticles produce Colour changes or aggregation-based signals. Various HMIs have utilised Ag, Au, and Pt nanoparticles for optical sensing. For instance, HMs like mercury (Hg^{2+}) are detectable by AgNPs. AgNPs assemble in the presence of Hg^{2+}, changing the material's hue from yellow to grey or black. And AuNPs are capable of finding heavy elements like lead. Red to blue colouration is evident when Pb^{2+} ions are bound. (Li et al. 2021; Tarekegn et al. 2022).

12.6.5 Paper-Based Microfluidic Devices

Paper-based analytical devices (PADs) have gained attention for on-site analysis, utilising lateral flow and photolithography techniques. These devices have been used for the semi-quantitative detection of various HMIs. These devices use paper as a substrate and rely on capillary action to transport fluids and react with target HM ions, resulting in visible or quantitative readout. For example, paper-based electrochemical sensors (PESs) offer simplicity and sensitivity, with electrodes printed on paper and connected to electrochemical readers. They are cost-effective and portable tools for HM detection (Devadhasan and Kim 2018).

12.7 CONCLUSION

Analytical techniques for HM analysis contribute a significant role in safeguarding human health and the environment. These techniques have evolved significantly, offering greater sensitivity, accuracy, and precision in detecting and quantifying HMs in various matrices. Researchers and regulatory authorities have access to many tools, from traditional methods like Atomic Absorption spectroscopy and ICP-MS to more recent innovations like XRF and voltammetry. The target HM, sample matrix, necessary detection limits, and cost considerations all play a part in the choice of analytical procedure. The dependability of outcomes must always be ensured, regardless of the method used, and strict quality control and quality assurance procedures must be followed. HM analysis must undergo continuous innovation and research to stay on top of new pollutants and successfully handle environmental and public health issues. In addition, international standards and regulations must be routinely updated to reflect the most recent developments in analytical methodologies to guarantee the preservation of ecosystems and the welfare of future generations.

In summary, HM analysis continues to evolve, and the application of cutting-edge analytical techniques is pivotal in addressing the challenges posed by HM contamination in our environment. As we move forward, a multidisciplinary approach, collaboration among scientists, and a commitment to sustainability will be essential in managing and mitigating the impact of HMs on our planet and its inhabitants.

ACKNOWLEDGEMENT

We appreciate the research possibilities provided by the Director of the CSIR-Indian Institute of Toxicology Research in Lucknow, India. Support from my Supervisor, Dr Nasreen Ghazi Ansari, is highly acknowledged for writing this chapter.

REFERENCES

Abah, James., Percy. Mashebe, and Onjefu Ameh Sylvanus Onjefu. 2016. "Preliminary assessment of some heavy metals pollution status of Lisikili River Water in Zambezi Region, Namibia." *International Journal of Environment and Pollution Research* 4 (2): 13–30.

Allegretta, Ignazio, Fabjola Bilo, Eva Marguí, Galina V Pashkova, and Roberto Terzano. 2022. "Development and application of X-rays in metal analysis of soil and plants." *Agronomy* 13 (1): 114. MDPI.

Anchidin-Norocel, Liliana, Wesley K. Savage, Gheorghe Gutt, and Sonia Amariei. 2021. "Development, optimization, characterization, and application of electrochemical biosensors for detecting nickel ions in food." *Biosensors (Basel)* 11 (12). https://doi.org/10.3390/bios11120519. https://www.ncbi.nlm.nih.gov/pubmed/34940276

Arı, Betül, Sezgin Bakırdere, and O Yavuz Ataman. 2020. "Development of sensitive analytical methods for the determination of thallium at trace levels by slotted quartz tube flame atomic absorption spectrometry." *Spectrochimica Acta Part B: Atomic Spectroscopy* 171: 105937.

Arjomandi, Mojtaba, and Hamid Shirkhanloo. 2019. "A review: Analytical methods for heavy metals determination in environment and human samples." *Analytical Methods in Environmental Chemistry Journal* 2 (03): 97–126.

Asher, Adedeji S., Kakulu E. Samuel, and Dauda S. Mary. 2020. "Analytical method for comparison of suitable wet digestion methods for heavy metal analysis in soil around a cement industry." *International Journal of Research and Scientific Innovation* 7: 41–47.

Atique, Usman, Sonia Iqbal, Noor Khan, Bainazir Qazi, Aqeel Javeed, Khalid Mahmood Anjum, Muhammad Sultan Haider, Tanveer Ali Khan, Sajid Mahmood, and Shahid Sherzada. 2020. "Multivariate assessment of water chemistry and metals in a river impacted by tanning industry." *Fresenius Environmental Bulletin* 29 (04A): 3013–3025.

Baldo, Maria Antonietta, Angela Maria Stortini, Paolo Oliveri, Riccardo Leardi, Ligia Maria Moretto, and Paolo Ugo. 2020. "Electrochemical preconcentration coupled with spectroscopic techniques for trace lead analysis in olive oils." *Talanta* 210: 120667. https://doi.org/10.1016/j.talanta.2019.120667. https://www.ncbi.nlm.nih.gov/pubmed/31987161.

Baluta, Sylwia, Francesca Meloni, Kinga Halicka, Adam Szyszka, Antonio Zucca, Maria Itria Pilo, and Joanna Cabaj. 2022. "Differential pulse voltammetry and chronoamperometry as analytical tools for epinephrine detection using a tyrosinase-based electrochemical biosensor." *RSC Advances* 12 (39): 25342–25353.

Beltrán, Blanca G, Victor Ramos-Sanchez, David Chávez-Flores, Rogelio Rodríguez-Maese, and Edwin Palacio. 2020. "Total reflection X-ray fluorescence spectroscopy (TXRF) method validation: Determination of heavy metals in dietary supplements." *Journal of Chemistry* 2020: 1–9.

Briffa, Jessica, Emmanuel Sinagra, and Renald Blundell. 2020. "Heavy metal pollution in the environment and their toxicological effects on humans." *Heliyon* 6 (9): e04691. https://doi.org/10.1016/j.heliyon.2020.e04691

Bustos, Daniel E., Juan A. Toro, Marisol Briceño, and Ricardo E. Rivas. 2022. "Use of slow atomization ramp in high resolution continuum source graphite furnace atomic absorption spectrometry for the simultaneous determination of Cd and Ni in slurry powdered chocolate samples." *Talanta* 247: 123547.

Cherkashina, Tatiana Yu, and Vadim A. Pellinen. 2021. "Applicability of X-ray fluorescence spectrometry for assessing geochemical features and heavy metal contamination of soils: Primary data." *International Journal of Environmental Analytical Chemistry* 101 (14): 2272–2287.

Clases, David, Maiken Ueland, Raquel Gonzalez de Vega, Philip Doble, and Daniel Pröfrock. 2021. "Quantitative speciation of volatile sulphur compounds from human cadavers by GC-ICP-MS." *Talanta* 221: 121424.

Costa, Cassiano L. S., Cláudia T. Prais, and Clésia C. Nascentes. 2022. "A simple method for glass analysis using total reflection X-ray fluorescence spectrometry." *Talanta* 243: 123354.

Das, Debashree Debasish, Nikita Sharma, and Pooja A Chawla. 2023. "Neutron activation analysis: An excellent nondestructive analytical technique for trace metal analysis." *Critical Reviews in Analytical Chemistry* 1–17. https://doi.org/10.1080/10408347.2023.2178841. https://www.ncbi.nlm.nih.gov/pubmed/36847238.

Devadhasan, Jasmine Pramila, and Jungkyu Kim. 2018. "A chemically functionalized paper-based microfluidic platform for multiplex heavy metal detection." *Sensors and Actuators B: Chemical* 273: 18–24.

Ding, Liang, Shui Wang, Bingjie Cai, Mancheng Zhang, and Changsheng Qu. 2018. "Application of portable X-ray fluorescence spectrometry in environmental investigation of heavy metal-contaminated sites and comparison with laboratory analysis." IOP Conference Series: Earth and Environmental Science.

El Hosry, Leina, Nicolas Sok, Rosalie Richa, Layal Al Mashtoub, Philippe Cayot, and Elias Bou-Maroun. 2023. "Sample preparation and analytical techniques in the determination of trace elements in food: A review." *Foods* 12 (4). https://doi.org/10.3390/foods12040895. https://www.ncbi.nlm.nih.gov/pubmed/36832970.

El Ouardi, Youssef, Alessio Giove, Markku Laatikainen, Catherine Branger, and Katri Laatikainen. 2021. "Benefit of ion imprinting technique in solid-phase extraction of heavy metals, special focus on the last decade." *Journal of Environmental Chemical Engineering* 9 (6): 106548.

Ene, Antoaneta, Ana Pantelică, Sorina-Simona Arbanaş, Violeta Pintilie, Florin Sloată, Florina Cristiana Căpriţă, Mihai Straticiuc, Dragoş Mirea, Andreea Şerban, and Claudia Stihi. 2019. "Development of analysis methodology using Proton Induced X-ray Emission (PIXE) as a complementary technique to determine trace elements in environmental matrices." *Analele Universităţii" Dunărea de Jos" din Galaţi. Fascicula II, Matematică, fizică, mecanică teoretică/Annals of the" Dunarea de Jos" University of Galati. Fascicle II, Mathematics, Physics, Theoretical Mechanics* 42 (2): 117–125.

Fabre, Cécile. 2020. "Advances in Laser-Induced Breakdown Spectroscopy analysis for geology: A critical review." *Spectrochimica Acta Part B: Atomic Spectroscopy* 166: 105799.

Faßbender, Sebastian, Marcus von der Au, Maren Koenig, Jürgen Pelzer, Christian Piechotta, Jochen Vogl, and Björn Meermann. 2021. "Species-specific isotope dilution analysis of monomethylmercury in sediment using GC/ICP-ToF-MS and comparison with ICP-Q-MS and ICP-SF-MS." *Analytical and Bioanalytical Chemistry* 413 (21): 5279–5289.

French, Amanda D., Hannah M. Ashbaugh, Gary Steinmetz, Melanie Barnes, Warren C. Conway, and David M. Klein. 2017. "The SMART (small mass, affordable, rapid, transfer-less) digestion method for heavy metal determinations." *International Journal of Environmental Analytical Chemistry* 97 (6): 499–507.

Furness, Robert W., David R. Thompson, and Paul M. Walsh. 2018. "Evidence from biological samples for historical changes in global metal pollution." In Robert W. Furness (Ed), *Heavy Metals in the Marine Environment*, 219–226. CRC Press.

Gandhi, Kamal, Neelima Sharma, Priyae Brath Gautam, Rajan Sharma, Bimlesh Mann, and Vanita Pandey. 2022. "Atomic absorption spectroscopy and flame photometry." In *Advanced Analytical Techniques in Dairy Chemistry*, 219–247. Springer.

Gao, Lingfeng, and Chiyang He. 2021. "Application of nanomaterials decorated with cyclodextrins as sensing elements for environment analysis." *Environmental Science and Pollution Research* 28 (42): 59499–59518.

Gong, Jian, and Pei Xie. 2020. "Research progress in sources, analytical methods, eco-environmental effects, and control measures of microplastics." *Chemosphere* 254: 126790.

Golui, Debasis, S. P. Datta, B. S. Dwivedi, M. C. Meena, V. K. Trivedi, S. Jaggi, and K. K. Bandyopadhya. 2020. "Effectiveness of toxicity characteristics leaching procedure for assessing metal hazards of polluted soils in relation to human health." *Soil and Sediment Contamination: An International Journal* 29 (3): 304–321.

Hechavarría-Hernández, Amauris, José Lucas Martins Viana, Laurent Barbiero, Ary Tavares Rezende-Filho, Celia Regina Montes, Adolpho José Melfi, and Anne Helene Fos. 2023. "Spatial and seasonal variation of arsenic speciation in Pantanal soda lakes." *Chemosphere*, 329, 138672.

Honeychurch, Kevin C., and Martina Piano. 2018. "Electrochemical (Bio) sensors for environmental and food analyses." *Biosensors (Basel)* 8 (3): 57. https://doi.org/10.3390/bios8030057. https://www.ncbi.nlm.nih.gov/pubmed/29932120.

Iftikhar, Ramsha, Iqra Parveen, Aqsa Mazhar, Muhammad Saqlain Iqbal, Ghulam Mustafa Kamal, Freeha Hafeez, Pang Ai Ling, and Mohsen Ahmadipour. 2022. "Small organic molecules as fluorescent sensors for the detection of highly toxic heavy metal cations in portable water." *Journal of Environmental Chemical Engineering* 11 (1): 109030.

Ingram, Peter, and Ann LeFurgey. 2017. "Analytical microscopy methods compared: EPXMA, XPXMA, and SIMS." *Ultrastructural Pathology* 41 (1): 115–117.

Ishii, Keizo. 2019. "PIXE and its applications to elemental analysis." *Quantum Beam Science* 3 (2): 12.

Juhászová, Lucie, Sebastian Burhenn, Linda Sagapova, Joachim Franzke, Jiří Dědina, and Jan Kratzer. 2019. "Hydride generation atomic absorption spectrometry with a dielectric barrier discharge atomizer: Method optimization and evaluation of analytical performance for tin." *Spectrochimica Acta Part B: Atomic Spectroscopy* 158: 105630.

Kinuthia, Geoffrey K., Veronica Ngure, Dunstone Beti, Reuben Lugalia, Agnes Wangila, and Luna Kamau. 2020. "Levels of heavy metals in wastewater and soil samples from open drainage channels in Nairobi, Kenya: Community health implication." *Scientific Reports* 10 (1): 8434.

Krishnan, Kumar, Elias B. Saion, C. K. Yap, Mee Yoke Chong, and A. S. Nadia. 2022. "Determination of trace elements in sediments samples by using neutron activation analysis." *Journal of Experimental Biology and Agricultural Science* 10 (1): 21–31.

La, Le Van, and Vishwas Joshi. 2021. "Neutron Activation Analysis of Heavy Elements in the Environment." CUNY Academic Works, CUNY New York City College of Technology, Publication and Research, 842.

Leon-Medina, Jersson X., Diego A. Tibaduiza, Juan C. Burgos, Martha Cuenca, and Dreidy Vasquez. 2022. "Classification of As, Pb and Cd heavy metal ions using square wave voltammetry, dimensionality reduction and machine learning." *IEEE Access* 10: 7684–7694.

Li, Yi-Kun, Ting Yang, Ming-Li Chen, and Jian-Hua Wang. 2021. "Recent advances in nanomaterials for analysis of trace heavy metals." *Critical Reviews in Analytical Chemistry* 51 (4): 353–372. https://doi.org/10.1080/10408347.2020.1736505. https://www.ncbi.nlm.nih.gov/pubmed/32182101.

López-Botella, Andrea, Raquel Sánchez, Raiza Paul, Jon Aizpurua, María José Gómez-Torres, and José-Luis Todolí-Torró. 2023. "Analytical determination of heavy metals in human seminal plasma—A systematic review." *Life* 13 (4): 925.

Madariaga, Juan Manuel. 2021. "X-ray fluorescence (XRF) techniques." In Juan Manuel Madariaga (Ed.), *Analytical Strategies for Cultural Heritage Materials and Their Degradation*, 23–44. The Royal Society of Chemistry.

Mastelaro, Valmor Roberto, and Edgar Dutra Zanotto. 2018. "X-ray absorption fine structure (XAFS) studies of oxide glasses—A 45-year overview." *Materials* 11 (2): 204.

Mayotha, Parita, Piyanete Chantiratikul, and Watsaka Siriangkhawut. 2023. "Sequential ultrasound-assisted digestion procedure for determination of cadmium and lead contaminants in sea grapes and some seaweed products." *Journal of Food Measurement and Characterization* 17 (1): 607–614.

Meermann, Björn, and Volker Nischwitz. 2018. "ICP-MS for the analysis at the nanoscale–A tutorial review." *Journal of Analytical Atomic Spectrometry* 33 (9): 1432–1468.

Nagao, Yuki, Kiyotaka Yoshii, Youichi Yoshida, and Yohey Hashimoto. 2023. "Characterization of trace elements in coal fly ash by extraction, micro-PIXE, TOF-SIMS, and XAFS." *Waste Management* 157: 18–24.

Nodehi, Marzieh, Mehdi Baghayeri, and Hojat Veisi. 2021. "Preparation of GO/Fe3O4@ PMDA/AuNPs nanocomposite for simultaneous determination of As^{3+} and Cu^{2+} by stripping voltammetry." *Talanta* 230: 122288.

Occupational Safety and Health Administration (OSHA). (1994). Permissible Exposure Limits (PELs). U.S. Department of Labor, Occupational Safety and Health Administration. 29 CFR 1910.1000.

Operacz, Agnieszka, Adrianna Bigaj, Karolina Hap, Tomasz Kotowski. 2022. The Effect of Sample Preparation and Measurement Techniques on Heavy Metals Concentrations in Soil: Case Study from Kraków, Poland, Europe. *Applied Sciences* 12 (4):2137–2154.

Oreste, Eliézer Quadro, Alexander Ossanes de Souza, Camila Corrêa Pereira, Daisa Hakbart Bonemann, Mariana Antunes Vieira, and Anderson Schwingel Ribeiro. 2020. "Evaluation of sample preparation methods for the determination of Cd, Cr and Pb in ceramic tableware by graphite furnace atomic absorption spectrometry." *Analytical Letters* 53 (3): 436–458.

Ortega, Richard. 2002. "Analytical methods for heavy metals in the environment: Quantitative determination, speciation, and microscopic analysis." In B. Sarkar (Ed.), *Heavy Metals in the Environment*, 35–38. New York, NY: Marcel Dekker, Inc.

Pasinszki, Tibor, Shilvee S Prasad, and Melinda Krebsz. 2023. "Quantitative determination of heavy metal contaminants in edible soft tissue of clams, mussels, and oysters." *Environmental Monitoring and Assessment* 195 (9): 1066. https://doi.org/10.1007/s10661-023-11686-9. https://www.ncbi.nlm.nih.gov/pubmed/37598134.

Pongrac, Paula, Mitja Kelemen, Primož Vavpetič, Katarina Vogel-Mikuš, Marjana Regvar, and Primož Pelicon. 2020. "Application of micro-PIXE (particle induced X-ray emission) to study buckwheat grain structure and composition." *Fagopyrum* 37 (1): 5–10.

Popmintchev, Dimitar, Benjamin R. Galloway, Ming-Chang Chen, Franklin Dollar, Christopher A. Mancuso, Amelia Hankla, Luis Miaja-Avila, Galen O'Neil, Justin M. Shaw, and Guangyu Fan. 2018. "Near-and extended-edge X-ray-absorption fine-structure spectroscopy using ultrafast coherent high-order harmonic supercontinua." *Physical Review Letters* 120 (9): 093002.

Rittirong, Anawat, and Kiadtisak Saenboonruang. 2018. "Quantification of aluminum and heavy metal contents in cooked rice samples from Thailand markets using inductively coupled plasma mass spectrometry (ICP-MS) and potential health risk assessment." *Emirates Journal of Food and Agriculture* 30 (5): 372–380.

Seidel, Peter, Doreen Ebert, Robert Schinke, Robert Möckel, Simone Raatz, Madlen Chao, Elke Niederschlag, Thilo Kreschel, Richard Gloaguen, and Axel D Renno. 2021. "Comparison of elemental analysis techniques for the characterization of commercial alloys." *Metals* 11 (5): 736.

Shah, Syed Kifayat Hussain, Javed Iqbal, Pervaiz Ahmad, Mayeen Uddin Khandaker, Sirajul Haq, and Muhammad Naeem. 2020. "Laser induced breakdown spectroscopy methods and applications: A comprehensive review." *Radiation Physics and Chemistry* 170: 108666.

Sharma, Hari Bhakta, Kumar Raja Vanapalli, Vikram Kumar Barnwal, Brajesh Dubey, and Jayanta Bhattacharya. 2021. "Evaluation of heavy metal leaching under simulated disposal conditions and formulation of strategies for handling solar panel waste." *Science of the Total Environment* 780: 146645.

Shen, Yujie, Mingcong Rong, Xiaodan Qu, Bolin Zhao, Jinhui Zou, Zhenbang Liu, Yu Bao, Ying He, Siyi Li, Xiaojin Wang, Meiyun Chen, Kaiying Chen, Yuwei Zhang, and Li Niu. 2022. "Graphene oxide-assisted synthesis of N, S Co-doped carbon quantum dots for fluorescence detection of multiple heavy metal ions." *Talanta* 241: 123224. https://doi.org/10.1016/j.talanta.2022.123224. https://www.ncbi.nlm.nih.gov/pubmed/35066284.

Song, Xiaochong, and Xiaojia Huang. 2022. "Recent developments in microextraction techniques for detection and speciation of heavy metals." *Advances in Sample Preparation* 2: 100019.

Tarekegn, Mekonnen Maschal, Raj Mohan Balakrishnan, Andualem Mekonnen Hiruy, Ahmed Hussen Dekebo, and Hema Susmitha Maanyam. 2022. "Nano-clay and iron impregnated clay nanocomposite for Cu^{2+} and Pb^{2+} ions removal from aqueous solutions." *Air, Soil and Water Research* 15: 11786221221094037.

U.S. Environmental Protection Agency (USEPA). 1994. *Methods for the Determination of Metals in Environmental Samples—Supplement I.* EPA/600/R-94/111. Office of Research and Development, Washington, DC

Vasanthi Sridharan, Nishanthi, and Mandal, Badal Kumar. 2022. "Simultaneous quantitation of lead and cadmium on an EDTA-reduced graphene oxide-modified glassy carbon electrode." *ACS Omega* 7 (49): 45469–45480. https://doi.org/10.1021/acsomega.2c06080

Wang, Zhe, Muhammad Sher Afgan, Weilun Gu, Yuzhou Song, Yun Wang, Zongyu Hou, Weiran Song, and Zheng Li. 2021. "Recent advances in laser-induced breakdown spectroscopy quantification: From fundamental understanding to data processing." *TrAC Trends in Analytical Chemistry* 143: 116385.

Watwe Varuna, Kulkarni Preeti. 2021. Evaluation of Cr(VI) adsorption on glutaraldehyde crosslinked chitosan beads using cyclic voltammetry employing gold electrode. *Journal of Analytical Science and Technology*. 12, 1–10.

Welz, Bernhard, and Maria Goreti R. Vale. 2019. "Atomic absorption spectrometry and related techniques." In Nelu Grinberg, Sonia Rodriguez (Eds.), *Ewing's Analytical Instrumentation Handbook, Fourth Edition*, 71–116. CRC Press.

Winterer, Markus. 2024. "xafsX: A program to process, analyse and reduce X-ray absorption fine structure spectra (XAFS)." *International Tables for Crystallography* (2024). Vol. I., Ch. 6.23, pp. 843–847. https://doi.org/10.1107/S1574870720003481

Xhoffer, Chris, René Van Grieken, and R. Van Grieken. 2018. "Environmental aerosol characterization by single particle analysis techniques." In Jacques Buffle, Herman P. van Leeuwen (Eds.), *Revival: Environmental Particles (1993)*, volume 2, 207–245. CRC Press.

Xinxin, MO, Matthew G Siebecker, GOU Wenxian, LI Ling, and LI Wei. 2021. "A review of cadmium sorption mechanisms on soil mineral surfaces revealed from synchrotron-based X-ray absorption fine structure spectroscopy: Implications for soil remediation." *Pedosphere* 31 (1): 11–27.

Yu, Chunhai. 2021. "Interference from soluble iron on mercury determination in water by cold vapor atomic absorption spectrometry (CV-AAS) with sodium borohydride as reductant." *Analytical Sciences* 37 (8): 1181–1184.

Zhang, Yinchao, Ting Li, He Chen, Siying Chen, Pan Guo, Xunbao Rui, and Baowei Li. 2019. "Assessment of high-purity quartz glass by laser induced fluorescence technique." Proc. SPIE 11063, Pacific Rim Laser Damage 2019: Optical Materials for High-Power Lasers. https://doi.org/10.1117/12.2539916

Zou, Zhirong, Yujia Deng, Jing Hu, Xiaoming Jiang, and Xiandeng Hou. 2018. "Recent trends in atomic fluorescence spectrometry towards miniaturized instrumentation – A review." *Analytica chimica acta* 1019: 25–37.

13 An Overview of Physical and Chemical Methods of Heavy Metal Removal from Wastewater

*Sushant Pathak, Asheesh Kumar, and Saif Ullah Khan**
Zakir Husain College of Engineering and Technology,
Aligarh Muslim University, Aligarh, India

13.1 INTRODUCTION

Due to the rapid increase in industrial and human activities (e.g., electroplating, tanning, painting, mining, batteries, pesticides, metal rinsing process, textile, petrochemical, paper industry, and electrolysis applications), the presence of heavy metals in wastewater has increased and constantly increasing. Heavy metals are commonly described as metals with remarkably high densities, atomic weights, or atomic numbers. Wastewater containing heavy metals enters the environment, endangering human life and the ecosystem. Heavy metals are not biodegradable and can be carcinogenic; the presence of these heavy metals above permissible limits leads to critical health problems for living beings. The most popular heavy metals are arsenic, mercury, chromium, lead, copper, and nickel. Even traces of these heavy metals are dangerous to human health, so removing these heavy metal ions from wastewater is very important for a clean environment and human health. Removal of heavy metals from inorganic wastewater can be achieved by conventional treatment processes such as ion exchange, chemical precipitation, electrochemical treatment. These processes have significant disadvantages, which are insufficient removal, high toxic sludge, and high energy consumption (Eccles, 1999).

Currently, many techniques have been studied to develop more efficient and cheaper technologies. Adsorption has emerged as one of the alternative therapies and in recent years the search for cheap adsorbents with the ability to bind metals has prevailed. Membrane separation is increasingly used for the treatment of inorganic wastewater due to its simplicity of use. Reverse osmosis, ultrafiltration, and nanofiltration are a few examples of different membrane filtration techniques (Kurniawan et al., 2006). Although different approaches can be used to treat inorganic wastewater, optimal treatment should be able to meet the specified maximum pollution level. Regulations, in addition to being reasonable, relevant, must be adaptable to local conditions. In this chapter, an overview of various physical and chemical methods for removing heavy metals from industrial wastewater is discussed along with their positives and limitations.

13.2 SOURCES OF HEAVY METAL CONTAMINATION

Heavy metal contamination can occur through various natural and anthropogenic (human-related) processes. These contaminants are toxic and can have detrimental effects on human health,

* Corresponding author: saifkhan@zhcet.ac.in

DOI: 10.1201/9781032685793-13

ecosystems, and the environment. One of the main sources of heavy metals in water is industrial activity. Many industries, such as mining, metallurgy, and chemical production, release heavy metals into the environment. For example, mining operations can introduce heavy metals such as lead, mercury, and cadmium into nearby water bodies through runoff or contaminated soil erosion. Similarly, metal production processes can release heavy metals into the atmosphere, which can eventually settle into water bodies. Improper disposal of industrial waste and lack of effective anti-pollution measures can cause problems and cause pollution of water resources. The burning of coal for energy production also releases heavy metals like mercury and lead into the atmosphere, which can then settle on land and water bodies.

Agricultural activities are another significant contributor to heavy metal contamination in water. Heavy metals can enter the soil through the use of chemical fertilizers, pesticides, and herbicides in agriculture. Over time, these metals can leach into groundwater or be washed into nearby streams and rivers by stormwater runoff. Moreover, application of organic and inorganic amendments, such as sewage sludge or industrial by-products, to agricultural fields can cause heavy metals to enter the soil. This can result in long-term soil contamination, affecting water quality in the surrounding area. Irrigation of crops with water containing heavy metals can exacerbate the problem.

Rapid urbanization and related municipal activities are also significant sources of heavy metals in water. Urban areas generate various pollutants, including heavy metals, through stormwater runoff, sewage disposal, and industrial activities within city limits. Roads, roofs, and parking lots can accumulate heavy metals from vehicle emissions, tire wear, and atmospheric degradation. When it rains, these metals are carried by stormwater into storm drains, often directly discharging into nearby water bodies, which become contaminated. In addition, municipal sewage treatment plants cannot always effectively remove heavy metals from wastewater, leading to the release of treated but still contaminated water into rivers and lakes. Many consumer products contain heavy metals, and when disposed of in landfills, they can leach into groundwater and soil over time. This combination of urbanization and inadequate wastewater treatment infrastructure can lead to the persistent presence of heavy metals in urban water systems.

Heavy metals purely arise within the Earth's crust, and weathering and erosion transfer these metals into surface waters and soils of another origin. Volcanic eruptions may also release heavy metals into the atmosphere. Activities like fishing, hunting, and the use of lead-based ammunition or fishing weights can introduce heavy metals into food chains, leading to bioaccumulation in aquatic and terrestrial organisms.

Effective environmental regulations and pollution control measures are essential to mitigate heavy metal contamination and protect human health and ecosystems from its harmful effects. Some of the heavy metals with their sources and impacts and permissible limits are listed in Table 13.1.

13.3 ENVIRONMENTAL IMPLICATIONS FROM THE HEAVY METALS

Heavy metals can accumulate in soil from many sources such as industrialization, mining, and agriculture. This reduces the fertility of the soil and makes it unsuitable for plant growth. Heavy metals can mix with groundwater and surface water and cause water pollution. Drinking dirty water harms water resources and human health. When plants or aquatic organisms absorb heavy metals from the environment, heavy metals enter food sources. As animals prey on small organisms, the amount of these metals increases, harming animals and humans at higher trophic levels. Heavy metals can damage aquatic ecosystems by damaging water supplies, reducing biodiversity, and altering the food supply. This could have an impact on the entire ecosystem. Some heavy metals, such as mercury, can be released into the air through industrial processes and coal burning. Once airborne, they can travel very far and eventually live in soil and water, causing environmental pollution.

To reduce the health and environmental risks associated with heavy metals, it is important to control industrial emissions and waste disposal, monitor water and soil quality, and take steps to

TABLE 13.1
Permissible Limits of Different Heavy Metal Ions in Drinking Water Reported by WHO, Their Common Sources, and Possible Effects on Human Health

Metals	WHO (mg/L)	Common Sources	Negative Impact above the Acceptable Limit
Lead	0.05	Batteries, PVC pipes in sanitation, paints, jewelry, recycled PVC, and agriculture.	Decreased kidney function, increased blood pressure, cardiovascular effects, and reproductive problems.
Arsenic	0.05	Pesticide, oxidation of pyrite and arsenic pyrite, arsenic-based preservatives on wooden utility poles, fertilizers, and dumping untreated effluent.	Disorder of the skin, liver, blood, and kidneys.
Cadmium	0.005	Photovoltaic cells, pigments, photographic and engraving process, electroplated parts, batteries, plastics, synthetic rubber, photoconductors, and paints.	Damage to the immune system, psychological disorders, and diarrhea.
Mercury	0.001	MSW incineration, combustion of coal, and volcanic emissions.	Damage to nervous system, kidneys, and gastrointestinal tract.
Nickel	0.1	Volcanic activity, weathering of rocks, forest fires, insoluble particles, and wind-blown dust.	Birth effects, respiratory failure, heart disorders, and lung cancer.
Chromium	0.05	Tanning, chrome plating industries, and leather industry.	Liver and kidney damages, stomach ulcer, respiratory problems, and skin rashes.
Copper	1.5	Photovoltaic cells, tanning, and fertilizers.	Anemia, and liver damage.

Source: Samavati et al., 2023.

reduce impact on the business and customer environment. Additionally, treatment methods such as phytoremediation (using plants to absorb heavy metals) and proper disposal of waste are important steps in controlling heavy metal pollution.

13.4 SIGNIFICANCE OF HEAVY METAL REMOVAL FROM WASTEWATER

The importance of heavy metal removal from wastewater cannot be overstated, as it is a critical component of environmental protection and public health. Heavy metals, such as lead, mercury, cadmium, and arsenic, are toxic to both the environment and human beings. When these metals find their way into our water sources, they can have devastating consequences, making effective removal processes indispensable. First and foremost, the significance of heavy metal removal from wastewater lies in safeguarding public health. Heavy metals are well-known for their adverse health effects, even at low concentrations. Prolonged exposure to these metals can lead to a myriad of health problems, including neurological disorders, kidney damage, and various cancers. Children and pregnant women are particularly vulnerable to the toxic effects of heavy metals, as these substances can interfere with the development of the nervous system and lead to irreversible damage. By removing heavy metals from wastewater before it is discharged into the environment or treated for drinking water, we can significantly reduce the risk of exposure to these harmful substances, thus protecting the well-being of communities and individuals.

Furthermore, the ecological importance of heavy metal removal cannot be ignored. Aquatic ecosystems, in particular, are incredibly sensitive to heavy metal contamination. When heavy metals enter water bodies, they can disrupt the delicate balance of these ecosystems, leading to a cascade of negative effects. For instance, heavy metals can accumulate in aquatic organisms, leading to biomagnification up the food chain, which ultimately affects the health of fish, amphibians, and even birds, that rely on aquatic habitats. This, in turn, can lead to a decline in biodiversity, disrupting the intricate web of life in freshwater environments. By removing heavy metals from wastewater, we can help protect these vital ecosystems, preserving not only the diverse array of species that inhabit them but also the ecosystem services they provide, such as water purification, nutrient cycling, and recreational opportunities.

Moreover, heavy metal removal is crucial for maintaining the quality of our drinking water sources. As industrial, agricultural, and urban activities continue to release heavy metals into the environment, these contaminants can find their way into groundwater and surface water supplies. Even at low concentrations, heavy metals can render water unsafe for consumption. It is essential to ensure that the water we use for drinking, cooking, and bathing is free from harmful levels of heavy metals. Implementing effective removal processes, such as coagulation, precipitation, adsorption, and ion exchange, can help guarantee that our drinking water meets safety standards and is free from the health risks associated with heavy metal exposure.

In addition, heavy metal removal is integral to compliance with environmental regulations. Governments and regulatory bodies worldwide have established stringent guidelines for the allowable levels of heavy metals in wastewater and drinking water. Failing to remove heavy metals adequately can result in severe legal consequences for industries and municipalities. Compliance with these regulations not only ensures the protection of public health and the environment but also helps maintain a responsible and sustainable approach to water management.

Last, the importance of heavy metal removal from wastewater cannot be understated, as it is intricately linked to the protection of public health, the preservation of ecosystems, the provision of safe drinking water, and compliance with environmental regulations. As we continue to face the challenges of industrialization, urbanization, and agricultural intensification, the significance of implementing and improving heavy metal removal processes remains paramount. Investing in research, technology, and infrastructure to enhance heavy metal rem oval is not just a matter of environmental responsibility; it is an essential step toward ensuring a healthier and more sustainable future for our planet and all its inhabitants.

13.5 PHYSIOCHEMICAL PROPERTIES OF HEAVY METALS

Heavy metals exhibit unique physicochemical properties that distinguish them from other elements in the periodic table. These properties play an important role in environmental behavior and toxicity potential. Here is a unique perspective on the physicochemical properties of heavy metals.

Heavy metals are characterized by high atomic mass and low electronegativity. This property causes a strong attraction between heavy metal ions and other atoms or molecules, forming stable complexes and bonds in water. Many heavy metals can exist in many oxidation states, which contributes to the versatility of chemical reactions. This property allows to participate in various redox processes in natural and industrial conditions. They also have a unique affinity for ligands such as organic molecules and inorganic anions; this property results in the formation of complex ions that can significantly affect mobility, solubility, and toxicity in the environment.

Heavy metal compounds generally show low solubility in water, especially at neutral alkaline pH levels. This behavior can lead to the precipitation of heavy metal ions as heavy phases or complexes, which can affect their deposition and bioavailability in aquatic ecosystems. Heavy metals are notorious for their ability to bioaccumulate in living organisms. Their unique physicochemical properties, including their high atomic mass and ability to form complexes, allow them to bind to biomolecules and accumulate in tissues over time, potentially reaching toxic levels. Some heavy metals, such as copper and zinc, are important trace elements in biological systems because of their catalytic role in enzymes. Although it is needed in normal amounts, it can be toxic in excess.

Heavy metal ions often give different colors to compounds and minerals. For example, copper can produce a green color, while iron can cause a red or yellow color. This color is often found in minerals and pigments used in various supplements. Heavy metals such as zinc, cadmium, and lead are known for their corrosion resistance and are valuable in various industrial products such as coatings and alloys. Some heavy metals, such as uranium and thorium, are naturally radioactive because their atomic nuclei are unstable. Their radioactivity is important for nuclear power and radiological safety.

Heavy metals are notorious for their toxicity to living organisms, especially for their ability to interfere with important biological processes and disrupt cell functions. These properties lead to strict regulations and controls on the concentration of heavy metals in the environment. In summary, heavy metals show a fascinating set of physicochemical properties that make them interesting and complex. Its unique properties contribute to its widespread use in various industrial applications and also highlight the importance of managing its presence in the environment to prevent adverse effects on ecosystems and human health.

13.6 PHYSICAL METHODS FOR HEAVY METAL REMOVAL

13.6.1 Membrane Filtration

Membrane filtration is a very effective method for removing heavy metals from wastewater. The process relies on specialized membranes with precise pore sizes that allow water molecules to pass through while blocking the passage of heavy metal ions and other contaminants. Membrane filtration for heavy metal removal exists in various types on the basis of the size of particle that can be retained such as microfiltration, nanofiltration, ultrafiltration, and reverse osmosis.

The process begins with the selection of an appropriate membrane based on the type and size of the heavy metals present in the wastewater. Different membranes have different pore sizes, which determine the size of particles or ions that can pass through. Before the wastewater enters the membrane filtration system, it often undergoes pretreatment to remove larger particles, solids, and any substances that could foul or damage the membrane. Pretreatment methods may include coagulation, flocculation, and sedimentation to aggregate and settle out these larger contaminants. The pretreated wastewater is then fed into the membrane filtration system. The wastewater flows over the surface of the membrane, and a pressure differential is applied across the membrane. This pressure forces the water through the membrane pores while retaining the heavy metal ions and other contaminants on the upstream side of the membrane.

The key to heavy metal removal lies in the selective permeability of the membrane. Membranes are engineered with pore sizes that are smaller than the size of heavy metal ions. As a result, heavy metal ions are physically blocked by the membrane and cannot pass through, effectively separating them from the purified water; while the purified water passes through the membrane and becomes part of the treated effluent, the heavy metal ions and other contaminants are concentrated on the membrane's surface. This concentrated solution, often referred to as the "concentrate" or "retentate," contains the captured heavy metals. The concentrate, which now contains a higher concentration of heavy metals, needs to be managed appropriately. To maintain the efficiency and longevity of the membrane filtration system, regular maintenance is essential. This includes cleaning the membranes to remove any accumulated fouling or scaling that may occur over time. The result of this process is purified water that is substantially free from heavy metal contaminants.

Membrane filtration offers several advantages for heavy metal removal; membranes are highly effective at removing heavy metal ions due to their selective permeability. Membrane filtration

TABLE 13.2
Heavy Metal Removal Using RO Method

Membrane	Heavy Metal	Initial Metal Conc.	Removal Efficiency	References
RO	Cu, Ni		99.5%	Mohsen-Nia et al. (2007)
RO	Ni, Zn	44–169 mg/L	99.2%, 98.8%	Ipek (2005)
RO	Cu	–	95%	Ozaki et al. (2002)
	Ni			
	Cr			
RO	Pb	–	100%	Dialynas and Diamadopoulos (2009)
	Ni		100%	
	Cr		89%	
	Cu		49%	

typically requires fewer chemicals compared to traditional chemical treatment methods. The process is energy-efficient, making it cost-effective and environmentally friendly. Membrane systems can be customized for various wastewater sources and heavy metal removal requirements. Overall, membrane filtration is a sustainable and efficient solution for the removal of heavy metals from wastewater, contributing to improved water quality and environmental protection (Table 13.2).

13.6.2 Coagulation and Flocculation

Heavy metals can be removed from solutions by flocculation and coagulation. Coagulation, also known as particle charge neutralization, is a process by which the net surface charges of colloidal particles are reduced to stabilize them through a process of electrostatic repulsion. Coagulation requires coagulants or organic materials that have been dissolved into larger forms. Coagulants used in this process are aluminum sulfate, polyaluminum chloride, magnesium chloride, aluminum hydroxide, polyethyleneimine, and others. Through successive collisions and interactions with dissolved organic polymers and inorganic polymers, flocculation gradually mixes the destabilized particles to increase the particle size. Filtration makes it easier to separate small particles from larger ones after they have clumped together.

13.6.3 Ion Exchange

Heavy metals can be effectively removed from wastewater through ion exchange, a process where undesirable metal ions in the water are exchanged for ions on a solid ion exchange resin. For instance, in the case of lead removal, a cation exchange resin with negatively charged sites can be employed. When the contaminated water passes through the resin bed, positively charged lead ions in the water are attracted to and adsorbed onto the negatively charged sites on the resin, effectively replacing other cations (e.g., calcium or sodium) that were initially attached to the resin. This exchange process results in the immobilization of lead ions within the resin, thus purifying the water by reducing its heavy metal content to acceptable levels. The ion exchange resin can be regenerated for reuse by washing it with a solution that has a higher concentration of the cations initially attached to the resin, releasing the captured heavy metal ions, and allowing the resin to continue its role in heavy metal removal. The aqueous phase's pH can affect this process. By using the ion exchange process, several kinds of resins are employed to remove heavy metals. Purolite C100 resin has been successfully employed to remove lead ions from the binary combination (Badawy et al., 2009). Thakare and Jana (2015) used the high-density ion exchange resin INDION 225H to get rid of copper from wastewater. Shaidan et al. (2012) used a strongly acidic cation exchange resin in a fixed bed to recover 97% of nickel from wastewater (Table 13.3).

TABLE 13.3
Heavy Metal Removal Using Ion Exchange Method

Heavy Metals	Initial Metal Concentration	Resin	Optimum pH	Removal Efficiency	References
Cu, Zn, Cd	–	Amberlite IR	–	–	Lee et al. (2007)
Pb	–	Purolite C100	–	99.17%	Badawy et al. (2009)
Cu	50–150 mg/L	INDION 225H	6.3–6.5		Thakare and Jana (2015)
Ni, Pb	800, 1000, 1250 mg/L	Amberjet 1200 Na	–	98% 99%	Zewail and Yousef (2015)
Ni	1800, 2800, 3800 mg/L	Acidic cation exchange resin	3–7	97%	Shaidan et al. (2012)
Cd, Pb	0.1–6 mmol/L	*Sargassum muticum* loaded with calcium	5	–	Carro et al. (2015)

13.6.4 Adsorption

Heavy metals can be effectively removed from wastewater through adsorption, a process in which heavy metal ions are attracted and bound to the surface of a solid adsorbent material. For example, activated carbon, zeolites, or certain metal oxides are commonly used adsorbents. When wastewater containing heavy metals encounters the adsorbent material, the heavy metal ions adhere to the surface of the adsorbent due to electrostatic forces, chemical bonding, or ion exchange. This results in the immobilization of heavy metals, reducing their concentration in the treated water to safe levels. The adsorption process is widely applied in water treatment and environmental remediation to mitigate heavy metal contamination, ensuring cleaner and safer water for various industrial and municipal applications (Tables 13.4 and 13.5).

13.7 CHEMICAL METHODS FOR HEAVY METAL REMOVAL

13.7.1 Neutralization

Heavy metals can be effectively removed from wastewater through a process known as neutralization, where the acidic or alkaline nature of the water is adjusted to precipitate the heavy metal ions, transforming them into insoluble forms that can be easily separated and removed. For example, let's

TABLE 13.4
Adsorption Capacities of Modified Natural Materials for Heavy Metal

Adsorbent	Adsorption Capacity (mg/g)						References
	Lead	Cadmium	Zinc	Copper	Chromium	Nickel	
Zeolite, clinoptilolite	1.6	2.4	0.5	1.64	0	0.4	Babel (2003)
Modified zeolite, MMZ	123					8	Nah et al. (2006)
HCl-treated clay			63.2	83.3			Vengris et al. (2001)
Calcined phosphate	155						Mouflih et al. (2005)
Activated phosphate	4						Pan et al. (2007)
Claypoly(methoxyethyl) acrylamide	81.02		20.6	29.8		80.9	Şölener et al. (2008)

TABLE 13.5
Adsorption Capacities of Some Agricultural and Biological Wastes for Heavy Metals

	Adsorption Capacity (mg/g)						
Adsorbent	**Lead**	**Cadmium**	**Zinc**	**Copper**	**Chromium**	**Nickel**	**References**
Maize cope and husk	456	493.7	495.9				Igwe and Abia (2005)
Orange peel						158	Ajmal et al. (2000)
Coconut shell charcoal					3.65		Babel and Kurniawan (2004)
Pecan shells activated carbon			13.9	31.7			Bansode et al. (2003)
Rice husk		2.0			0.79		Bishnoi et al. (2004)
Modified rice hull					23.4		Tang et al. (2003)
Bacillus – bacterial biomass	467	85.3	418	381	39.9		Ahluwalia and Goyal (2007)

consider the case of removing lead (Pb) from industrial wastewater. Lead is a toxic heavy metal commonly found in industrial effluents. To neutralize and remove lead, an alkali such as sodium carbonate (Na_2CO_3) can be added to the acidic wastewater. The hydroxide ions (OH–) from the sodium hydroxide react with the lead ions (Pb^{2+}) present in the water to form insoluble lead carbonate ($PbCO_3$) precipitates. These precipitates settle out of the water, allowing for their easy removal through sedimentation or filtration. The resulting treated water is significantly less contaminated with lead, meeting environmental discharge standards and ensuring the protection of ecosystems and human health. Neutralization is a cost-effective method for heavy metal removal, but it requires careful monitoring of pH levels and appropriate chemical dosing to achieve optimal results. Here are some chemical agents and the compound formed defined for removal of particular heavy metal (Table 13.6).

13.7.2 Solvent Extraction

Solvent extraction is a powerful technique employed for the selective removal of heavy metals from various solutions, offering an efficient means of environmental remediation and metal recovery. This process involves the transfer of metal ions from an aqueous phase into an organic solvent phase, facilitated by specific chemical extractants. One illustrative example is the removal of copper

TABLE 13.6
Heavy Metal Removal Using Neutralization

Heavy Metal	Heavy Metal Chemical Agent for Neutralization	Insoluble Compound Formed
Lead (Pb)	Sodium carbonate (Na_2CO_3)	Lead carbonate ($PbCO_3$)
Cadmium (Cd)	Sodium sulfide (Na_2S)	Cadmium sulfide (CdS)
Mercury (Hg)	Sodium sulfide (Na_2S)	Mercuric sulfide (HgS)
Copper (Cu)	Sodium hydroxide (NaOH)	Copper hydroxide ($Cu(OH)_2$)
Chromium (Cr)	Sodium hydroxide (NaOH)	Chrom. hydroxide ($Cr(OH)_3$)
Arsenic (As)	Calcium hydroxide ($Ca(OH)_2$)	Calcium arsenate ($Ca_3(AsO_4)_2$)
Nickel (Ni)	Sodium hydroxide (NaOH)	Nickel hydroxide ($Ni(OH)_2$)

ions from industrial wastewater using an organic phase containing a chelating agent like LIX 984N. When the wastewater, containing copper ions, comes into contact with the organic phase, the chelating agent selectively binds to the copper ions, forming a stable complex. This complex is then separated from the aqueous phase by settling or centrifugation. Subsequently, the copper-loaded organic phase can undergo a stripping process to recover the copper ions as a concentrated solution.

13.7.3 Chemical Precipitation

Chemical precipitation is a widely used method for removing heavy metals from wastewater through the formation of insoluble metal compounds, allowing for their subsequent separation from the treated water. In this process, specific chemical reagents are added to the contaminated wastewater, inducing a chemical reaction that results in the precipitation of heavy metal ions as solid particles. An example of this process is the removal of lead (Pb^{2+}) from industrial wastewater containing elevated levels of this toxic heavy metal. To achieve this, a commonly employed reagent is sodium sulfide (Na_2S), which reacts with lead ions to form insoluble lead sulfide (PbS). The balanced chemical equation for this reaction is as follows:

$$Pb_2 + + Na_2S \rightarrow PbS + 2Na +$$

As lead sulfide precipitates out of the solution, it forms solid particles that can be separated using various methods like settling, sedimentation, or filtration. Once removed, the treated water is significantly less contaminated with lead, rendering it safer for discharge into the environment or for further processing. Chemical precipitation is effective for the removal of many heavy metals, but it does have some drawbacks, including the generation of sludge, the need for careful pH control, and the potential for re-solubilization of precipitates under certain conditions, which must be carefully managed in industrial wastewater treatment processes (Table 13.7).

13.7.4 Electrochemical Treatment

Electrochemical treatment is a powerful method for removing heavy metals from wastewater, harnessing the principles of electrochemistry to transform and separate these toxic contaminants. In this process, electrodes are immersed in the contaminated water, and an electric current is applied, inducing a series of redox reactions that lead to the precipitation or deposition of heavy metal ions, effectively separating them from the water. For example, let's consider the removal of lead (Pb) from industrial wastewater. In an electrochemical cell, lead ions in the wastewater migrate toward the cathode (negative electrode), where they gain electrons and are reduced to solid lead metal,

TABLE 13.7
Heavy Metal Removal Using Chemical Precipitation

Heavy Metal	Chemical Used for Precipitation	Insoluble Compound Formed
Lead (Pb)	Sodium sulfide (Na_2S)	Lead sulfide (PbS)
Cadmium (Cd)	Sodium sulfide (Na_2S)	Cadmium sulfide (CdS)
Mercury (Hg)	Sodium sulfide (Na_2S)	Mercuric sulfide (HgS)
Copper (Cu)	Sodium hydroxide (NaOH)	Copper hydroxide ($Cu(OH)_2$)
Chromium (Cr)	Sodium hydroxide (NaOH)	Chromium hydroxide ($Cr(OH)_3$
Arsenic (As)	Ferric chloride ($FeCl_3$)	Ferric arsenate ($FeAsO_4$)
Nickel (Ni)	Sodium hydroxide (NaOH)	Nickel hydroxide ($Ni(OH)_2$)

TABLE 13.8
Heavy Metal Removal Using Electrochemical Treatment

Heavy Metal	Electrode or Material Used	Electrochemical Process/Outcome
Lead (Pb)	Lead dioxide (PbO_2) or platinum (Pt) electrode	Lead ions (Pb^{2+}) are reduced to solid lead (Pb) at the cathode.
Cadmium (Cd)	Cadmium alloy or titanium (Ti) electrode	Cadmium ions (Cd^{2+}) are reduced to solid cadmium (Cd) at the cathode.
Copper (Cu)	Copper or platinum (Pt) electrode	Copper ions (Cu^{2+}) are reduced to solid copper (Cu) at the cathode.
Chromium (Cr)	Graphite or platinum (Pt) electrode	Hexavalent chromium ions (Cr^{6+}) are reduced to trivalent chromium ions (Cr^{3+}) or deposited as chromium hydroxide at the cathode.
Arsenic (As)	Iron or aluminum electrode	Arsenic ions (As^{3+} and As^{5+}) are reduced and adsorbed onto iron or aluminum hydroxide precipitates formed at the cathode.
Nickel (Ni)	Nickel or platinum (Pt) electrode	Nickel ions (Ni^{2+}) are reduced to solid nickel (Ni) at the cathode.

effectively removing them from the solution. Simultaneously, water is oxidized at the anode (positive electrode), generating oxygen gas. The precipitated lead particles can then be easily separated, leaving behind purified water. This electrochemical method offers several advantages, including high efficiency, minimal chemical usage, adaptability to various heavy metals, and the potential for metal recovery, making it a promising solution for addressing heavy metal pollution in industrial and environmental settings (Table 13.8).

13.8 RECENT ADVANCEMENTS IN HEAVY METAL REMOVAL

As of today, there are several recent advances and emerging technologies in the field of heavy metal removal from wastewater. However, please note that developments in scientific and technological fields continue to evolve rapidly, so it's essential to check for the most current information. The use of nanomaterials, such as graphene oxide, carbon nanotubes, and various nanocomposites, has shown promise in efficiently adsorbing heavy metals from wastewater due to their high surface area and reactivity. Researchers are exploring the use of bio-based materials like agricultural waste, fruit peels, and biomass-derived materials as low-cost and sustainable adsorbents for heavy metal removal.

Electrochemical techniques, including electrocoagulation, electroflotation, and electrochemical reduction, have gained attention for their effectiveness in heavy metal removal and recovery. Advancements in adsorption processes involve the development of modified adsorbents, such as functionalized nanomaterials and magnetic adsorbents, which can be easily separated and reused. Improved ion exchange resins with higher selectivity and capacity for heavy metals are being developed for efficient removal. Bioremediation methods, including the use of microorganisms and plants, are being studied to treat heavy metal-contaminated sites and wastewater. Advances in membrane filtration technologies, including nanofiltration and reverse osmosis, are improving the removal of heavy metals from water sources. Researchers are exploring the integration of multiple treatment methods, such as combining adsorption with membrane filtration or electrochemical processes, to enhance heavy metal removal efficiency.

It's important to stay updated with the latest research and technological advancements in heavy metal removal as the field continues to evolve to meet the challenges of environmental pollution and water quality management. Researchers, industries, and governments are working together to develop innovative and sustainable solutions for this critical environmental issue.

13.9 FACTOR AFFECTING HEAVY METAL REMOVAL EFFICIENCY

The pH plays a crucial role in the removal of heavy metals from wastewater through various treatment processes. By adjusting the pH, we can control the solubility and precipitation of heavy metal ions. In some cases, lowering the pH facilitates the formation of metal hydroxides or sulfides, which then settle out of the water, allowing for easy removal. Conversely, raising the pH can help complex ions of heavy metals, turning them into less toxic forms that can be more efficiently separated from water. Thus, manipulation of pH is a key strategy in the successful removal of heavy metals from wastewater, contributing to clean and safe water resources. The initial concentration of heavy metals in the wastewater is a critical factor. Higher concentrations may require more extensive treatment processes and longer contact times for efficient removal. The choice of adsorbent or precipitant material plays a vital role. Different materials have varying affinities for specific heavy metals. Selecting the right adsorbent or precipitant is essential for effective removal. The duration of contact between the wastewater and the treatment material (e.g., adsorbent, precipitant) affects removal efficiency. Longer contact times often lead to better removal. Temperature can influence the kinetics of chemical reactions involved in heavy metal removal. In some cases, higher temperatures can enhance removal rates, while in others, lower temperatures may be favorable.

The presence of other ions in the wastewater can compete with heavy metal ions for binding sites on adsorbents or precipitants. High ionic strength can reduce removal efficiency. Some heavy metals exist in multiple oxidation states (e.g., chromium, iron). The redox conditions of the wastewater can affect which oxidation state of the metal is present and, consequently, its removal behavior for solid adsorbents; the particle size and surface area of the material can impact removal efficiency. Smaller particles and higher surface areas often result in better adsorption. Adequate mixing or agitation of the wastewater with the treatment material is essential to ensure even contact and maximize removal efficiency.

For some treatment methods, the ability to regenerate or reuse the adsorbent or precipitant is a crucial factor in terms of cost-effectiveness and sustainability. The choice of treatment technology, whether it's chemical precipitation, adsorption, membrane filtration, or electrochemical methods, can significantly impact removal efficiency. Each method has its own advantages and limitations. Meeting regulatory requirements for heavy metal discharge limits is essential. Regulations can vary by region and industry and may dictate the level of removal needed.

13.10 COMPARATIVE ANALYSIS OF REMOVAL METHODS

Comparative analysis of different methods for removing heavy metals from wastewater can help in selecting the most suitable approach based on specific requirements and constraints. Here's a comparative analysis of some common heavy metal removal methods (Table 13.9).

13.11 REGULATORY FRAMEWORK AND GUIDELINES

The regulatory framework and guidelines for heavy metals in wastewater and environmental protection vary from one country to another. However, there are common elements and international standards that many nations follow to address heavy metal pollution and protect public health and the environment. In the United States, for example, the Environmental Protection Agency (EPA) sets guidelines and regulations for heavy metal discharge into water bodies under the Clean Water Act (CWA). The EPA establishes maximum allowable concentrations for various heavy metals in wastewater and surface waters. Companies and municipalities must adhere to these limits and

TABLE 13.9
Comparative Analysis of Some Common Heavy Metal Removal Methods

Removal Method	Advantages	Disadvantages	Applicability
Chemical precipitation	Effective for a wide range of metals.	Generates large volumes of sludge.	Municipal and industrial wastewater with moderate metal concentrations.
Adsorption	High removal efficiency, tunable for specific metals.	Adsorbent saturation and regeneration required.	Various applications, especially for metal removal from dilute solutions.
Ion exchange	High selectivity and efficiency.	Limited capacity for some metals.	Low to moderate metal concentrations in industrial effluents.
Membrane filtration	Effective for particulate and dissolved metals.	Membrane fouling may occur.	Treatment of low to moderate metal concentrations in various industries.
Electrochemical methods	Efficient and selective removal, potential for recovery.	Electricity and maintenance required.	Suitable for various metal ions, especially in electroplating and mining wastewater.

regularly monitor their effluents to ensure compliance. Similarly, the European Union (EU) has directives such as the Water Framework Directive (WFD) and the Industrial Emissions Directive (IED) that set standards for heavy metal emissions and discharge limits. EU member states are responsible for implementing these directives into their national laws.

Internationally, the World Health Organization (WHO) provides guidelines for drinking water quality, which include permissible levels of heavy metals in drinking water to protect public health. The WHO collaborates with countries to establish and update these guidelines. The United Nations' Stockholm Convention and the Minamata Convention address specific heavy metals, such as mercury, and aim to reduce their production, use, and release into the environment on a global scale. In addition to governmental regulations, there are often industry-specific standards and guidelines. For example, the mining and metallurgical industries have specific guidelines.

13.12 FUTURE SCOPE

The future goals associated with heavy metal removal are aligned with the broader objectives of environmental protection, public health, sustainable resource management, and addressing complex wastewater scenarios where heavy metals coexist with other contaminants, such as organic pollutants and nutrients. Integrated treatment approaches that can remove multiple contaminants simultaneously will be essential. As we look ahead, a lot of research and development needs to be done with efforts aiming at improved efficiency and selectivity of heavy metal removal. This includes the development of novel adsorbents, membrane materials, and treatment processes that can effectively target specific heavy metal ions, even at low concentrations. Moreover, there is a growing emphasis on sustainable practices in heavy metal removal. Future goals include minimizing waste generation, maximizing the recovery and recycling of heavy metals, and reducing the environmental footprint of treatment processes. This aligns with the principles of the circular economy. Furthermore, advances in sensor technology and data analytics are expected to enable real-time monitoring and control of heavy metal removal processes. This will enhance the ability to respond promptly to fluctuations in metal concentrations and optimize treatment efficiency.

13.13 CONCLUSION

Heavy metals can be removed by physical and chemical methods; the choice of method depends on the technical application, economic factors, and the simplicity of the approach. All methods are important in some situations, such as severe water removal, but some methods are more important than others. At low and high metal concentrations, the approach can be very useful. Membrane filtration is also important because it uses fewer chemicals, produces very solid waste, and is very effective at removing heavy metals even at low flow rates. Advantages of the process of thickening and flocculation include settling and dewatering sludge, but this approach requires more chemicals and money. With the ion exchange method, the pH of the wastewater does not change, and the process is chemically safe, stable, and reliable. However, the cost of the membrane is very expensive, and the rubber needs to be regenerated to regenerate the rubber. Adsorption is an effective method to remove heavy metals from wastewater; both artificial and natural adsorbents remove heavy metals based on the phenomenon of deposition on the surface of the adsorbent. Removal of heavy metals from aqueous solutions can be done economically and efficiently by using waste products as adsorbents. The type of adsorbent used to remove heavy metals affects how effective adsorption is. The choice of technique is mainly influenced by the amount of heavy metals present, environmental impact, cost, and ease of use.

ACKNOWLEDGMENT

Authors are thankful to the Aligarh Muslim University, Aligarh, for providing necessary support for this work.

REFERENCES

Ahluwalia, S. S., & Goyal, D. (2007). Microbial and plant derived biomass for removal of heavy metals from wastewater. *Bioresource Technology*, *98*(12), 2243–2257. https://doi.org/10.1016/j.biortech.2005.12.006

Ajmal, M., Rao, R. A. K., Ahmad, R., & Ahmad, J. (2000). Adsorption studies on Citrus reticulata (fruit peel of orange): Removal and recovery of Ni(II) from electroplating wastewater. *Journal of Hazardous Materials*, *79*(1–2), 117–131. https://doi.org/10.1016/S0304-3894(00)00234-X

Babel, S. (2003). Low-cost adsorbents for heavy metals uptake from contaminated water: A review. *Journal of Hazardous Materials*, *97*(1–3), 219–243. https://doi.org/10.1016/S0304-3894(02)00263-7

Babel, S., & Kurniawan, T. A. (2004). Cr(VI) removal from synthetic wastewater using coconut shell charcoal and commercial activated carbon modified with oxidizing agents and/or chitosan. *Chemosphere*, *54*(7), 951–967. https://doi.org/10.1016/j.chemosphere.2003.10.001

Badawy, N. A., El-Bayaa, A. A., Abdel-Aal, A. Y., & Garamon, S. E. (2009). Chromatographic separations and recovery of lead ions from a synthetic binary mixtures of some heavy metal using cation exchange resin. *Journal of Hazardous Materials*, *166*(2–3), 1266–1271. https://doi.org/10.1016/j.jhazmat.2008.12.044

Bansode, R. R., Losso, J. N., Marshall, W. E., Rao, R. M., & Portier, R. J. (2003). Adsorption of metal ions by pecan shell-based granular activated carbons. *Bioresource Technology*, *89*(2), 115–119. https://doi.org/10.1016/S0960-8524(03)00064-6

Bishnoi, N. R., Bajaj, M., Sharma, N., & Gupta, A. (2004). Adsorption of Cr(VI) on activated rice husk carbon and activated alumina. *Bioresource Technology*, *91*(3), 305–307. https://doi.org/10.1016/S0960-8524(03)00204-9

Carro, L., Barriada, J. L., Herrero, R., & Sastre de Vicente, M. E. (2015). Interaction of heavy metals with Ca-pretreated *Sargassum muticum* algal biomass: Characterization as a cation exchange process. *Chemical Engineering Journal*, *264*, 181–187. https://doi.org/10.1016/j.cej.2014.11.079

Dialynas, E., & Diamadopoulos, E. (2009). Integration of a membrane bioreactor coupled with reverse osmosis for advanced treatment of municipal wastewater. *Desalination*, *238*(1–3), 302–311. https://doi.org/10.1016/j.desal.2008.01.046

Eccles, H. (1999). Treatment of metal-contaminated wastes: Why select a biological process? *Trends in Biotechnology*, *17*(12), 462–465. https://doi.org/10.1016/S0167-7799(99)01381-5

Igwe, J., & Abia, A. (2005). Competitive adsorption of Zn (II), Cd (II) AND Pb (II) ions from aqueous and non-aqueous solution by maize cob and husk. *African Journal of Biotechnology*, *4*, 1113–1116.

Ipek, U. (2005). Removal of Ni(II) and Zn(II) from an aqueous solution by reverse osmosis. *Desalination, 174*(2), 161–169. https://doi.org/10.1016/j.desal.2004.09.009

Kurniawan, T. A., Chan, G. Y. S., Lo, W.-H., & Babel, S. (2006). Physico–chemical treatment techniques for wastewater laden with heavy metals. *Chemical Engineering Journal, 118*(1–2), 83–98. https://doi.org/10.1016/j.cej.2006.01.015

Lee, I.-H., Kuan, Y.-C., & Chern, J.-M. (2007). Equilibrium and kinetics of heavy metal ion exchange. *Journal of the Chinese Institute of Chemical Engineers, 38*(1), 71–84. https://doi.org/10.1016/j.jcice.2006.11.001

Mohsen-Nia, M., Montazeri, P., & Modarress, H. (2007). Removal of Cu2+ and Ni2+ from wastewater with a chelating agent and reverse osmosis processes. *Desalination, 217*(1–3), 276–281. https://doi.org/10.1016/j.desal.2006.01.043

Mouflih, M., Aklil, A., & Sebti, S. (2005). Removal of lead from aqueous solutions by activated phosphate. *Journal of Hazardous Materials, 119*(1–3), 183–188. https://doi.org/10.1016/j.jhazmat.2004.12.005

Nah, I. W., Hwang, K.-Y., Jeon, C., & Choi, H. B. (2006). Removal of Pb ion from water by magnetically modified zeolite. *Minerals Engineering, 19*(14), 1452–1455. https://doi.org/10.1016/j.mineng.2005.12.006

Ozaki, H., Sharma, K., & Saktaywin, W. (2002). Performance of an ultra-low-pressure reverse osmosis membrane (ULPROM) for separating heavy metal: Effects of interference parameters. *Desalination, 144*(1–3), 287–294. https://doi.org/10.1016/S0011-9164(02)00329-6

Pan, B. C., Zhang, Q. R., Zhang, W. M., Pan, B. J., Du, W., Lv, L., Zhang, Q. J., Xu, Z. W., & Zhang, Q. X. (2007). Highly effective removal of heavy metals by polymer-based zirconium phosphate: A case study of lead ion. *Journal of Colloid and Interface Science, 310*(1), 99–105. https://doi.org/10.1016/j.jcis.2007.01.064

Samavati, Z., Samavati, A., Goh, P. S., Fauzi Ismail, A., & Sohaimi Abdullah, M. (2023). A comprehensive review of recent advances in nanofiltration membranes for heavy metal removal from wastewater. *Chemical Engineering Research and Design, 189*, 530–571. https://doi.org/10.1016/j.cherd.2022.11.042

Shaidan, N. H., Eldemerdash, U., & Awad, S. (2012). Removal of Ni(II) ions from aqueous solutions using fixed-bed ion exchange column technique. *Journal of the Taiwan Institute of Chemical Engineers, 43*(1), 40–45. https://doi.org/10.1016/j.jtice.2011.06.006

Şölener, M., Tunali, S., Özcan, A. S., Özcan, A., & Gedikbey, T. (2008). Adsorption characteristics of lead(II) ions onto the clay/poly(methoxyethyl)acrylamide (PMEA) composite from aqueous solutions. *Desalination, 223*(1–3), 308–322. https://doi.org/10.1016/j.desal.2007.01.221

Tang, P. L., Lee, C. K., Low, K. S., & Zainal, Z. (2003). Sorption of Cr(VI) and Cu(II) in aqueous solution by ethylenediamine modified rice hull. *Environmental Technology, 24*(10), 1243–1251. https://doi.org/10.1080/09593330309385666

Thakare, Y. N., & Jana, A. K. (2015). Performance of high density ion exchange resin (INDION225H) for removal of Cu(II) from waste water. *Journal of Environmental Chemical Engineering, 3*(2), 1393–1398. https://doi.org/10.1016/j.jece.2015.01.002

Vengris, T., Binkien, R., & Sveikauskait, A. (2001). Nickel, copper and zinc removal from waste water by a modified clay sorbent. *Applied Clay Science, 18*(3–4), 183–190. https://doi.org/10.1016/S0169-1317(00)00036-3

Zewail, T. M., & Yousef, N. S. (2015). Kinetic study of heavy metal ions removal by ion exchange in batch conical air spouted bed. *Alexandria Engineering Journal, 54*(1), 83–90. https://doi.org/10.1016/j.aej.2014.11.008

14 Green Guardians
Harnessing Nature's Wisdom for Heavy Metal Detoxification through Biological and Eco-Friendly Solutions

*Ashish Kumar Mishra, Shivani Gupta, Parvati Madheshiya, Gereraj Sen Gupta, and Supriya Tiwari**
Laboratory of Ecotoxicology, Institute of Science, Banaras Hindu University, Varanasi, India

14.1 INTRODUCTION

The indiscriminate exploitation of global mineral resources has engendered copious amounts of mining waste, imposing direct harm upon the environment and jeopardizing human well-being (Ajani et al., 2022; Asim and Nageswara Rao, 2021; Wang et al., 2023). Heavy metal contamination, a prevalent menace in modern times, stems chiefly from industrial processes, mining activities, agricultural practices, and rapid urbanization. Heavy metals such as lead, cadmium, mercury, arsenic, and chromium, emanating from these sources, have permeated diverse environmental realms, including soil, water bodies, sediments, and even the atmosphere. Their persistent presence and inherent toxicity pose grave threats not only to ecosystem health but also to human well-being, carrying potential long-term consequences ranging from neurological disorders to cancer (Ali et al., 2021; Bhat et al., 2022a; Briffa et al., 2020). As society marches forward with industrialization and urban expansion, coupled with advancements in living standards (Li et al., 2022; Mukherjee et al., 2022; Samarajeewa, 2022), the prevalence of soil heavy metal pollution escalates. This insidious form of contamination, akin to water pollution and land degradation, infiltrates the food chain, triggering a domino effect of environmental degradation and human health ailments (Ali et al., 2021; Bhat et al., 2022b; Mohiuddin et al., 2022).

Cadmium (Cd) stands out among the toxic heavy metals causing soil contamination due to its high mobility, toxicity, and ability to accumulate significantly. Exposure to Cd can lead to systemic toxicity and potentially contribute to the development of cancerous conditions (Cheng et al., 2023; Filippini et al., 2020; Tabelin et al., 2018). Lead (Pb) induces various health complications, including cardiovascular, central nervous system, renal, and reproductive issues (Bouida et al., 2022; Kumar et al., 2020; Olufemi et al., 2022). Copper (Cu), while essential in appropriate amounts, can be harmful when imbalanced, potentially leading to disorders like Menkes disease (Ingle et al., 2018; Liang et al., 2023). Chromium (Cr), existing primarily as Cr^{3+} or Cr^{6+}, poses risks to both plants and humans, affecting enzyme activity, causing stunted growth in plants, and leading to poisoning and cancer in humans (Fendorf, 1995). Nickel (Ni) overconsumption can result in allergic reactions and damage various organs, including the nervous system (Das and Büchner, 2007;

* Corresponding author: supriyabhu@gmail.com

DOI: 10.1201/9781032685793-14

Das et al., 2010; Šulinskienė et al., 2019). Consequently, effective environmental management and innovative, sustainable approaches to heavy metal removal are crucial to mitigate these risks.

Traditional methods for remediating soil contaminated by heavy metals, such as soil washing, encapsulation, precipitation, ion exchange, adsorption, membrane filtration, solidification, excavation, and landfilling have been employed to address soil pollution to a certain extent (Hoang et al., 2022; Li et al., 2019; Luptakova et al., 2012; Raffa et al., 2021). However, these approaches face notable limitations that impede their widespread use. Many conventional techniques are invasive and disturb ecosystems, resulting in habitat destruction (Priya et al., 2023). Notably, excavation carries the risk of secondary contamination, potentially spreading pollutants to nearby areas or releasing them into the atmosphere (Blenis et al., 2023). Furthermore, traditional methods often struggle to completely extract tightly bound metals from the soil matrix, sometimes necessitating the removal of large volumes of soil, which can reduce available land for agricultural or other purposes (Lee et al., 2021). Despite their partial effectiveness, these techniques are associated with high costs, the generation of secondary pollutants, and limited suitability for large-scale remediation projects. Moreover, they may not align with the increasing focus on sustainability and environmental stewardship in modern society.

In response to these challenges, there has been a growing interest in exploring alternative remediation strategies that harness the power of nature in a sustainable and eco-friendly manner. Biological and eco-friendly methods have emerged as promising avenues for addressing heavy metal contamination, leveraging the inherent capabilities of living organisms and natural processes to sequester, transform, or degrade heavy metal pollutants (Aryal 2024; Hu et al., 2024). Biological approaches to heavy metal removal incorporate the application of living organisms to plant or modify the metals present in contaminated environments. These approaches are frequently considered cost-effective, sustainable, and environmentally suitable alternatives to more traditional chemical or physical processes (Tripathi and Gaur, 2021). Bioremediation has enhanced efficiency, cost, and social acceptability (Alaira et al., 2021). These methods encompass a wide array of approaches, including phytoremediation, microbial remediation, and mycoremediation, each utilizing specific organisms such as plants, bacteria, fungi, or algae. These organisms possess unique biochemical mechanisms that enable them to uptake, immobilize, or detoxify heavy metals through processes such as bioaccumulation, biosorption, biotransformation, and rhizofiltration (Bala et al., 2022; Kumar et al., 2021; Raklami et al., 2022; Tekere, 2020). Moreover, eco-friendly methods emphasize the use of natural materials and processes to achieve heavy metal remediation goals (Figure 14.1). Examples include the application of organic amendments, biochar, compost, and green sorbents derived from agricultural waste or natural minerals. These materials not only exhibit a high affinity for heavy metal ions but also promote soil health, microbial activity, and overall ecosystem resilience (Liang et al., 2021; Paz-Ferreiro et al., 2014; Rana et al., 2021).

This chapter aims to provide a comprehensive exploration of the various biological and eco-friendly methods employed for the removal of heavy metals from contaminated environments. Through an in-depth analysis of their mechanisms, applications, advantages, limitations, and case studies, we seek to elucidate the potential of these methods as sustainable solutions for mitigating the impacts of heavy metal contamination on both environmental and human health. By bridging the gap between scientific research and practical implementation, we aspire to contribute to the advancement of environmentally sound remediation strategies that promote the well-being of our planet and its inhabitants (Figure 14.1).

14.2 BIOLOGICALS METHODS

14.2.1 Microbial Bioremediation

While heavy metals cannot be decomposed, they can change from one inorganic complex oxidation state to another. A microbial remediation technique may be applied to achieve this transformation.

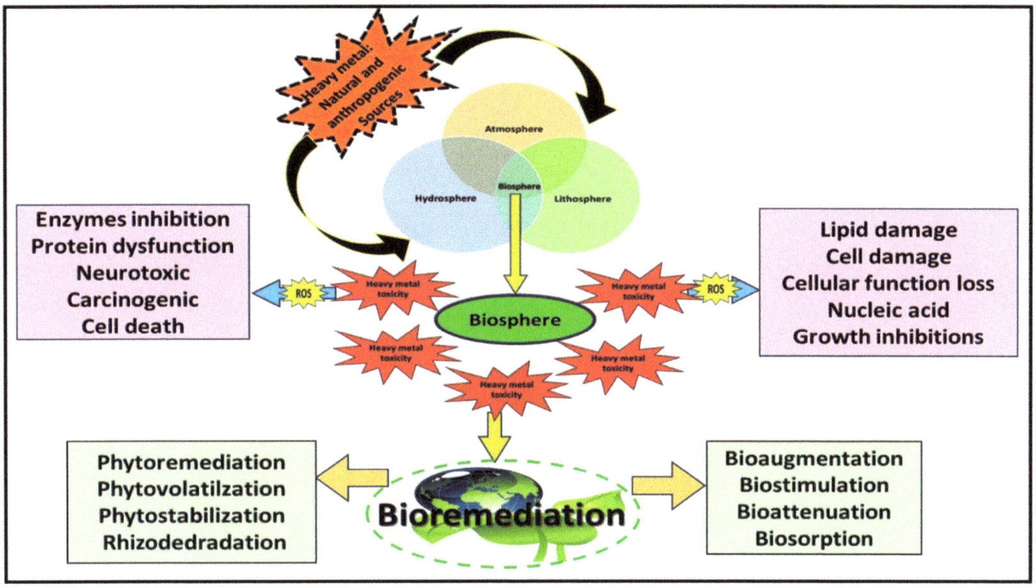

FIGURE 14.1 Illustration depicting the intricate interplay between heavy metal toxicity and various biological remediation strategies, highlighting the potential of natural processes in mitigating environmental contamination.

This process utilizes microorganisms to decontaminate organic and inorganic heavy metals from the environment. Bacteria and fungi are the predominant microorganisms utilized for the biological remediation of contaminated soil, such as the bacterium *Bacillus* sp. L14 (EB L14) was used to remove copper, cadmium, and lead, and fungi are effective organisms for biosorption of cadmium (II), copper (II), mercury (II), magnesium (II), lead (II), and zinc (II) from contaminated soil (Bano et al., 2018; Verma et al., 2021). After incubating for 14 days, the fungi were able to eliminate 83–66% of the total amount of heavy metals. Different species of *Aspergillus* are used to remove mercury and lead from the soil (Khan et al., 2019). It is an environmentally friendly, sustainable technology to overcome heavy metal contamination. This method is a remediation of the mixture that is growing in popularity and shows considerable potential. It has been studied and utilized for pollutant cleanup in increasing quantities (Dadrasnia et al., 2015). The availability and concentration of metal ions influence how microbial communities respond to heavy metals. The mechanism through which microbes react to metals is highly complex and depends upon a variety of variables, including the variety of metal, the type of media, and the species of microbes involved either in the passive (adsorption) or active (bioaccumulation) uptake of heavy metals (Figure 14.2) Several bio-indicators are utilized to express the heavy metal contamination (Madheshiya et al., 2022). Heavy metal ions can be attached by several functional groups that exist in microbial cell walls, which are primarily made up of polysaccharides, lipids, and proteins. These chains include carboxylate, hydroxyl, amino, and phosphate. The bioremediation process depends on microbial viability, cell death, competition, predation, pH, temperature, and moisture (Emenike et al., 2018). The mechanism involved in microbial remediation of heavy metals typically includes several key processes: Bioaccumulation (adsorb heavy metals onto their cell surfaces or accumulate them within their cells), Chelation (Microorganisms produce chelating agents, such as organic acids and peptides, which bind to heavy metals, forming stable complexes), Biotransformation (microorganisms possess enzymes capable of transforming toxic forms of heavy metals into less harmful or more easily removable forms through processes such as reduction, oxidation, or methylation) and Synergistic Interactions (Microbial communities often work synergistically, with different species contributing

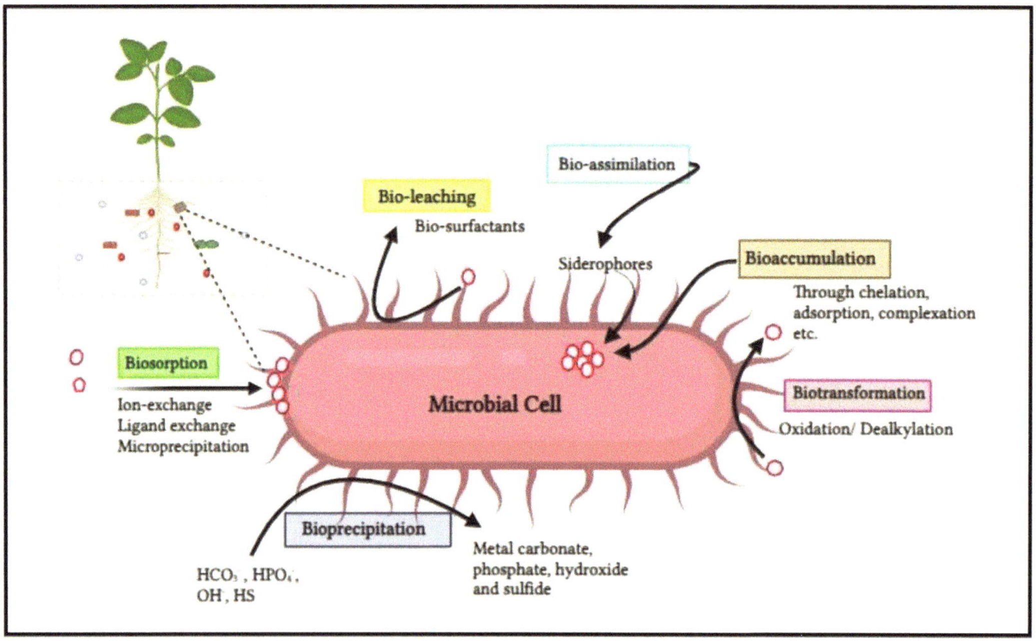

FIGURE 14.2 Revealing the elegant interaction of microbes with heavy metals.

unique capabilities to the overall remediation process). Overall, microbial remediation of heavy metals involves a combination of physical, chemical, and biological processes that work together to reduce the concentration and toxicity of heavy metals in the environment (Iravani and Varma, 2022; Wu et al., 2010). According to, Emenike et al. (2016) there are various methods of eliminating heavy metals via microbial remediation including bioaugmentation, biostimulation, and bioattenuation:

14.2.2 Bioaugmentation

Bioaugmentation, a strategy to enhance the degradation activities at contaminated sites, by the introduction of specific microbes or microbial consortia. The efficacy of bioaugmentation is subject to the influence of numerous abiotic and biotic factors. This method typically harnesses indigenous bacteria, applying them to contaminated sites to expedite the removal of undesirable compounds. By leveraging the inherent metabolic capabilities of these microbial communities, bioaugmentation promotes the breakdown of heavy metals (Emenike et al., 2016). Through targeted microbial interventions, bioaugmentation represents a promising approach in the realm of environmental bioremediation, offering a tailored solution for mitigating contamination challenges. Since nonbiological methods are more costly compared to bioaugmentation, they are primarily used for soils with significantly lower concentrations of metal-degrading microbes and those comprising compounds that require multi-process remediation, including processes that are harmful or toxic to microbes that are suitable for small-scale sites (Emenike et al., 2016, 2018).

14.2.3 Biostimulation

Biostimulation is a strategic approach aimed at enhancing bioremediation processes by modifying the environmental conditions to stimulate the activity of existing bacteria. This method involves the addition of specific nutrients such as phosphorus, nitrogen, oxygen, or carbon to the contaminated site. The primary objective of nutrient supplementation is to bolster the population or metabolic activity of naturally occurring microorganisms, thereby amplifying their capacity for bioremediation. One of the key advantages of biostimulation lies in its reliance on indigenous microbial communities, which have adapted to the unique conditions of the subsurface environment. These native microbes possess an inherent ability to degrade contaminants and are optimally distributed throughout the soil matrix, facilitating efficient remediation. However, several factors can influence the effectiveness of biostimulation in soil remediation, particularly for heavy metal removal. These factors include nutrient availability, pH levels, temperature, moisture content, oxygen availability, soil properties, and the type of contaminant present. Addressing these variables is essential for optimizing the biostimulation process and ensuring successful remediation outcomes. By carefully managing these factors, biostimulation can be a highly effective and environmentally sustainable approach for addressing soil contamination challenges (Emenike et al., 2018).

14.2.4 Bioattenuation

This approach makes utilization of natural processes to control the chemical spill's dispersion and reduce the concentration of pollutants at affected sites. As a result, pollutants in the environment are allowed to naturally degrade, decrease, or transform without intervention. In the process of cleaning up a place, natural attenuation is frequently used in conjunction with the control or removal of the source of contamination. The effectiveness of bioattenuation is generally dependent on the conditions available at the sites of contamination. The physical, chemical, and biological properties of the soil and groundwater, as well as the various types of pollutants present, significantly influence the rate of bioattenuation. Pollutant concentrations can be lowered via dilution or dispersion, biodegradation and chemical transformations can reduce the bulk of contaminants, and adsorption can immobilize pollutants by binding them to soil particles. It utilizes biological transformations and natural processes for dispersing harmful substances, making it a feasible and cost-effective remedy for remediating environmental problems (Emenike et al., 2018) (Figure 14.2).

By harnessing the diverse metabolic capabilities of microbes, microremediation strategies offer a promising approach for mitigating heavy metal contamination in various environmental matrices. However, the effectiveness of these strategies depends on factors such as microbial community composition, environmental conditions, and the specific contaminants involved. Ongoing research is focused on optimizing microremediation approaches and understanding the complex interactions between microbes and heavy metals in contaminated environments.

14.2.5 Phytoremediation

Phytoremediation, stemming from the Greek prefix "phyto" meaning "plant" and the Latin root "remedium" meaning "to correct or remove evil", involves using plants to alleviate or eliminate environmental contaminants. It draws upon research from diverse fields such as constructed wetlands, oil spill management, and agricultural practices focusing on the accumulation of heavy metals by plants. Since its introduction, the term has gained widespread usage, encompassing various specific applications and interpretations. Plants are utilized in phytoremediation procedures to either convert heavy metals in the roots into less toxic elements or to absorb heavy metals from polluted soil in the roots, stems, and leaves (Table 14.1) The modern technologies of phytoremediation are based on different uptake mechanisms which include phytoextraction, phytostabilization, phytovolatization, and rhizodegradation.

14.2.6 Phytoaccumulation or Phytoextraction

Plants develop to store pollutants in their aboveground components, such as shoots, leaves, and roots, a process called phytoextraction. It comprises the absorption, transportation, and accumulation of

TABLE 14.1
Plant Species Investigated for Heavy Metal Phytoremediation Studies

S. No.	Plant Species	Metals Associated	References
1.	*Agrostis capillaris*	Cu, Pb	Galal et al. (2017)
2.	*Pteris vittata*	As	Datta et al. (2017)
3.	*Eichhornia crassipes*	Pb, Zn, Hg, Ni, Cd	Rezania et al. (2016)
4.	*Epilobium dodonaei*	Cu, Zn, Pb	Ranđelović et al. (2016)
5.	*Arundo donax*	Ni, Cd	
6.	*Medicago sativa* L.	Cd	Ghnaya et al. (2015)
7.	*Cynodon dactylon*	Mn, Cu, Fe, Pb	Islam et al. (2015)
8.	*Micranthemum umbrosum*	As, Cd	Islam et al. (2015)
9.	*Brassica juncea*	Se	Dixit et al. (2015)
10.	*Astragalus bisulcatus*		
11.	*Latuca sativa* L.	Cu, Fe, Mn, Zn, Ni, Cd, Pb, Co, As	Quainoo et al. (2015); Rashid et al. (2014)
12.	*Pistia stratiotes* L.	Cd, Cr	Akter et al. (2014); Das et al. (2014); Farnese et al. (2014)
13.	*Astragalus bisulcatus*	Se, Pb	Ali et al. (2013)
14.	*Spartina argentinensis*	Cr	Nalla et al. (2012)
15.	*Ricinus communis*	Cd	Huang et al. (2011)
16.	*Pisum sativum*	Pb, Cu, Zn, Fe, Cd, Ni, As, Cr	Hegedűsová et al. (2009); Małecka et al. (2008); Sharma et al. (2010); Wani et al. (2008)
17.	*Jatropha curcas* L.	Cd, Cu, Ni, Pb	Abhilash et al. (2009)
18.	*Liriodendron tulipifera*	Ni, Hg	Ruiz and Daniell (2009)
19.	*Oryza sativa* L.	Cd, Cu	Liu et al. (2022a)
20.	*Tagetes* sp.		Hernández-Allica et al. (2008)
21.	Giant duckweed	Fe, Cr, Cd, Cu	Ko et al. (2008)
22.	*Populus canescens*	Zn	Bittsánszky et al. (2005)
23.	*Salix viminalis, Salix fragilis*	Cd, Cu, Pb, Zn	Pulford and Watson (2003)
24.	*Cyperus alternifolius*	Fe, Pd, Cr, Cu	Mench and Martin (1991)

heavy metals along with nutrients. To remove any remaining traces of the heavy metal, these plants have been grown in heavily polluted regions where the plant parts usually growing above the ground accumulate these metals (Sharma, 2012). Several plants such as *Haumaniastrum robertii, Pteris vittata* and *Thlaspi caerulescens* accumulate heavy metals (copper, arsenic, and zinc) therefore, these plants can reduce the bioaccumulation of metal ions in the food chain (Meena et al., 2021). A broad root system, high growth rate, large biomass, and resistance to heavy metals comprise a benefit of phytoextraction. Based on plant accumulation properties, phytoaccumulation has been categorized into two methods; the first method is the utilization of naturally existing hyperaccumulator plants, which can accumulate greater heavy metals than plants of other species. Such plants can accumulate heavy metals at a rate that is 10–500 times greater than that of common plants. However, in the second method to increase the high-biomass plant's ability to store heavy metals from the atmosphere, chelates or soil amendments are used (Meena et al., 2021; Suman et al., 2018).

14.2.7 Phytostabilization

Natural methods for the phytostabilization of heavy metals comprise metal valence reduction, precipitation, and absorption. Although the entire elimination of heavy metals from the environment might not be possible, this phytostabilization is more effective for the rapid immobilization of heavy metals from soil and groundwater (Senthil Kumar and Gunasundari, 2018). Phytostabilization is a basic economical approach that employs plants to regulate the dispersion of contaminants containing heavy metals in the rhizosphere within the root tissues. Despite the physical elimination of the pollutants, this process involves the deactivation and immobilization of contaminants, consequently ceasing the subsequent movement of contaminants to the same food chain. Phytostabilization through decreasing soil pH significantly increases heavy metal availability. The microbial population associated with plant roots and rhizospheres is involved in multiple processes, involving nutrient recycling, potentially dangerous pollutant detoxification, and soil management (Ali et al., 2020).

14.2.8 Phytovolatilization

Some heavy metals, such as selenium and mercury can be removed by phytovolatilization. The pollutants could not be removed using this approach. It only changed the nontoxic forms of pollutants from their lethal form. The resulting accumulation of more substances that could re-accumulate in the water bodies is the method's disadvantage. Genetically engineered plants are primarily utilized in this method for absorbing pollutants; in particular, *Brassica juncea* and *Brassica napus* are used to phytovolatilize mercury and selenium from soil. Mercury and selenium are considered the most suitable to be remediated through phytovolatilization (Meena et al., 2021). According to Ali et al. (2020), Indian mustard and canola varieties are very helpful in the phytovolatilization of selenium. Other toxic metals like mercury have been phytoremediated by using the phytovolatilization process Bisht et al. (2020).

14.2.9 Rhizodegradation

This method comprises the degradation of heavy metal by using a root system. Degradation of substances found in plant- and microbe-derived proteins and enzymes has been attributed to specialized activity in the rhizospheric region. The consequence of bacteria and plants working together symbiotically is rhizodegradation. Using the root system, rhizofiltration is a form of phytoremediation that is used to clean extracted water with a reduced concentration of pollutants. Rhizospheric microflora-stimulated pollutant decomposition is also known as phytostimulation (Lee et al., 2020). For other metals that are heavily concentrated in the roots, such as cadmium, copper, lead, zinc, nickel, and chromium, this technique can be applied. Toxin removal from water or soil is commonly achieved by the use of sunflowers, rye, tobacco, corn, Indian mustard, spinach, and pulses (Luciano et al., 2013; Meena et al., 2021).

14.2.10 Genetically Engineered Organisms (GEOs)

Genetically engineered organisms (GEOs) have undergone genetic engineering to enhance their ability to efficiently remove toxic heavy metals, surpassing other treatment methods. GEOs, including bacteria, algae, and fungi, and evaluate their potential in heavy metal removal. In recent times, GEOs have emerged as a prominent solution for removing toxic heavy metals, offering both economic feasibility and environmental safety (Han et al., 2022; Saravanan et al., 2022). Numerous factors can impact the efficacy of bioremediation employing GEOs. These include reaction conditions, the chemical makeup of heavy metals, redox potential, and nutritional status, among others. Critical processes such as chelation, biotransformation, oxidative stress response, metal regulation, and transportation, as well as the modification of cell surfaces, are pivotal in the remediation of heavy metals (Fasani et al., 2018). In phytoremediation using genetically modified plants, the choice of appropriate plant species, plant-microorganism interactions, translocation mechanisms, tolerance mechanisms, metal properties, and environmental factors play crucial roles. Additionally, molecular alterations have been employed to express metal-binding proteins on cell surfaces by overexpressing genes or introducing exogenous DNA, resulting in the creation of transgenic algae with heightened selectivity and effectiveness in adsorbing heavy metals (Iravani and Varma, 2022). Despite their potential effectiveness, the utilization of genetically modified organisms (GMOs) for metal removal in phytoremediation presents several challenges including regulatory hurdles, ecological risks, public perception, gene stability and expression, and ethical considerations as well.

14.3 ECO FRIENDLY MATERIALS

14.3.1 Biosorbents

Biosorption stands out as a key method for heavy metal detoxification, particularly at lower concentrations. It is a naturally occurring physiochemical process observed in certain biomass, where contaminants are passively concentrated and bound to the cellular structure (Nascimento et al., 2023; Sao et al., 2014). In the biosorption process, two distinct phases are involved: a solid phase, known as the biosorbent, which typically comprises materials such as plants, bacteria, and fungi, and a liquid phase, typically aqueous, containing the metal ions to be removed, referred to as the sorbate. The fundamental concept of this technology lies in the interaction between a sorbate (an atom, ion, or molecule) and a biosorbent (the surface of the biological agent) (Ali Redha, 2020). This interaction leads to the sorption and accumulation of the sorbate at the sorbate-biosorbent interface, thereby decreasing the concentration of the sorbate in the solution. Various mechanisms facilitate biosorption, including complexation, chelation, reduction, precipitation, and ion exchange (Calderon et al., 2020). A diverse array of biological materials, both living and non-living, have been utilized as biosorbents, including algal biomass, fungi, bacteria, agricultural waste, and more (Abdi and Kazemi, 2015). Typically, biomass chosen for biosorption should be abundant in organic ligands or functional groups. An optimal biosorbent should be readily accessible, environmentally benign, possess high biosorption capacity, and be suitable for large-scale usage and repeated use. The production of intracellular enzymes and extracellular polysaccharides on the surface of microbes aids in the bioaccumulation and detoxification of toxic metals (Yaashikaa et al., 2021). Biopolymers such as cellulose, chitosan, alginate, and pectin are favored adsorbents due to their high adsorption capacities, compatibility with living systems, and ability to be reused. The interactions between metals and biosorbents vary depending on the microbial species and the characteristics of the metals involved (Kayan and Kayan, 2021). Biosorbents offer valuable advantages over conventional methods of heavy metal detoxification (Gadd, 2009; Sameera et al., 2011):

- *Versatile*: Biosorption is effective across a wide range of conditions, including temperature, pH, and the presence of other metal ions.
- *Cost-effective*: Biosorbents are inexpensive as they are derived from waste biological materials and can be easily produced using simple nutrient media.
- *Selective for metals*: Different types of biomass exhibit varying degrees of selectivity for different solutes. Through physico-chemical treatment, biomass can be made capable of removing multiple heavy metals simultaneously during the sorption process.
- *Regenerative*: Biosorbents can be reused in the sorption process once the metal has been extracted from the biomass.
- *Reduced waste generation*: Biosorption eliminates secondary issues such as sludge formation commonly associated with traditional methods.

14.3.1.1 Examples of Effective Biosorbents

14.3.1.1.1 Bacteria

Bacteria are categorized as either gram-positive or gram-negative, with the primary distinction lying in their cell wall composition and thickness. Gram-positive bacteria have thicker peptidoglycan layers in their cell walls, connected by amino acid bridges (De Benedetti et al., 2021). The initial step of biosorption involves metal ions attaching to functional groups (such as amine, carboxyl, hydroxyl, phosphate, sulfate, and amine) present on the bacterial cell wall. The general process of metal uptake includes the binding of metal ions to reactive groups on the bacterial cell wall, followed by the internalization of heavy metal ions into the cell (Priyadarshanee and Das, 2021). Gram-positive bacteria exhibit a higher potential for the removal of heavy metal cations due to their significant electronegative charge density, which is attributed to the presence of teichoic and teichuronic acids (polyalcohol) linked by phosphodiester bonds and attached to the peptidoglycan of the cell wall (Tsezos et al., 2006). A scientific study demonstrated that filamentous Gram-positive bacteria known as actinomycetes exhibited a cadmium removal capacity of approximately 95% (Gahlout et al., 2021). Various genera of bacteria, including *Bacillus*, *Pseudomonas*, *Micrococcus*, *Streptococcus*, *Staphylococcus*, and *Enterobacter*, have been effectively utilized as biosorbents for heavy metals in both soil and water (Rana et al., 2021). Bacterial products such as exopolysaccharides (EPS) are predominantly acidic and contain hydroxyl, phosphoryl, sulfide, carboxyl, and amide groups, which enable them to bind metals in soil (Bramhachari and Nagaraju, 2017). For example, the cyanobacterium *Anabaena doliolum* Ind1, isolated from a coal mining site, demonstrated the ability to adsorb 69%–92% of cadmium from soil (Goswami et al., 2015). A strain of *Providencia vermicola* exhibited Pb(II) tolerant properties, absorbing up to 155 mg/g biomass through an internalization mechanism (Rana et al., 2021). In arsenic-contaminated environments, arsenic binds with exopolysaccharide (EPS) released by *Rhizobium radiobacter*. The uptake of arsenate by whole cell pellet and EPS was reported to be 0.045 and 0.068 mg/g biomass, respectively (Deepika et al., 2016). Additionally, metal-tolerant rhizospheric bacteria *Bacillus cereus* were employed for the removal of Cr(VI), Ni(II), and Cd(II), achieving accumulations ranging from 90 to 122, 128 to 146, and 120 to 135 mg/g, respectively (Akhter et al., 2017).

14.3.1.1.2 Fungi

Fungal biomasses have gained significant interest as biosorbent materials for heavy metal removal, primarily due to their ease of cultivation and availability as industrial waste products (Ghosh et al., 2021). The biosorption of elements by fungi occurs within the cell wall, which is composed of complex macromolecular structures predominantly consisting of polysaccharides (80–90%), along with proteins, lipids, and pigments as minor constituents. Chitin is also a common component of fungal cell walls. This diversity of structural elements provides numerous functional groups capable of binding metal ions of toxicological significance (Javaid et al., 2011). Utilizing fungi as biosorbents offers several advantages: fungi have the ability to utilize a wide range of nutrients and possess unique metabolic processes that enable them to utilize toxic waste products and organic compounds as nutrients. Additionally, certain fungi can degrade recalcitrant compounds by secreting specialized enzymes, which are later absorbed inside the cells as nutrients. Fungi are also capable of surviving in diverse habitats that may be unsuitable for other eukaryotes (Ayele et al., 2021). Research conducted over the years has demonstrated that prominent fungal species like *S. cerevisiae, Aspergillus flavus, Aspergillus tamarii,* and *Trichoderma atroviride* possess innate abilities to absorb a wide range of heavy metals including copper, lead, nickel, iron, cobalt, arsenic, cadmium, and chromium (Yan and Viraraghavan, 2008). Moreover, the accumulation of zinc and copper by *Acremonium persicinum* and *Penicillium simplicissimum*, respectively, demonstrates the potential of these fungal species in sequestering heavy metals in soil (Abdel-Wareth, 2023). *Cunninghamella elegans*-derived chitin and chitosan were utilized for the removal of metals from aqueous solutions, with chitosan exhibiting superior biosorption capacity for Cu(II) (75%), while chitin demonstrated maximum biosorption of Fe(II) (56%) at pH 4 (Rana et al., 2021). A combination of *Aspergillus fumigatus* and *Aspergillus flavus* was employed to remove Cd(II) and Cr(VI) from industrial effluent (Talukdar et al., 2020). *Penicillium janthinillum* Conidia immobilized using poly-vinyl alcohol were effective in removing Cu, Pb, and Cd from an aqueous medium (Sen et al., 2021). A study also investigated the functional groups present in *Mucor rouxii* and their role in the biosorption of lead, cadmium, nickel, and zinc (Yan and Viraraghavan, 2008).

14.3.1.1.3 Algae

The utilization of algae as biosorbents is also a good alternative for heavy metal detoxification due to their minimal nutrient requirements, high sorption capacity, abundant availability, high surface area to volume ratio, reduced volume of sludge generated for disposal, and potential for metal regeneration and recovery (Escudero et al., 2018). Algae have been reported to biosorb approximately 15.3–84.6% of heavy metals, exhibiting higher capacities compared to other microbial Biosorbents (Kanamarlapudi et al., 2018). Different groups of algae exhibit variations in cell wall composition. Brown algae primarily comprise cellulose for structural support, alginic acid (composed of mannuronic and guluronic acid), and sulfated polysaccharides with high carboxyl group contents, which play roles in metal biosorption. Red algae are noted for their sulfated polysaccharides made of galactans, containing abundant hydroxyl and carboxyl groups, contributing to their biosorption potential. Green algae contain cellulose along with polysaccharides bound to proteins, featuring functional groups like amino, sulfate, hydroxyl, and carboxyl (González et al., 2011).

The polysaccharide components present in algal biomass, including xylans, alginic acid, mannans, chitosan, glucans, chitin, melanins, and hyaluronic acid, offer functional sites facilitating heavy metal biosorption. The –OH groups in algal polysaccharides develop a negative charge at higher pH levels and exhibit secondary metal binding properties at pH > 10 (Prasher et al., 2021). Algae-based biosorbents, including seaweeds and various micro/macroalgae species, possess numerous functional groups and biomacromolecules. Research investigated the biosorption capacity of green algal species *Spirogyra* sp. and *Cladophora* sp. for removing Pb(II) and Cu(II) from aqueous solutions. *Spirogyra* exhibited capacities of 87.2 mg/g for Pb(II) and 38.2 mg/g for Cu(II), while *Cladophora* showed capacities of 45.4 mg/g for Pb(II) and 13.7 mg/g for Cu(II) (Kanamarlapudi et al., 2018). Furthermore, a study assessed the removal of cadmium (II) and lead (II) from aqueous solutions using the blue-green alga *Anabaena sphaerica*, revealing relatively high biosorption capacities of 111.1 mg/g for cadmium (II) and 121.95 mg/g for lead (II). Additionally, *Sargassum* biomass demonstrated the capability to biosorb cadmium and lead cations through partial or complete esterification of carboxylic sites present on its cell wall (Kleinübing et al., 2011).

14.3.2 Green Adsorbents

Green adsorbents for heavy metal detoxification refer to environment-friendly materials derived from natural sources that are utilized to remove heavy metal contaminants from various matrices, such as water, soil, and industrial effluents. These adsorbents are considered "Green" due to their sustainable origin and low environmental impact compared to conventional methods (Crini et al., 2018).

Primarily derived from agricultural sources and their by-products, as well as through the pyrolysis of agricultural waste, these materials are enriched in cellulose, offering significant biosorption capacity for contaminants. Other constituents of agricultural waste biomass include hemicellulose, lignin, lipids, proteins, simple sugars, water hydrocarbons, and starch, all of which harbor diverse functional groups facilitating complexation with contaminants, aiding in heavy metal removal from polluted environments (Gupta et al., 2021; Karić et al., 2022). The proportions of these chemicals vary among different plant species, with crop residue typically containing 35–50% cellulose, 20–30% lignin, and 15–30% hemicellulose. Lignocellulosic materials, being a major biomass source, present a complex structure that poses challenges for degradation. However, the presence of aliphatic and phenolic hydroxyl groups, methoxyl groups, and carbonyl groups within lignin can significantly enhance its adsorption efficiency. Consequently, these waste materials possess enriched surface functional groups, a porous structure, and a large specific area, rendering them promising adsorbents for heavy metal removal (Jain et al., 2016).

Numerous agricultural wastes and forest products, including sugarcane bagasse, pine bark, oak wood and bark biochars, corn straw, and fruit and vegetable peels, have been identified as effective adsorbents for heavy metals (Sabir et al., 2021) (Table 14.2). Grapefruit peel, for instance, has

TABLE 14.2
Green Adsorbents Explored in Heavy Metal Phytoremediation Studies

S. No.	Material	Heavy Metal	Sorption Capacity (mg/g)	Reference
1.	Peach and apricot stones	Cu(II)	4.45	Kobya et al. (2005)
2.	Sludge (paper mill)	Ni	13.7	Suryan and Ahluwalia (2012)
		Cu	5.5	
		Pb	14.1	
		Cd	14.8	
3.	Garden Grass	Cu(II)	58.34	Hossain et al. (2012)
4.	Lignin	Cr(III)	17.97	Wu et al. (2008)
5.	Coconut tree sawdust	Pb	25	Putra et al. (2014)
6.	Chitosan	Cr(VI)	7.94	Aydın and Aksoy (2009)
7.	Rice husk	Ni	3.88	Hegazi (2013)
		Cd	16.18	
		Cu	29	
		Pb	4.23	
		Zn	5.38	
8.	Banana peel	Zn	5.80	Annadurai et al. (2003)
		Pb	7.97	
		Ni	6.88	
9.	Waste tea leaves	Pb	8.6	Ahluwalia and Goyal (2005)
10.	Egg shell	Pb	19.9	Putra et al. (2014)
11.	Almond shells	Pb(II)	9	Maharshi et al. (2009)
12.	Treated agave bagasse	Pb	54.3	Velazquez-Jimenez et al. (2013)
13.	Cellulose-chitosan composite	Cu(II)	75.82	Peng et al. (2014)
14.	Raw corn cob	Pb	16.2	Tan et al. (2010)

been found to biosorb cadmium and nickel with capacities of 42.09 and 46.13 mg/g, respectively, from aqueous solutions. Additionally, the bark powder of *Acacia leucocephala* has been utilized as a low-cost biosorbent for the removal of Cu(II), Cd(II), and Pb(II) with capacities of 147.1, 167.7, and 185.2 mg/g, respectively, from aqueous solutions. Agroforestry waste mixtures composed of fern, rice husk, and oak leaves have also shown promising results, with the mixture exhibiting 10% greater efficiency in removing heavy metals compared to single waste sources (Rosales et al., 2015).

Rice husk, a widely used material for heavy metal adsorption from soil and water, is composed of the protective layer of the paddy grain and contains a high silica content (92–95%), making it suitable for heavy metal adsorption (Ullah et al., 2021). Raw sugarcane bagasse and its primary functional constituents exhibit substantial adsorption capability for heavy metals. The notable heavy metal adsorption capacity of sugarcane bagasse is attributed to its highly porous structure and the presence of carbon-oxygen functional groups. These functional groups facilitate the process by either filling the pores or binding heavy metals through mechanisms such as exchanging hydrogen ions for metal ions or forming complexes with the metal ions through the donation of an electron pair (Khoo et al., 2018).

Moreover, modified agricultural solid waste has been extensively employed as an effective adsorbent for removing various contaminants from wastewater, with improvements attributed to surface properties modifications. Chemical treatments, such as acid or alkaline solutions, are commonly used to modify raw biomass, removing natural fats, waxes, and lignin compounds and enhancing adsorption properties (Liu et al., 2022b). Recently, the production of activated carbon, biochar, and charcoal from agricultural residues has emerged as a cost-effective adsorbent option due to their high selectivity, porosity, and surface area. Biochar, derived from the pyrolysis of agricultural and forest residual biomass, has demonstrated effectiveness in heavy metal removal from water and soil, owing to its low cost, easy availability, high stability, and eco-friendly nature (Biswal and Balasubramanian, 2023). For instance, corncob biochar has been found to aid in the sequestration of Mn (II), Pb(II), and Cd(II) in diesel oil-contaminated soil, while biochar derived from rice husk has also been used for heavy metal removal. Various raw and chemically activated agricultural wastes, including jackfruit, pecan shell, bamboo, eucalyptus bark, maize cob, and castor hull, have likewise shown promise as effective adsorbents for aqueous phase heavy metal ion removal (Chaiyaraksa and Phumcharoen, 2021).

Additionally, these cost-effective biosorbents derived from agricultural waste offer a dual benefit in reducing environmental pollution. Firstly, they effectively remove heavy metal contamination at a minimal expense, and secondly, they help mitigate the volume of waste generated from agricultural activities. Such initiatives contribute to achieving climate neutrality by promoting industrial practices aligned with circular economy principles and fostering sustainable waste management strategies (Kainth et al., 2024). Moreover, this approach aligns with a zero-pollution objective through the implementation of pollution prevention and remediation measures. Globally, a significant amount of agricultural waste is either disposed of through dumping or incineration, leading to resource wastage and environmental degradation. Utilizing these agricultural byproducts to produce biosorbents presents an attractive alternative for heavy metal detoxification, thus making biosorption a more sustainable solution (Younas et al., 2024).

14.4 CASE STUDIES

14.4.1 Case Study 1

Annually, a significant volume of biological waste is generated, necessitating its conversion into effective materials to mitigate environmental pollution. This not only diminishes the ecological risks associated with disposing of these wastes in natural habitats but also reduces waste management costs. Utilizing natural biological residues as sorbent materials has emerged as a promising avenue of research due to their renewability, high yield, low cost, wide availability, and eco-friendly attributes. Recent studies have highlighted the potential of this approach (Ahsan et al., 2018; Mo et al., 2018; Ranasinghe et al., 2018; Villen-Guzman et al., 2019). In a notable case study conducted by Deniz and Ersanli (2021), a renewable biosorbent was developed from the biological residue of the common coastal plant *Zostera marina*, which is abundant in the coastal regions of the northern hemisphere. This plant generates substantial leafy waste annually. The biosorbent was utilized for environmentally friendly removal of heavy metal pollution from water, with manganese serving as a standard heavy metal solution for evaluating the decontamination efficiency. The World Health Organization (WHO) has set the maximum acceptable concentration of manganese in drinking water at 0.05 mg L-1 due to its detrimental physiological, biochemical, and morphological effects on living organisms. Additionally, manganese is listed in the Hazardous Substances Priority List by the Toxic Substances and Disease Registry (ATSDR) (Amorim et al., 2018; Queiroz et al., 2018).

The batch biosorption behavior of the biosorbent was investigated through characterization, parameter assessment, and kinetic and equilibrium studies. Notably, prior to this study, there had been no documented instances of utilizing waste from *Z. marina* for purging heavy metal pollution from aquatic environments.

14.5 METHODOLOGY AND OUTCOMES: HEAVY METAL SOLUTION AND RENEWABLE BIOSORBENT MATERIAL

Manganese (II) chloride dihydrate served as the heavy metal agent to assess the efficacy of the biosorbent treatment. A manganese stock solution was initially prepared at a concentration of 1,000 mg/L using distilled water, and the desired concentrations were subsequently derived from this stock through dilution. The initial pH of these solutions was adjusted using 0.1 mol/L solutions of sodium hydroxide and hydrochloric acid. The waste material from the common coastal plant, *Zostera marina*, sourced from the Black Sea region (Sinop, Turkey), underwent multiple rinses with normal and distilled water to eliminate impurities. Subsequently, it was dried in a laboratory oven at 80°C for 24 hours to remove moisture. The dried waste was then ground using a laboratory mill and sieved to obtain particles smaller than 0.5 mm. The resulting biosorbent material was stored in a glass bottle for use in the heavy metal biosorption experiments. In this study, a renewable biosorbent was derived from the biowaste of the *Z. marina* plant and applied to eliminate heavy metal (manganese) contamination from water environments. This biosorbent features a rough surface and a variety of functional groups conducive to heavy metal ion retention. Through experimentation, it was determined that the maximum capacity of the biosorbent was 58.426 mg/g, indicating its potential for effective green purification of heavy metals.

14.5.1 Case Study 2

This study presents the creation of a highly effective biosorbent, termed sulfonated spent coffee waste (SCW) (SCW-SO_3H), achieved by altering surface chemistry and incorporating sulfonic acid functionalities onto the SCW. The adsorption capabilities of SCW-SO_3H were evaluated for various environmental pollutants, including methylene blue (MB), tetracycline (TC), and chromium Cr(VI). The biosorbent derived from SCW exhibited notable adsorption capacity and rapid adsorption kinetics for the removal of MB, TC, and Cr(VI) owing to the surface modifications implemented on the SCW.

14.6 METHODOLOGY AND OUTCOMES: HEAVY METAL SOLUTION AND RENEWABLE BIOSORBENT MATERIAL

The gathered SCW underwent a series of treatments: initially washed with Milli-Q water, then boiled to eliminate any coloring agents such as caffeine, tannin, and other dyes. Following this, the solid was filtered, washed, and dried in a vacuum oven at 70°C for 12 hours. Subsequently, 3 g of the dried SCW was combined with 45 ml of sulfuric acid. After dilution, the mixture underwent vacuum filtration and extensive washing with water to remove excess acid. The resulting sample was then subjected to heating in a vacuum oven at 80°C for 12 hours before being ground into a fine powder for subsequent use.

The study revealed that the developed biosorbent, sulfonated SCW (SCW-SO_3H), synthesized by incorporating sulfonic acid polar functionalities onto the polymeric biomass (cellulose and lignin) of the SCW, exhibited notable efficiency as a biosorbent in removing various environmental pollutants such as MB, TC, and Cr(VI), displaying higher adsorption capacities.

14.7 LIMITATIONS AND DRAWBACKS

While biological and eco-friendly methods for heavy metal detoxification offer several advantages, they also come with certain limitations and drawbacks:

- *Selectivity*: Biological methods may not be highly selective toward specific heavy metals, leading to the potential removal of essential trace metals along with the target contaminants.
- *Efficiency*: The efficiency of biological methods can vary depending on factors such as the type of biomass used, environmental conditions, and the concentration of heavy metals. In some cases, these methods may not achieve complete removal of contaminants.
- *Kinetics*: Biological processes typically operate at slower rates compared to chemical methods, resulting in longer treatment times and potentially higher costs.

- *Biomass regeneration*: Regenerating biomass for reuse in heavy metal detoxification processes can be challenging and may require additional treatment steps, leading to increased complexity and costs.
- *Environmental factors*: Biological methods may be influenced by environmental factors such as temperature, pH, and the presence of other contaminants, which can affect their effectiveness.
- *Scale-up challenges*: Scaling up biological methods from laboratory-scale to industrial-scale applications may pose technical challenges and require significant investment in infrastructure and resources.
- *Potential ecological impacts*: Introducing foreign biomass into natural ecosystems for heavy metal remediation purposes may have unintended ecological consequences, such as alterations to microbial communities or disruptions to ecosystem dynamics.
- *Regulatory considerations*: Compliance with regulatory standards and guidelines for the use of biological methods in heavy metal detoxification may pose logistical challenges and require extensive documentation and monitoring.

Overall, while biological and eco-friendly methods offer promising alternatives for heavy metal detoxification, addressing these limitations and drawbacks is essential to ensure their practical applicability and effectiveness in real-world scenarios.

14.8 FUTURE PERSPECTIVES

The future outlook for utilizing biological and eco-friendly solutions for heavy metal detoxification appears promising, with ongoing advancements poised to address current limitations and enhance their effectiveness. Emerging biotechnological approaches, such as the genetic engineering of microorganisms for enhanced metal uptake or the development of novel biosorbents with tailored properties, hold considerable potential for improving the efficiency and specificity of metal removal processes. Additionally, the integration of bioremediation techniques with sustainable agricultural practices, such as phytoremediation and biochar application, could offer synergistic benefits for soil and water remediation while promoting soil health and fertility. Furthermore, the adoption of innovative technologies, such as nanobiotechnology and bioreactor systems, may enable more efficient and cost-effective treatment of heavy metal-contaminated sites on a larger scale. As awareness of environmental sustainability grows and regulatory frameworks evolve, there is increasing momentum toward the adoption of biological and eco-friendly solutions as integral components of comprehensive heavy metal pollution management strategies. Continued research and collaboration across multidisciplinary fields are essential to unlock the full potential of these approaches and ensure their successful implementation in real-world scenarios.

ACKNOWLEDGMENT

Authors are grateful to the Department of Botany, Institute of Science, Banaras Hindu University, Varanasi, for providing necessary support for this work.

REFERENCES

Abdel-Wareth, M.T.A., 2023. Sequestration and Detoxification of Heavy Metals by Fungi. In *Sustainable Industrial Wastewater Treatment and Pollution Control* (pp. 185–209). Singapore: Springer Nature Singapore.

Abdi, O. and Kazemi, M., 2015. A review study of biosorption of heavy metals and comparison between different biosorbents. *Journal of Materials and Environmental Science, 6*(5), pp. 1386–1399.

Abhilash, P.C., Jamil, S. and Singh, N., 2009. Transgenic plants for enhanced biodegradation and phytoremediation of organic xenobiotics. *Biotechnology Advances*, *27*(4), pp. 474–488.

Ahluwalia, S.S. and Goyal, D., 2005. Removal of heavy metals by waste tea leaves from aqueous solution. *Engineering in Life Sciences*, *5*(2), pp. 158–162.

Ahsan, M.A., Jabbari, V., Islam, M.T., Kim, H., Hernandez-Viezcas, J.A., Lin, Y., Díaz-Moreno, C.A., Lopez, J., Gardea-Torresdey, J. and Noveron, J.C., 2018. Green synthesis of a highly efficient biosorbent for organic, pharmaceutical, and heavy metal pollutants removal: Engineering surface chemistry of polymeric biomass of spent coffee waste. *Journal of Water Process Engineering*, *25*, pp. 309–319.

Ajani, M.B., Maleka, P.P., Penabei, S. and Usman, I.T., 2022. Health risk assessment of heavy metals concentration from soil; a case study of the Mayo-Dallah in Southern area of Chad. *Journal of Radiation Research and Applied Sciences*, *15*(1), pp. 130–138.

Akhter, K., Ghous, T., Andleeb, S., Ejaz, S., Khan, B.A. and Ahmed, M.N., 2017. Bioaccumulation of heavy metals by metal-resistant bacteria isolated from *Tagetes minuta* rhizosphere, growing in soil adjoining automobile workshops. *Pakistan Journal of Zoology*, *49*(5).

Akter, S., Afrin, R., Mia, M.Y. and Hossen, M.Z., 2014. Phytoremediation of chromium (Cr) from tannery effluent by using water lettuce (*Pistia stratiotes*). *ASA University Review*, *8*(2).

Alaira, S., Padilla, C., Alcantara, E. and Aggangan, N., 2021. Social acceptability of the bioremediation technology for the rehabilitation of an abandoned mined-out area in Mogpog, Marinduque, Philippines. *Journal of Environmental Science and Management*, *24*(1).

Ali Redha, A., 2020. Removal of heavy metals from aqueous media by biosorption. *Arab Journal of Basic and Applied Sciences*, *27*(1), pp. 183–193. https://doi.org/10.1080/25765299.2020.1756177

Ali, H., Khan, E. and Sajad, M.A., 2013. Phytoremediation of heavy metals—Concepts and applications. *Chemosphere*, *91*(7), pp. 869–881.

Ali, M.M., Hossain, D., Al-Imran, A., Khan, M.S., Begum, M. and Osman, M.H., 2021. Environmental pollution with heavy metals: A public health concern. In *Heavy Metals—Their Environmental Impacts and Mitigation*, pp. 771–783.

Ali, S., Abbas, Z., Rizwan, M., Zaheer, I.E., Yavaş, İ., Ünay, A., Abdel-Daim, M.M., Bin-Jumah, M., Hasanuzzaman, M. and Kalderis, D., 2020. Application of floating aquatic plants in phytoremediation of heavy metals polluted water: A review. *Sustainability*, *12*(5), p. 1927.

Amorim, S.S., Ruas, F.A.D., Barboza, N.R., de Oliveira Neves, V.G., Leão, V.A. and Guerra-Sa, R., 2018. Manganese (Mn2þ) tolerance and biosorption by *Meyerozyma guilliermondii* and *Meyerozyma caribbica* strains. *Journal of Environmental Chemical Engineering*, *6*(4), pp. 4538–4545. doi:10.1016/j.jece.2018.06.061.

Annadurai, G., Juang, R.S. and Lee, D.J., 2003. Adsorption of heavy metals from water using banana and orange peels. *Water Science and Technology*, *47*(1), pp. 185–190.

Aryal, M., 2024. Rhizomicrobiome dynamics: A promising path towards environmental contaminant mitigation through bioremediation. *Journal of Environmental Chemical Engineering*, *12*(2), p. 112221.

Asim, M. and Nageswara Rao, K., 2021. Assessment of heavy metal pollution in Yamuna River, Delhi-NCR, using heavy metal pollution index and GIS. *Environmental Monitoring and Assessment*, *193*(2), p. 103.

Aydın, Y.A. and Aksoy, N.D., 2009. Adsorption of chromium on chitosan: Optimization, kinetics and thermodynamics. *Chemical Engineering Journal*, *151*(1–3), pp. 188–194.

Ayele, A., Haile, S., Alemu, D. and Kamaraj, M., 2021. Comparative utilization of dead and live fungal biomass for the removal of heavy metal: A concise review. *The Scientific World Journal*, *2021*, p. 5588111.

Bala, S., Garg, D., Thirumalesh, B.V., Sharma, M., Sridhar, K., Inbaraj, B.S. and Tripathi, M., 2022. Recent strategies for bioremediation of emerging pollutants: A review for a green and sustainable environment. *Toxics*, *10*(8), p. 484.

Bano, A., Hussain, J., Akbar, A., Mehmood, K., Anwar, M., Hasni, M.S., Ullah, S., Sajid, S. and Ali, I., 2018. Biosorption of heavy metals by obligate halophilic fungi. *Chemosphere*, *199*, pp. 218–222.

Bhat, S.A., Bashir, O., Haq, S.A.U., Amin, T., Rafiq, A., Ali, M., Américo-Pinheiro, J.H.P. and Sher, F., 2022a. Phytoremediation of heavy metals in soil and water: An eco-friendly, sustainable and multidisciplinary approach. *Chemosphere*, *303*, p. 134788.

Bhat, S.A., Sher, F., Hameed, M., Bashir, O., Kumar, R., Vo, D.V.N., Ahmad, P. and Lima, E.C., 2022b. Sustainable nanotechnology based wastewater treatment strategies: Achievements, challenges and future perspectives. *Chemosphere*, *288*, p. 132606.

Bisht, R., Chanyal, S. and Srivastava, R.K., 2020. A systematic review on phytoremediation technology: Removal of pollutants from waste water and soil. *International Journal of Research in Engineering, Science and Management*, *3*, pp. 54–59.

Biswal, B.K. and Balasubramanian, R., 2023. Use of biochar as a low-cost adsorbent for removal of heavy metals from water and wastewater: A review. *Journal of Environmental Chemical Engineering*, *11*(5), p. 110986.

Bittsánszky, A., Kömives, T., Gullner, G., Gyulai, G., Kiss, J., Heszky, L., Radimszky, L. and Rennenberg, H., 2005. Ability of transgenic poplars with elevated glutathione content to tolerate zinc (2+) stress. *Environment International*, *31*(2), pp. 251–254.

Blenis, N., Hue, N., Maaz, T.M. and Kantar, M., 2023. Biochar production, modification, and its uses in soil remediation: A review. *Sustainability*, *15*(4), p. 3442.

Bouida, L., Rafatullah, M., Kerrouche, A., Qutob, M., Alosaimi, A.M., Alorfi, H.S. and Hussein, M.A., 2022. A review on cadmium and lead contamination: Sources, fate, mechanism, health effects and remediation methods. *Water*, *14*(21), p. 3432.

Bramhachari, P.V. and Nagaraju, G.P., 2017. Extracellular polysaccharide production by bacteria as a mechanism of toxic heavy metal biosorption and biosequestration in the marine environment. In Marine Pollution and Microbial Remediation (pp. 67–85). Singapore: Springer.

Briffa, J., Sinagra, E. and Blundell, R., 2020. Heavy metal pollution in the environment and their toxicological effects on humans. *Heliyon*, *6*(9), p. e04691.

Chaiyaraksa, C. and Phumcharoen, K., 2021. Reduction of heavy metal movement in soil contaminated with diesel using corncob-biochar. *EnvironmentAsia*, *14*(1), pp. 41–51.

Cheng, Y., Qiu, L., Shen, P., Wang, Y., Li, J., Dai, Z., Qi, M., Zhou, Y. and Zou, Z., 2023. Transcriptome studies on cadmium tolerance and biochar mitigating cadmium stress in muskmelon. *Plant Physiology and Biochemistry*, *197*, p. 107661.

Crini, G., Lichtfouse, E., Wilson, L.D. and Morin-Crini, N., 2018. Green adsorbents for pollutant removal. *Environmental Chemistry for a Sustainable World*, *18*, pp. 23–71.

Dadrasnia, A., Salmah, I., Emenike, C.U. and Shahsavari, N., 2015. Remediation of oil contaminated media using organic material supplementation. *Petroleum Science and Technology*, *33*(9), pp. 1030–1037.

Das, K.K. and Büchner, V., 2007. Effect of nickel exposure on peripheral tissues: Role of oxidative stress in toxicity and possible protection by ascorbic acid. *Reviews on Environmental Health*, *22*(2), pp. 157–173.

Das, K.K., Das, S.N. and Dhundasi, S.A., 2010. Nickel: Molecular Diversity, Application, Essentiality and Toxicity in Human Health. In *Biometals, Molecular Structures, Binding Properties and Applications* (pp. 33–58). New York: Nova Science Publishers Inc.

Das, S., Goswami, S. and Talukdar, A.D., 2014. A study on cadmium phytoremediation potential of water lettuce, *Pistia stratiotes* L. *Bulletin of Environmental Contamination and Toxicology*, *92*, pp. 169–174.

Datta, R., Das, P., Tappero, R., Punamiya, P., Elzinga, E., Sahi, S., Feng, H., Kiiskila, J. and Sarkar, D., 2017. Evidence for exocellular arsenic in fronds of *Pteris vittata*. *Scientific Reports*, *7*(1), p. 2839.

De Benedetti, S., Fisher, J.F. and Mobashery, S., 2021. Bacterial Cell Wall: Morphology and Biochemistry. In *Practical Handbook of Microbiology* (pp. 167–204). Boca Raton: CRC Press.

Deepika, K.V., Raghuram, M., Kariali, E. and Bramhachari, P.V., 2016. Biological responses of symbiotic *Rhizobium radiobacter* strain VBCK1062 to the arsenic contaminated rhizosphere soils of mung bean. *Ecotoxicology and Environmental Safety*, *134*, pp. 1–10.

Deniz, F. and Ersanli, E.T., 2021. A renewable biosorbent material for green decontamination of heavy metal pollution from aquatic medium: A case study on manganese removal. *International Journal of Phytoremediation*, *23*(3), pp. 231–237. doi:10.1080/15226514.2020.1807905.

Dixit, R., Wasiullah, X., Malaviya, D., Pandiyan, K., Singh, U.B., Sahu, A., Shukla, R., Singh, B.P., Rai, J.P., Sharma, P.K. and Lade, H., 2015. Bioremediation of heavy metals from soil and aquatic environment: An overview of principles and criteria of fundamental processes. *Sustainability*, *7*(2), pp. 2189–2212.

Emenike, C.U., Agamuthu, P. and Fauziah, S.H., 2016. Blending *Bacillus* sp., *Lysinibacillus* sp. and *Rhodococcus* sp. for optimal reduction of heavy metals in leachate contaminated soil. *Environmental Earth Sciences*, *75*, pp. 1–8.

Emenike, C.U., Jayanthi, B., Agamuthu, P. and Fauziah, S.H., 2018. Biotransformation and removal of heavy metals: A review of phytoremediation and microbial remediation assessment on contaminated soil. *Environmental Reviews*, *26*(2), pp. 156–168.

Escudero, L.B., Quintas, P.Y., Wuilloud, R.G. and Dotto, G.L., 2018. Biosorption of Metals and Metalloids. In Green Adsorbents for Pollutant Removal: Innovative Materials (pp. 35–86). Cham: Springer.

Farnese, F.D.S., Oliveira, J.A., Lima, F.S., Leão, G.A., Gusman, G.S. and Silva, L.C., 2014. Evaluation of the potential of *Pistia stratiotes* L. (water lettuce) for bioindication and phytoremediation of aquatic environments contaminated with arsenic. *Brazilian Journal of Biology*, *74*, pp. S108–S112.

Fasani, E., Manara, A., Martini, F., Furini, A. and DalCorso, G., 2018. The potential of genetic engineering of plants for the remediation of soils contaminated with heavy metals. *Plant, Cell & Environment*, *41*(5), pp. 1201–1232.

Fendorf, S.E., 1995. Surface reactions of chromium in soils and waters. *Geoderma*, *67*(1–2), pp. 55–71.

Filippini, T., Torres, D., Lopes, C., Carvalho, C., Moreira, P., Naska, A., Kasdagli, M.I., Malavolti, M., Orsini, N. and Vinceti, M., 2020. Cadmium exposure and risk of breast cancer: A dose-response meta-analysis of cohort studies. *Environment International*, *142*, p. 105879.

Gadd, G.M., 2009. Biosorption: Critical review of scientific rationale, environmental importance and significance for pollution treatment. *Journal of Chemical Technology & Biotechnology*, *84*(1), pp. 13–28.

Gahlout, M., Prajapati, H., Tandel, N. and Patel, Y., 2021. Biosorption: An Eco-Friendly Technology for Pollutant Removal. In Microbial Rejuvenation of Polluted Environment (Volume 2, pp. 207–227). Singapore: Springer.

Galal, T.M., Gharib, F.A., Ghazi, S.M. and Mansour, K.H., 2017. Phytostabilization of heavy metals by the emergent macrophyte *Vossia cuspidata* (Roxb.) Griff.: A phytoremediation approach. *International Journal of Phytoremediation*, *19*(11), pp. 992–999.

Ghnaya, T., Mnassri, M., Ghabriche, R., Wali, M., Poschenrieder, C., Lutts, S. and Abdelly, C., 2015. Nodulation by *Sinorhizobium meliloti* originated from a mining soil alleviates Cd toxicity and increases Cd-phytoextraction in *Medicago sativa* L. *Frontiers in Plant Science*, *6*, p. 863.

Ghosh, S., Selvakumar, G., Ajilda, A.A.K. and Webster, T.J., 2021. Microbial Biosorbents for Heavy Metal Removal. In *New Trends in Removal of Heavy Metals from Industrial Wastewater* (pp. 213–262). Elsevier.

González, F., Romera, E., Ballester, A., Blázquez, M.L., Muñoz, J.Á. and García-Balboa, C., 2011. Algal Biosorption and Biosorbents. In Microbial Biosorption of Metals (pp. 159–178). Dordrecht: Springer.

Goswami, S., Diengdoh, O.L., Syiem, M.B., Pakshirajan, K. and Kiran, M.G., 2015. Zn (II) and cu (II) removal by *Nostoc muscorum*: A cyanobacterium isolated from a coal mining pit in Chiehruphi, Meghalaya, India. *Canadian Journal of Microbiology*, *61*(3), pp. 209–215.

Gupta, A., Sharma, V., Sharma, K., Kumar, V., Choudhary, S., Mankotia, P., Kumar, B., Mishra, H., Moulick, A., Ekielski, A. and Mishra, P.K., 2021. A review of adsorbents for heavy metal decontamination: Growing approach to wastewater treatment. *Materials*, *14*(16), p. 4702.

Han, P., Teo, W.Z. and Yew, W.S., 2022. Biologically engineered microbes for bioremediation of electronic waste: Wayposts, challenges and future directions. *Engineering Biology*, *6*(1), pp. 23–34.

Hegazi, H.A., 2013. Removal of heavy metals from wastewater using agricultural and industrial wastes as adsorbents. *HBRC Journal*, *9*(3), pp. 276–282.

Hegedűsová, A., Jakabová, S., Vargová, A., Hegedűs, O. and Pernyeszi, T.J., 2009. Use of phytoremediation techniques for elimination of lead from polluted soils. *Nova Biotechnologica*, *9*(2), p. 125.

Hernández-Allica, J., Becerril, J.M. and Garbisu, C., 2008. Assessment of the phytoextraction potential of high biomass crop plants. *Environmental Pollution*, *152*(1), pp. 32–40.

Hoang, A.T., Nižetić, S., Cheng, C.K., Luque, R., Thomas, S., Banh, T.L. and Nguyen, X.P., 2022. Heavy metal removal by biomass-derived carbon nanotubes as a greener environmental remediation: A comprehensive review. *Chemosphere*, *287*, p. 131959.

Hossain, M.A., Ngo, H.H., Guo, W.S. and Setiadi, T., 2012. Adsorption and desorption of copper (II) ions onto garden grass. *Bioresource Technology*, *121*, pp. 386–395.

Huang, H., Yu, N., Wang, L., Gupta, D.K., He, Z., Wang, K., Zhu, Z., Yan, X., Li, T. and Yang, X.E., 2011. The phytoremediation potential of bioenergy crop Ricinus communis for DDTs and cadmium co-contaminated soil. *Bioresource Technology*, *102*(23), pp. 11034–11038.

Hu, Y., Wang, J., Yang, Y., Li, S., Wu, Q., Nepovimova, E., Zhang, X. and Kuca, K., 2024. Revolutionizing soil heavy metal remediation: Cutting-edge innovations in plant disposal technology. *Science of The Total Environment*, *918*, p. 170577.

Ingle, A.P., Paralikar, P., Shende, S., Gupta, I., Biswas, J.K., da Silva Martins, L.H. and Rai, M., 2018. Copper in Medicine: Perspectives and Toxicity. In Biomedical Applications of Metals (pp. 95–112). Cham: Springer.

Iravani, S. and Varma, R.S., 2022. Genetically engineered organisms: Possibilities and challenges of heavy metal removal and nanoparticle synthesis. *Clean Technologies*, *4*(2), pp. 502–511.

Islam, M.A., Hossain, M.T., Khatun, M. and Hossen, M.S., 2015. Environmental impact assessment on frequency of pesticide use during vegetable production. *Progressive Agriculture*, *26*(2), pp. 97–102.

Jain, C.K., Malik, D.S. and Yadav, A.K., 2016. Applicability of plant based biosorbents in the removal of heavy metals: A review. *Environmental Processes*, *3*, pp. 495–523.

Javaid, A., Bajwa, R., Shafique, U. and Anwar, J., 2011. Removal of heavy metals by adsorption on *Pleurotus ostreatus*. *Biomass and Bioenergy*, *35*(5), pp. 1675–1682.

Kainth, S., Sharma, P. and Pandey, O.P., 2024. Green sorbents from agricultural wastes: A review of sustainable adsorption materials. *Applied Surface Science Advances*, *19*, p. 100562.

Kanamarlapudi, S.L.R.K., Chintalpudi, V.K. and Muddada, S., 2018. Application of biosorption for removal of heavy metals from wastewater. *Biosorption*, *18*(69), pp. 70–116.
Karić, N., Maia, A.S., Teodorović, A., Atanasova, N., Langergraber, G., Crini, G., Ribeiro, A.R. and Đolić, M., 2022. Bio-waste valorisation: Agricultural wastes as biosorbents for removal of (in) organic pollutants in wastewater treatment. *Chemical Engineering Journal Advances*, *9*, p. 100239.
Kayan, G.Ö. and Kayan, A., 2021. Composite of natural polymers and their adsorbent properties on the dyes and heavy metal ions. *Journal of Polymers and the Environment*, *29*(11), pp. 3477–3496. https://doi.org/10.1007/s10924-021-02154-x.
Khan, I., Ali, M., Aftab, M., Shakir, S., Qayyum, S., Haleem, K.S. and Tauseef, I., 2019. Mycoremediation: A treatment for heavy metal-polluted soil using indigenous metallotolerant fungi. *Environmental Monitoring and Assessment*, *191*, pp. 1–15.
Khoo, R.Z., Chow, W.S. and Ismail, H., 2018. Sugarcane bagasse fiber and its cellulose nanocrystals for polymer reinforcement and heavy metal adsorbent: A review. *Cellulose*, *25*, pp. 4303–4330.
Kleinübing, S.J., Da Silva, E.A., Da Silva, M.G.C. and Guibal, E., 2011. Equilibrium of Cu (II) and Ni (II) biosorption by marine alga *Sargassum filipendula* in a dynamic system: Competitiveness and selectivity. *Bioresource Technology*, *102*(7), pp. 4610–4617.
Ko, B.G., Anderson, C.W., Bolan, N.S., Huh, K.Y. and Vogeler, I., 2008. Potential for the phytoremediation of arsenic-contaminated mine tailings in Fiji. *Soil Research*, *46*(7), pp. 493–501.
Kobya, M., Demirbas, E., Senturk, E. and Ince, M., 2005. Adsorption of heavy metal ions from aqueous solutions by activated carbon prepared from apricot stone. *Bioresource Technology*, *96*(13), pp. 1518–1521.
Kumar, A., Kumar, A., MMS, C.P., Chaturvedi, A.K., Shabnam, A.A., Subrahmanyam, G., Mondal, R., Gupta, D.K., Malyan, S.K., Kumar, S.S. and A. Khan, S., 2020. Lead toxicity: Health hazards, influence on food chain, and sustainable remediation approaches. *International Journal of Environmental Research and Public Health*, *17*(7), p. 2179.
Kumar, M., Seth, A., Singh, A.K., Rajput, M.S. and Sikandar, M., 2021. Remediation strategies for heavy metals contaminated ecosystem: A review. *Environmental and Sustainability Indicators*, *12*, p. 100155.
Lee, B.X.Y., Hadibarata, T. and Yuniarto, A., 2020. Phytoremediation mechanisms in air pollution control: A review. *Water, Air, & Soil Pollution*, *231*(8), p. 437.
Lee, S.H., Kim, S.O., Lee, S.W., Kim, M.S. and Park, H., 2021. Application of soil washing and thermal desorption for sustainable remediation and reuse of remediated soil. *Sustainability*, *13*(22), p. 12523.
Liang, M., Lu, L., He, H., Li, J., Zhu, Z. and Zhu, Y., 2021. Applications of biochar and modified biochar in heavy metal contaminated soil: A descriptive review. *Sustainability*, *13*(24), p. 14041.
Liang, Y., Pan, Z., Zhu, M., Gao, R., Wang, Y., Cheng, Y. and Zhang, N., 2023. Exposure to essential and non-essential trace elements and risks of congenital heart defects: A narrative review. *Frontiers in Nutrition*, *10*, p. 1121826.
Liu, A., Wang, W., Zheng, X., Chen, X., Fu, W., Wang, G., Ji, J., Jin, C. and Guan, C., 2022a. Improvement of the Cd and Zn phytoremediation efficiency of rice (*Oryza sativa*) through the inoculation of a metal-resistant PGPR strain. *Chemosphere*, *302*, p. 134900.
Liu, G., Dai, Z., Liu, X., Dahlgren, R.A. and Xu, J., 2022b. Modification of agricultural wastes to improve sorption capacities for pollutant removal from water—A review. *Carbon Research*, *1*(1), p. 24.
Li, P., Wang, X., Zhang, T., Zhou, D. and He, Y., 2008. Effect of several amendments on rice growth and uptake of copper and cadmium from a contaminated soil. *Journal of Environmental Sciences*, *20*, pp. 449–455.
Li, F.J., Yang, H.W., Ayyamperumal, R. and Liu, Y., 2022. Pollution, sources, and human health risk assessment of heavy metals in urban areas around industrialization and urbanization-Northwest China. *Chemosphere*, *308*, p. 136396.
Li, C., Zhou, K., Qin, W., Tian, C., Qi, M., Yan, X. and Han, W., 2019. A review on heavy metals contamination in soil: Effects, sources, and remediation techniques. *Soil and Sediment Contamination: An International Journal*, *28*(4), pp. 380–394.
Luciano, A., Viotti, P., Torretta, V. and Mancini, G., 2013. Numerical approach to modelling pulse-mode soil flushing on a Pb-contaminated soil. *Journal of Soils and Sediments*, *13*, pp. 43–55.
Luptakova, A., Ubaldini, S., Macingova, E., Fornari, P. and Giuliano, V., 2012. Application of physical–chemical and biological–chemical methods for heavy metals removal from acid mine drainage. *Process Biochemistry*, *47*(11), pp. 1633–1639.
Madheshiya, P., Gupta, G.S., Sahoo, A. and Tiwari, S., 2022. Biomonitoring Tools and Bioprogramming: An Overview. In *New Paradigms in Environmental Biomonitoring Using Plants* (pp. 341–366). Elsevier.
Małecka, A., Piechalak, A., Morkunas, I. and Tomaszewska, B., 2008. Accumulation of lead in root cells of *Pisum sativum*. *Acta Physiologiae Plantarum*, *30*, pp. 629–637.

Meena, M., Sonigra, P. and Yadav, G., 2021. Biological-based methods for the removal of volatile organic compounds (VOCs) and heavy metals. *Environmental Science and Pollution Research*, *28*, pp. 2485–2508.

Mehrasbi, M.R., Farahmandkia, Z., Taghibeigloo, B. and Taromi, A., 2009. Adsorption of lead and cadmium from aqueous solution by using almond shells. *Water, Air, and Soil Pollution*, *199*, pp. 343–351.

Mench, M. and Martin, E., 1991. Mobilization of cadmium and other metals from two soils by root exudates of *Zea mays* L., *Nicotiana tabacum* L. and *Nicotiana rustica* L. *Plant and Soil*, *132*, pp. 187–196.

Mohiuddin, M., Hossain, M.B., Ali, M.M., Hossain, M.K., Habib, A., Semme, S.A., Rakib, M.R.J., Rahman, M.A., Yu, J., Al-Sadoon, M.K. and Gulnaz, A., 2022. Human health risk assessment for exposure to heavy metals in finfish and shellfish from a tropical estuary. *Journal of King Saud University-Science*, *34*(4), p. 102035.

Mo, J., Yang, Q., Zhang, N., Zhang, W., Zheng, Y. and Zhang, Z., 2018. A review on agro-industrial waste (AIW) derived adsorbents for water and wastewater treatment. *Journal of Environmental Management*, *227*, pp. 395–405. doi:10.1016/j.jenvman.2018.08.069.

Mukherjee, R., Datta, S., Sinha, D., Maurya, A.K. and Roy, S., 2022. Heavy Metal Contamination in Plants: An Overview. In Heavy Metals in Plants: Physiological to Molecular Approach (pp. 16–49). Boca Raton: CRC Press.

Nalla, S., Hardaway, C.J. and Sneddon, J., 2012. Phytoextraction of selected metals by the first and second growth seasons of *Spartina alterniflora*. *Instrumentation Science & Technology*, *40*(1), pp. 17–28.

Nascimento, J.M.D., Otaviano, J.J.S., Sousa, H.S.D. and Oliveira, J.D.D., 2023. Biological Method of Heavy Metal Management: Biosorption and Bioaccumulation. In Heavy Metals in the Environment: Management Strategies for Global Pollution (pp. 315–360). American Chemical Society. doi:10.1021/bk-2023-1456.ch016.

Olufemi, A.C., Mji, A. and Mukhola, M.S., 2022. Potential health risks of lead exposure from early life through later life: Implications for public health education. *International Journal of Environmental Research and Public Health*, *19*(23), p. 16006.

Paz-Ferreiro, J., Lu, H., Fu, S., Mendez, A. and Gasco, G., 2014. Use of phytoremediation and biochar to remediate heavy metal polluted soils: A review. *Solid Earth*, *5*(1), pp. 65–75.

Peng, S., Meng, H., Ouyang, Y. and Chang, J., 2014. Nanoporous magnetic cellulose–chitosan composite microspheres: Preparation, characterization, and application for Cu(II) adsorption. *Industrial & Engineering Chemistry Research*, *53*(6), pp. 2106–2113.

Prasher, P., Mudila, H. and Sharma, M., 2021. Biosorption and bioaccumulation of pollutants for environmental remediation. *Microbial Rejuvenation of Polluted Environment: Volume*, *2*, pp. 379–405.

Priyadarshanee, M. and Das, S., 2021. Biosorption and removal of toxic heavy metals by metal tolerating bacteria for bioremediation of metal contamination: A comprehensive review. *Journal of Environmental Chemical Engineering*, *9*(1), p. 104686.

Priya, A.K., Muruganandam, M., Ali, S.S. and Kornaros, M., 2023. Clean-up of heavy metals from contaminated soil by phytoremediation: A multidisciplinary and eco-friendly approach. *Toxics*, *11*(5), p. 422.

Pulford, I.D. and Watson, C., 2003. Phytoremediation of heavy metal-contaminated land by trees—a review. *Environment International*, *29*(4), pp. 529–540.

Putra, W.P., Kamari, A., Yusoff, S.N.M., Ishak, C.F., Mohamed, A., Hashim, N. and Isa, I.M., 2014. Biosorption of Cu (II), Pb (II) and Zn (II) ions from aqueous solutions using selected waste materials: Adsorption and characterisation studies. *Journal of Encapsulation and Adsorption Sciences, 2014*.

Quainoo, A.K., Konadu, A. and Kumi, M., 2015. The potential of shea nut shells in phytoremediation of heavy metals in contaminated soil using lettuce (Lactuca sativa) as a test crop.

Queiroz, P.S., Barboza, N.R., Cordeiro, M.M., Leão, V.A. and Guerra-Sa, R., 2018. Rich growth medium promotes an increased on Mn(II) removal and manganese oxide production by *Serratia marcescens* strains isolates from wastewater. *Biochemical Engineering Journal*, *140*, 148–156. doi:10.1016/j.bej.2018.09.018.

Raffa, C.M., Chiampo, F. and Shanthakumar, S., 2021. Remediation of metal/metalloid-polluted soils: A short review. *Applied Sciences*, *11*(9), p. 4134.

Raklami, A., Meddich, A., Oufdou, K. and Baslam, M., 2022. Plants—Microorganisms-based bioremediation for heavy metal cleanup: Recent developments, phytoremediation techniques, regulation mechanisms, and molecular responses. *International Journal of Molecular Sciences*, *23*(9), p. 5031.

Rana, A., Sindhu, M., Kumar, A., Dhaka, R.K., Chahar, M., Singh, S. and Nain, L., 2021. Restoration of heavy metal-contaminated soil and water through biosorbents: A review of current understanding and future challenges. *Physiologia Plantarum*, *173*(1), pp. 394–417.

Ranasinghe, S.H., Navaratne, A.N. and Priyantha, N., 2018. Enhancement of adsorption characteristics of Cr(III) and Ni(II) by surface modification of jackfruit peel biosorbent. *Journal of Environmental Chemical Engineering* *6*(5), pp. 5670–5682. doi:10.1016/j.jece.2018.08.058.

Ranđelović, D., Gajić, G., Mutić, J., Pavlović, P., Mihailović, N. and Jovanović, S., 2016. Ecological potential of *Epilobium dodonaei* Vill. for restoration of metalliferous mine wastes. *Ecological Engineering, 95*, pp. 800–810.

Rashid, A., Mahmood, T., Mehmood, F., Khalid, A., Saba, B., Batool, A. and Riaz, A., 2014. Phytoaccumulation, competitive adsorption and evaluation of chelators-metal interaction in lettuce plant. *Environmental Engineering & Management Journal (EEMJ), 13*(10), pp. 2583–2592.

Rezania, S., Taib, S.M., Din, M.F.M., Dahalan, F.A. and Kamyab, H., 2016. Comprehensive review on phytotechnology: Heavy metals removal by diverse aquatic plants species from wastewater. *Journal of Hazardous Materials, 318*, pp. 587–599.

Rosales, E., Ferreira, L., Sanromán, M.Á., Tavares, T. and Pazos, M., 2015. Enhanced selective metal adsorption on optimised agroforestry waste mixtures. *Bioresource Technology, 182*, pp. 41–49.

Ruiz, O.N. and Daniell, H., 2009. Genetic engineering to enhance mercury phytoremediation. *Current Opinion in Biotechnology, 20*(2), pp. 213–219.

Sabir, A., Altaf, F., Batool, R., Shafiq, M., Khan, R.U. and Jacob, K.I., 2021. Agricultural Waste Absorbents for Heavy Metal Removal. In Green Adsorbents to Remove Metals, Dyes and Boron from Polluted Water (pp. 195–228). Cham: Springer.

Samarajeewa, U., 2022. Heavy metals and food safety in Sri Lanka: A review. *Journal of the National Science Foundation of Sri Lanka, 50*(3), p. 541.

Sameera, V., Naga Deepthi, C.H., Srinu Babu, G. and Ravi Teja, Y., 2011. Role of biosorption in environmental cleanup. *Journal of Microbial & Biochemical Technology 1*, pp.1–8. doi:10.4172/1948-5948.R1-001.

Sao, K., Khan, F., Pandey, P.K. and Pandey, M., 2014. A review on heavy metals uptake by plants through biosorption. *International Proceedings of Economics Development and Research, 75*, p. 78.

Saravanan, A., Kumar, P.S., Ramesh, B. and Srinivasan, S., 2022. Removal of toxic heavy metals using genetically engineered microbes: Molecular tools, risk assessment and management strategies. *Chemosphere, 298*, p. 134341.

Sen, R., Jha, P. and Jobby, R., 2021. Immobilized Fungal Technology: a New Perspective for Bioremediation of Heavy Metals. In *Bioremediation for Environmental Sustainability* (pp. 541–559). Elsevier.

Senthil Kumar, P. and Gunasundari, E., 2018. Bioremediation of Heavy Metals. In Bioremediation: Applications for Environmental Protection and Management, pp. 165–195. Singapore: Springer.

Sharma, S., 2012. Bioremediation: Features, strategies and applications. *Asian Journal of Pharmacy and Life Science, 2231*, p. 4423.

Sharma, S., Sharma, P. and Mehrotra, P., 2010. Bioaccumulation of heavy metals in *Pisum sativum* L. growing in fly ash amended soil. *Journal of American Science, 6*(6), pp. 43–50.

Šulinskienė, J., Bernotienė, R., Baranauskienė, D., Naginienė, R., Stanevičienė, I., Kašauskas, A. and Ivanov, L., 2019. Effect of zinc on the oxidative stress biomarkers in the brain of nickel-treated mice. *Oxidative Medicine and Cellular Longevity, 2019*, p. 8549727.

Suman, J., Uhlik, O., Viktorova, J. and Macek, T., 2018. Phytoextraction of heavy metals: A promising tool for clean-up of polluted environment? *Frontiers in Plant Science, 9*, p. 1476.

Suryan, S. and Ahluwalia, S.S., 2012. Biosorption of heavy metals by paper mill waste from aqueous solution. *International Journal of Environmental Sciences, 2*(3), pp. 1331–1343.

Tabelin, C.B., Igarashi, T., Villacorte-Tabelin, M., Park, I., Opiso, E.M., Ito, M. and Hiroyoshi, N., 2018. Arsenic, selenium, boron, lead, cadmium, copper, and zinc in naturally contaminated rocks: A review of their sources, modes of enrichment, mechanisms of release, and mitigation strategies. *Science of the Total Environment, 645*, pp. 1522–1553.

Talukdar, D., Sharma, R., Jaglan, S., Vats, R., Kumar, R., Mahnashi, M.H. and Umar, A., 2020. Identification and characterization of cadmium resistant fungus isolated from contaminated site and its potential for bioremediation. *Environmental Technology & Innovation, 17*, p. 100604.

Tan, G., Yuan, H., Liu, Y. and Xiao, D., 2010. Removal of lead from aqueous solution with native and chemically modified corncobs. *Journal of Hazardous Materials, 174*(1–3), pp. 740–745.

Tekere, M., 2020. Biological Strategies for Heavy Metal Remediation. In Methods for Bioremediation of Water and Wastewater Pollution (pp. 393–413). Cham: Springer.

Tripathi, M. and Gaur, R., 2021. Bioactivity of Soil Microorganisms for Agriculture Development. In *Microbes in Land Use Change Management* (pp. 197–220). Elsevier.

Tsezos, M., Remoundaki, E. and Hatzikioseyian, A. 2006, October. Biosorption-principles and applications for metal immobilization from waste-water streams. In Proceedings of EU-Asia Workshop on Clean Production and Nanotechnologies (pp. 23–33). Seoul.

Ullah, A., Tahir, A., Rashid, H.U., ur Rehman, T., Danish, S., Hussain, B. and Akca, H., 2021. Strategies for reducing Cd concentration in paddy soil for rice safety. *Journal of Cleaner Production, 316*, p. 128116.

Velazquez-Jimenez, L.H., Pavlick, A. and Rangel-Mendez, J.R., 2013. Chemical characterization of raw and treated agave bagasse and its potential as adsorbent of metal cations from water. *Industrial Crops and Products*, *43*, pp. 200–206.

Verma, S., Bhatt, P., Verma, A., Mudila, H., Prasher, P. and Rene, E.R., 2021. Microbial technologies for heavy metal remediation: Effect of process conditions and current practices. *Clean Technologies and Environmental Policy*, *25*, pp. 1485–1507.

Villen-Guzman, M., Gutierrez-Pinilla, D., Gomez-Lahoz, C., VeredaAlonso, C., Rodriguez-Maroto, J.M. and Arhoun, B., 2019. Optimization of Ni (II) biosorption from aqueous solution on modified lemon peel. *Environmental Research. 179*(Pt B), 108849. doi:10.1016/j.envres.2019.108849.

Wang, N., Wang, Y., Li, B., Huang, F., Sun, C., Li, X., Zhao, R. and Wang, Y., 2023. Characteristics of a copper-cadmium tolerant strain screened from tailings and its potential in remediation of heavy metal contaminated soil. *Water, Air, & Soil Pollution*, *234*(4), p. 277.

Wani, P.A., Khan, M.S. and Zaidi, A., 2008. Effects of heavy metal toxicity on growth, symbiosis, seed yield and metal uptake in pea grown in metal amended soil. *Bulletin of Environmental Contamination and Toxicology*, *81*, pp. 152–158.

Wu, G., Kang, H., Zhang, X., Shao, H., Chu, L. and Ruan, C., 2010. A critical review on the bio-removal of hazardous heavy metals from contaminated soils: Issues, progress, eco-environmental concerns and opportunities. *Journal of Hazardous Materials*, *174*(1–3), pp. 1–8.

Wu, Y., Zhang, S., Guo, X. and Huang, H., 2008. Adsorption of chromium (III) on lignin. *Bioresource Technology*, *99*(16), pp. 7709–7715.

Yaashikaa, P.R., Kumar, P.S., Saravanan, A. and Vo, D.V.N., 2021. Advances in biosorbents for removal of environmental pollutants: A review on pretreatment, removal mechanism and future outlook. *Journal of Hazardous Materials*, *420*, p.126596. https://doi.org/10.1016/j.jhazmat.2021.126596.

Yan, G. and Viraraghavan, T., 2008. Mechanism of biosorption of heavy metals by Mucor rouxii. *Engineering in Life Sciences*, *8*(4), pp. 363–371.

Younas, F., Younas, S., Bibi, I., Farooqi, Z.U.R., Hameed, M.A., Mohy-Ud-Din, W., Shehzad, M.T., Hussain, M.M., Shakil, Q., Shahid, M. and Niazi, N.K., 2024. A critical review on the separation of heavy metal(loid)s from the contaminated water using various agricultural wastes. *International Journal of Phytoremediation*, *26*(3), pp. 349–368.

15 Green Solutions for Heavy Metal Remediation

Unveiling the Potential of Agricultural Waste through Techno-Economic Analysis and Life Cycle Assessment

*Priyanka Singh, Umesh Kumar, Akanksha Dwivedi, Prince Kumar Singh, Indrajeet Kumar, Saumya Mishra, Avanish Prajapati, Supriya Tiwari, and Rajesh Kumar Sharma**
Laboratory of Ecotoxicology, Department of Botany, Institute of Science, Banaras Hindu University, Varanasi, India

15.1 INTRODUCTION

The escalating global concern over the contamination of the environment with heavy metals reflects the critical awareness of their deleterious effects on ecosystems, human health, and the broader environmental landscape (Abadi et al., 2023). Heavy metals, including lead, cadmium, chromium, mercury, arsenic, and others, are pervasive in diverse environmental realms such as soil,[1] water, and air, primarily attributable to anthropogenic endeavors such as industrialization, mining activities, agricultural practices, wastewater disposal, and improper waste handling (Sharma et al., 2007; Singh et al., 2022). Due to their non-biodegradable characteristics, heavy metals tend to accumulate in the environment, posing significant threats to living organisms through the processes of bioaccumulation and biomagnification within the food web (Singh and Sharma, 2021). Conventional approaches for addressing heavy metal pollution, involving physical, chemical, and biological methods such as chemical precipitation, ion exchange, electro-coagulation, reverse osmosis, adsorption, aerobic and anaerobic biofilm filtration, etc., have been extensively utilized in remediation practices (Choudhary et al., 2022). However, these methodologies often incur high expenditures, lead to the production of secondary pollutants, and might not be appropriate for large-scale remediation initiatives. In recent years, the exploration of nature-based, sustainable, or eco-friendly methods has placed particular emphasis on utilizing agricultural waste for the remediation of heavy metals (Datta et al., 2023). Adopting sustainable practices in heavy metal remediation approaches not only addresses the urgent challenges posed by heavy metal pollution but also promotes the resilience and regeneration of ecosystems, safeguarding their integrity for future generations (Yang et al., 2024).

Agricultural production has tripled in the last five decades, due to primarily driven by the expansion of agricultural land for farming, the transformative impact of the green revolution on enhancing productivity, and the escalating demands due to population growth (Food and Agriculture Organization of the United Nations [FAO], 2017a; Organization for Economic Co-operation and

* **Corresponding author:** rajeshbot78@bhu.ac.in

DOI: 10.1201/9781032685793-15

Development [OECD], 2019). Agriculture sustains the production of an estimated 23.7 million tons food per day worldwide (Food and Agriculture Organization of the United Nations [FAO], 2017b). At global level, rapid growth in agricultural practices has led to a significant increase in agricultural waste generation, resulting in intensifying environmental pressure and negative impacts on soil integrity, air quality, and water systems (Food and Agriculture Organization of the United Nations [FAO], 2017c). There are several types of agricultural waste that have become a public concern and pose a threat to the sustainability of agricultural regimes, including crop residues (such as leaf litter, stems, straws, husks, and weeds), livestock wastes (comprising dung, residual milk, and waste feed), poultry waste (including spilled feed, feathers, and bedding material), slaughterhouse waste (encompassing blood, hair, hides, flesh, and bones), agro-industrial waste (such as bagasse, molasses, and peels), and oil-seed cakes (including palm kernel cake, groundnut, soybean, mustard, and coconut) (Bustillo-Lecompte and Mehrvar, 2015; Duque-Acevedo et al., 2020; Seidavi et al., 2019; Thangaraju et al., 2020; Tripathi et al., 2019).

The transformation of agricultural waste materials into value-added products, capitalizing on their potent ability to immobilize heavy metals, plays a pivotal role in mitigating environmental contamination (Tang et al., 2023). Agricultural waste exhibits various functional groups, including carboxyl, hydroxyl, carbonyl, and amino groups, facilitating heavy metal binding through complexation and chelation processes (Younas et al., 2024). Furthermore, advancements in technology such as chemical treatment, heat treatment, and nanoparticle deposition have been explored to optimize the efficacy of agricultural waste-based adsorbents (Yadav et al., 2023). Moreover, the integration of agricultural waste into remediation strategies aligns with the principles of circular economy, promoting the reuse and recycling of waste materials to minimize resource utilization and waste generation (Yang et al., 2023). The application of techno-economic analysis (TEA) presents a methodical framework for evaluating the economic viability of incorporating agricultural waste into heavy metal remediation processes. By assessing the costs and benefits associated with different remediation methods, TEA empowers informed decision-making regarding investment, technology selection, and process optimization. The economic viability of agricultural waste-based remediation strategies is analyzed by considering factors such as raw material acquisition costs, processing expenses, and the market value of remediated products (Ferdous et al., 2023).

Life cycle assessment (LCA) complements TEA by analyzing the environmental footprint of agricultural waste-based remediation processes across their entire life cycle (Saba et al., 2024). LCA emerges as an indispensable tool for analyzing the environmental footprint of products and processes, spanning from the extraction of raw materials through production, utilization, and disposal or recycling. Utilizing the inherent attributes of agricultural waste materials and integrating advanced analytical approaches such as TEA and LCA empowers us to fully exploit their potential as cost-effective, highly efficient, and environmentally friendly remediation technologies (Tushar et al., 2023). Fostering interdisciplinary research, innovation, and stakeholder involvement can pave the path toward a cleaner, healthier, and more sustainable future, wherein agricultural waste transforms into a valuable resource for environmental preservation and resource conservation.

15.2 AGRICULTURAL WASTE: A PROMISING RESOURCE FOR HEAVY METAL REMEDIATION

Agricultural and food industry waste arises from agricultural activities and the processing of agricultural and food products. Waste from food and agricultural activities is either used to extract value-added goods or left unused. Multiple articles have examined the adsorption of emerging pollutants (heavy metals) using agricultural waste such as fruit peels (e.g., citrus, banana, mango, and pineapple) (Tee et al., 2022), wood (e.g., stem and cobs) (Shaheen et al., 2019), and cereals (e.g., rice, wheat, and corn) (Karić et al., 2022). Agricultural waste biomass is expected to expand at a pace of 10% annually in the next years, driven by rising worldwide demand for crops and high consumption

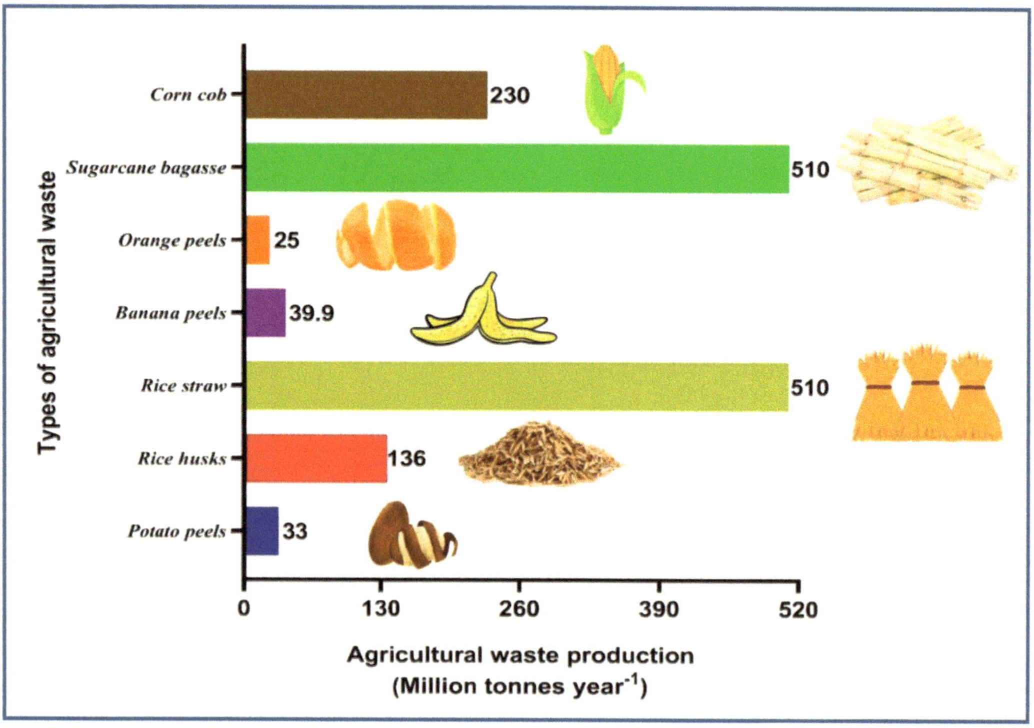

FIGURE 15.1 The worldwide annual production of agricultural waste. (Modified from Chong et al., 2023.)

of agricultural goods. Around 150 billion metric tons of agricultural waste are produced annually. Figure 15.1 illustrates the annual worldwide production of various categories of agricultural waste.

In recent times, transforming various agricultural waste varieties into biochar has demonstrated enhanced abilities for adsorption and mechanisms targeting a range of pollutants like dyes and heavy metals (Figure 15.2) (Zhao et al., 2021).

Biochar, characterized by its stability, fine texture, carbon richness, and porous structure, is derived from biomass feedstock through high-temperature pyrolysis (300–700°C) under restricted or anaerobic conditions (Meng et al., 2019). Increasing the porosity of activated carbon involves undergoing physical processes such as steam, CO_2, or O_2 activation; chemical processes like acid, base, or salt activation; and biological processes such as phytoremediation, phytostabilization, phytoextraction, and phytovolatilization (Figure 15.2). Biochar, characterized by its loose and porous structure, possesses a significant specific surface area and abundant functional groups such as carboxyl, phenolic, carbonyl, and hydroxyl groups (Shang et al., 2020).

These attributes enhance its ion exchange sites and soil cation exchange capacity, facilitating the generation of processes like heavy metal adsorption, precipitation, or ion exchange (Al-Wabel et al., 2015). However, the majority of biochar tends to be alkaline, thereby elevating soil pH and reducing the mobility of heavy metals within it (Ali et al., 2017). Additionally, biochar contains specific nutrients that, when integrated into the soil, can enhance organic matter levels, soil fertility, and crop yields (Zhao et al., 2019). According to Cui et al. (2016), during a five-year trial period, the bioavailability of Cd and Pb decreased by 8%–45% and 14%–50%, respectively. A field trial demonstrated that applying maize straw-derived biochar at a rate of 15 t ha^{-1} resulted in a 20.0% reduction in DTPA-Cd under mild Cd pollution and a 34.5% reduction under moderate Cd pollution (Shen et al., 2016). In another investigation, the introduction of woody biochar to a contaminated site in the UK's chemical industry sector led to a decrease in extraction concentrations of nickel (Ni(II)) and zinc

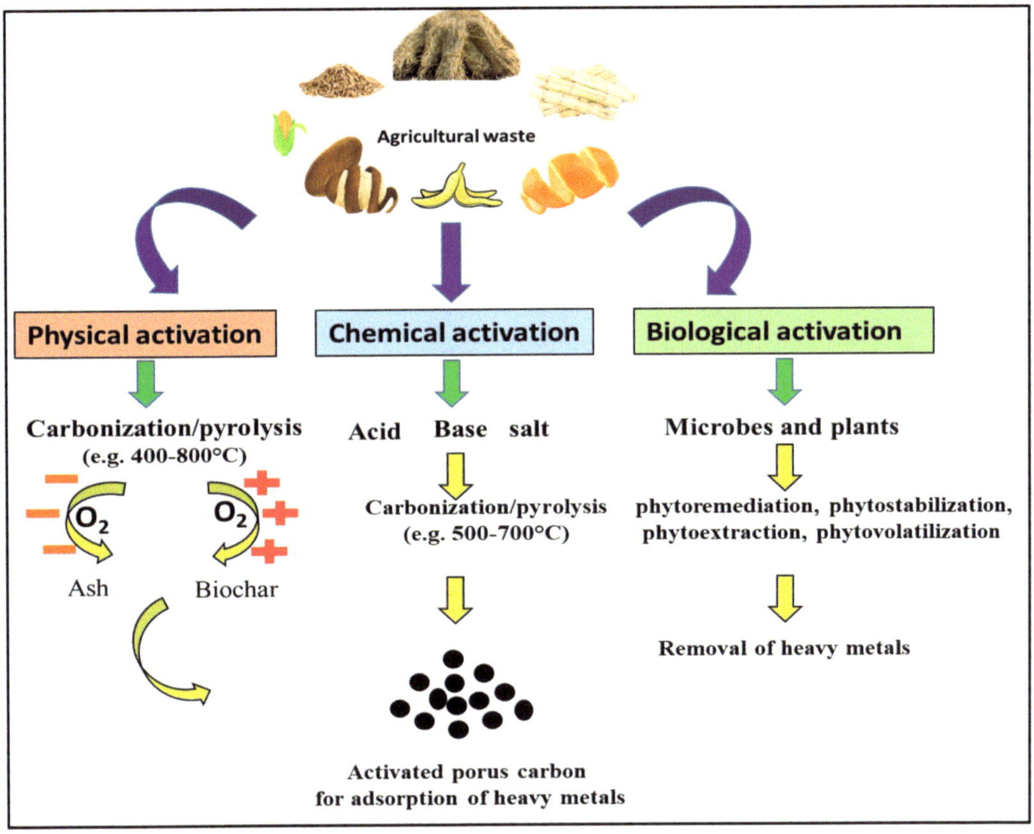

FIGURE 15.2 A schematic illustrating the stages involved in physical activation, chemical activation, and biological activation of raw materials.

(Zn(II)) by 83%–98% in carbonic acid leaching tests over a span of three years (Yang et al., 2017). Furthermore, treating tobacco stalk biochar resulted in a decrease of 64.2% in extractable Cd and 94.9% in Zn content at a 5% application rate, while tobacco plant shoot and root biomass increased by 30.3% and 36.2%, respectively (Saravanan et al., 2022). The following examples illustrate how agricultural waste has been harnessed to extract heavy metals from polluted soil.

15.2.1 Exploring Rice Husk as a Viable Solution for Heavy Metal Immobilization

Rice husk, a by-product of rice cultivation predominantly found in Asian countries, constitutes 10%–20% of the world's annual rice production, which totals around 500 million metric tons. It comprises 70%–85% organic matter, including lignin, cellulose, and sugars, with the remaining portion being silica. Due to its rapid uptake and high adsorption capacity, rice husk is an appealing material for adsorption purposes. Remarkable removal efficiencies have been observed for lead(II) and cadmium(II), reaching 99.8% and 99.2%, respectively, when treated with red dyestuff, and 100% for lead(II) and 93.3% for mercury(II) when treated with yellow dyestuff. Under optimal conditions, the removal rates for chromium, zinc, copper, and cadmium ions from aqueous solutions are 79%, 85%, 80%, and 85%, respectively (Belete, 2017). Kinetic removal experiments conducted in batch setups reveal that the uptake percentages of lead, cadmium, copper, and zinc were 54.3%, 8.24%, 51.4%, and 56.7%, respectively, using rice husk carbon, while with rice husk ash at the same dosages, the figures varied to 74.04%, 43.4%, 70.08%, and 77.2%, respectively (Nhapi et al., 2011).

15.2.2 Sugarcane Bagasse: A Promising Resource for Heavy Metal Immobilization

Sugarcane bagasse, a by-product of the sugar refining industry, exhibits increased removal efficiency for Cd(II) and Zn(II) as the pH surpasses 2, reaching 100% uptake at pH levels exceeding 8.0 (Mohan and Singh, 2002). Moreover, research indicates that the sorption capacity of the resulting activated carbon toward Cd(II) and Zn(II) is comparable to or even superior to that of other available adsorbents. Khan et al. (2001) reported that a dosage of 0.8 g/50 mL of adsorbent is adequate to achieve 80%–100% removal of Cr(VI) from aqueous solutions with an initial metal concentration of 20 mg L^{-1}, especially effective at pH 1, with removal efficacy diminishing to less than 15% at pH 3.

15.2.3 Orange Peels: Unveiling Their Potential for Heavy Metal Immobilization

Orange peel, discarded globally by the soft drink and orange juice industry without treatment, has garnered attention from numerous scientists due to its abundance in the orange processing industry, with many exploring its potential as an adsorbent for various water pollutants. Researchers have investigated the efficacy of orange peel in adsorbing Zn, Cu, Pb, Ni, and Cr from aqueous solutions (Boumediene et al., 2018). The adsorption sequence was found to be Ni > Cu > Pb > Zn > Cr. The adsorption of Ni was influenced by pH, with maximum removal occurring at pH 6. The extent of Ni(II) removal was observed to be dependent on factors such as sorbent dose, initial concentration, pH, and temperature. Adsorption kinetics followed a first-order model. The process was determined to be endothermic, indicating monolayer adsorption of Ni(II) with a maximum adsorption efficiency of 96% at 50°C for an initial concentration of 50 mg L^{-1} at pH 6 (Ajmal et al., 2000).

15.2.4 Peanut Shell: A Sustainable Solution for Heavy Metal Immobilization

Wilson et al. (2006) introduced peanut shell-based activated carbon for eliminating various heavy metal ions. Furthermore, Chamarthy et al. (2001) investigated activated peanut shell for removing five heavy metal ions, noting its superior efficiency compared to numerous commercial adsorbents. Furthermore, Kannan and Sundaram (2001) analyzed peanut shell carbon, rice, and coconut shell carbons for methylene blue removal, with peanut shell demonstrating the lowest loading capacity compared to rice and coconut shells. Malik et al. (2007) conducted a comprehensive evaluation to assess the efficacy of activated carbon derived from peanut shell in the removal of malachite green, a commonly used synthetic dye known for its environmental persistence and potential health risks. Through meticulous experimentation and rigorous analysis, they systematically evaluated the adsorption capacity of peanut shell-based activated carbon for malachite green removal from aqueous solutions.

15.2.5 Tea Waste: Exploring Its Potential for Heavy Metal Immobilization

The higher concentrations of aromatic, carboxylate, hydroxyl, and phenolic groups in tea leaves result in enhanced ion exchange behavior in the related bio-waste, leading to improved adsorption capacity (Hussain et al., 2018). Tea waste has demonstrated significant adsorption capabilities for several metals such as Cu(II), Pb(II), Cr(VI), Ni(II), and As(III). For instance, almost 99% of Zn(II) (50 mg dm^{-3} starting concentration) were adsorbed at pH 4.2 onto tea manufacturing waste in batch settings (Wasewar et al., 2009). Additionally, Cherdchoo et al. (2019) revealed that when subjected to the same experimental conditions—250 mg L^{-1} Cr(VI) concentration and pH 2—the removal rate of Cr(VI) was significantly higher with mixed waste tea (94 mg g^{-1}) than with coffee grounds (87 mg g^{-1}). Malkoc and Nuhoglu (2006) conducted column studies to assess the efficacy of tea waste in adsorbing Ni(II). The highest adsorption of Ni(II) and the longest breakthrough time were observed at pH 4. Adsorption reduced as the flow rate increased, with adsorption values of 11, 10, and 7 mg g^{-1} observed at flow rates of 5, 10, and 20 mL min^{-1}, respectively. Additionally, adsorption rose when particle size decreased from 1–3 to 0.15–0.25 mm.

15.2.6 Unveiling the Potential of Potato Peel Waste for Heavy Metal Remediation

Potato peels are a globally abundant solid waste that can be used as a biosorbent. Guechi and Hamdaoui (2016) investigated the use of untreated dried potato peels for removing Cu(II). Aman et al. (2008) utilized potato peels to create charcoal and investigated its adsorption capacity for Cu(II) in industrial wastewater samples. The potato peels were cleaned, dried, and then subjected to heat treatment in a furnace at 700°C (973K) for 2 hours. The potato peel charcoal, referred to as PPC, was pulverized and particles of 0.2 mm were obtained through sifting. One gram of PPC removed 99.8% of Cu(II) (150 mg L^{-1}, 100 mL) at pH 6 after being shaken for 20 minutes at 30°C. Mahale et al. (2016) investigated the removal of nickel using a powder with a particle size of 0.147 mm generated by treating potato peels with HCl, rinsing with distilled water, drying at 60°C, and sieving. The removal efficiency was highly influenced by pH, with the highest adsorption efficiency occurring at pH 9. The material showed a removal effectiveness of around 90% in a 250 mg L^{-1} Ni(II) solution after 50 minutes of contact time at the optimal dosage of 1.4 g L^{-1}.

The incorporation of agricultural waste into heavy metal remediation strategies offers valuable and versatile solutions that are cost-effective, sustainable, and environmentally beneficial in combating pollution. By capitalizing on the inherent properties of agricultural residues, innovative techniques can be devised to address heavy metal contamination in various environmental contexts (Table 15.1).

TABLE 15.1
Evaluation of Heavy Metal Immobilization Potential Using Different Raw and Modified Agricultural Waste Materials

S. No.	Agricultural Waste	Production Methods	Heavy Metals	References
1.	Palm fuel ash	Combustion	Cu, Pb, Zn, Ni, Cr	Bello et al., 2010; Chowdhury et al., 2011; Chu and Hashim, 2002; Hashim et al., 2003
2.	Coconut shell fibers	Natural form	Cu, Pb, Zn, Ni, Cr	Abdulrasaq and Basiru, 2010; Israel and Eduok, 2012; Okafor et al., 2012
3.	Coconut husk	Natural form	Cu, Ni, Pb, Zn	Malik and Dahiya, 2017
4.	Mangosteen	Carbonization and activation	Cu, Pb, Ni, Cr	Huang et al., 2013; Zein et al., 2010
5.	Rice husk	Carbonization and activation	Cu, Pb, Zn, Ni, Cr, As	Zakir et al., 2013
6.	Durian peel	Carbonization and activation	Pb, Ni, Cr	Lesmana et al., 2009; Narayanasamy et al., 2022
7.	Corncob	Carbonization and activation	Cu, Pb, Zn, Ni, Cr, Fe	Olayebi and Adebayo, 2017; Sallau et al., 2012
8.	Corn straw	Natural form	Cd, Pb	Yousef et al., 2020
9.	Coir pith	Carbonization and activation	Pb, Hg, Cd, Ni, Cu	Kadirvelu et al., 2001
10.	Raw rice bran	Natural form and combustion	Cd, Cu, Pb, Zn, Cr, Ni	Daifullah et al., 2003; Montanher et al., 2005; Oliveira et al., 2005
11.	Rice husk	Modified with water and HCL	Cr, Cu, Zn, Cd	Hegazi et al., 2013
	Rice husk	Phosphate treated	Cd	
	Rice husk	Tartaric acid treated	Cu	Wong et al., 2003
12.	Rice husk ash	Combustion	Cd, Cr, Cu, Zn, As, Pb	Bansal et al., 2009; Daifullah et al., 2003

(Continued)

TABLE 15.1 *(Continued)*
Evaluation of Heavy Metal Immobilization Potential Using Different Raw and Modified Agricultural Waste Materials

S. No.	Agricultural Waste	Production Methods	Heavy Metals	References
13.	Rice Straw	Natural form	Pb, Cu, Cd, Zn, Ni	Johns et al., 1998
14.	Lemon residues	Modified with $FeCl_3$	As	Pérez et al., 2023
15.	Orange peel	Modified with grafted copolymerization	Cd, Ni, Pb	Feng et al., 2011
16.	Orange peel	Modified with NaOH 0.8 M and $CaCl_2$ 0.8 M	Cu, Pb, Zn	Feng and Guo, 2012; Sulyman et al., 2017
17.	Orange peel	Natural form	Ni, Hg	Chinyelu et al., 2015; Gönen and Serin, 2012
18.	Sugarcane bagasse and its activated carbon	Carbonization and activation	Ni, Cr, Zn, Cd	Garg et al., 2008; Khan et al., 2001
19.	Sugarcane bagasse	Natural form	Mn, Cd, Hg, Zn, Cu	Krishnan et al., 2016; Moubarik and Grimi, 2015
20.	Rubber wood saw dust	Natural form	Cr	Karthikeyan et al., 2005
21.	Saw dust of *Pinus sylvestris*	Natural form	Pb, Cd	Taty-Costodes et al., 2003
22.	Maple saw dust	Natural form	Pb, Cu	Yu et al., 2000
23.	Mango saw dust	Natural form	Cu	Ajmal et al., 1998
24.	Charred saw dust	Chemically treated	As, Cr	Akolgo et al., 2021
25.	Groundnut shells and saw dust	Natural form	Ni, Cu, Zn	Shukla and Pai, 2005
26.	Saw dust of oak and black locust hard wood	Modified	Ni, Cu, Zn	Sciban et al., 2006
27.	Saw dust of Cedrus deodar wood	Modified with sodium hydroxide	Cd	Memon et al., 2007
28.	Walnut saw dust	Modified with formaldehyde in sulfuric acid	Pb	Bulut et al., 2007
29.	Orange Saw dust		Cu, Ni, Cr	Shukla and Pai, 2005
30.	Oak tree saw dust	Modified with hydrochloric acid	Pb	Argun and Dursun, 2007
31.	Pineapple peel fiber	Modified with succinic anhydride	Cu, Cd, Pb	Hu et al., 2011
32.	Pineapple peel	Natural form	Cr	Shakya and Agarwal, 2019; Yousef et al., 2020

(Continued)

TABLE 15.1 (Continued)
Evaluation of Heavy Metal Immobilization Potential Using Different Raw and Modified Agricultural Waste Materials

S. No.	Agricultural Waste	Production Methods	Heavy Metals	References
33.	Wheat bran/ Wheat straw	Modified with sulfuric acid	Cd, Cr, Hg, Cu, Ni, Pb	Alalwan et al., 2020; Farooq et al., 2011
34.	Almond shell biochar	Pyrolysis	Ni, Cd	Kılıc et al., 2013
35.	Banana peel	Natural form	Pb, Ni	Ayob et al., 2021
36.	Mango kernel	Bio composite	Cr	Akram et al., 2017
37.	Jackfruit peel	Natural form	Mn, Pb, Cd, Cu	Ayob et al., 2021; Ibrahim et al., 2020
38.	Jackfruit leaves powder	Natural form	Ni	Boruah et al., 2015
39.	Avocado peel	Natural form	Pb, Ni	Ahmad and Danish, 2022; Mallampati et al., 2015
40.	Banana peel biochar	Pyrolysis	Pb	Zhou et al., 2017
41.	*Brassica campestris*	Natural form	Ni, Pb	Baby Shaikh et al., 2018
42.	Pistachio hull waste	Natural form	Hg	Rajamohan, 2014
43.	Flax fiber tows	Natural form	Cu, Pb, Zn	Abbar et al., 2017

15.3 TECHNO-ECONOMIC ANALYSIS OF AGRICULTURAL WASTE SOLUTIONS

Using agricultural waste for heavy metal remediation supports circular economy concepts and sustainable agriculture by improving soil health and agricultural productivity while managing waste. Agricultural waste materials efficiently decrease heavy metal levels in soil by methods such as adsorption, ion exchange, and microbial activity. For instance, biochar and composting have strong sorption abilities and enhance microbial activity, converting heavy metal contaminants into less harmful forms. From an economic standpoint, using agricultural waste for remediation provides a cost-efficient option compared to traditional approaches, taking into account elements like material accessibility, processing expenses, and expandability. Utilizing local sourcing decreases transportation costs and bolsters regional economies. TEA combines technological feasibility and economic viability to make educated decisions based on cost-benefit evaluation. This method evaluates the financial expenses of agricultural waste clean-up compared to the advantages of enhancing soil quality, increasing crop output, and decreasing environmental pollution. This analysis offers a thorough framework for assessing the viability and advantages of sustainable remediation approaches.

15.3.1 Cost Considerations

Utilizing agricultural waste for heavy metal decontamination raises multiple economic factors to consider. The collection of agricultural waste, including crop leftovers, straw, and husks, is influenced by factors such as local abundance, transportation expenses, and the type of trash (Shabani et al., 2013; Shaheen et al., 2019). In regions where agriculture is flourishing, there may be a lot of garbage that is both plentiful and inexpensive. However, in areas where there is a shortage of waste or when it is being used for other purposes such as animal feed or bioenergy generation, costs can increase. Enhancing the effectiveness of agricultural waste for heavy metal decontamination is achieved through processes such as shredding, drying, grinding, or chemical treatment (Dai et al., 2018). The prices are influenced by factors such as operational size, treatment method, and equipment used. Enhancing efficiency through processing increases total costs. Costs associated with applying agricultural waste depend on elements such as the amount of garbage, personnel, equipment, and site details (Koul et al., 2022). Specialized equipment and knowledge can increase expenses, along with continuous monitoring and maintenance to ensure effectiveness and reduce environmental risks. Comparing agricultural waste-based solutions to traditional remediation techniques such as chemical extraction or thermal treatment shows potential cost benefits, especially in areas where waste materials are abundant and cost-effective (Hu et al., 2024). Additionally, these technologies frequently result in reduced environmental impacts and provide extra advantages such as improving soil quality and increasing organic matter content. Therefore, comprehensive cost-benefit analyses are required to determine the feasibility and potential for expansion of the project. Moreover, agricultural waste-based solutions offer long-lasting benefits by enhancing soil fertility and ecosystem health, requiring minimal maintenance costs. In contrast, traditional methods may demand continuous maintenance and costly interventions over time. Utilizing agricultural waste for remediation promotes environmental sustainability by repurposing waste materials and involving local communities, unlike traditional approaches that may result in environmental drawbacks like energy consumption and chemical contamination (Habineza et al., 2017; Sharma et al., 2018). Overall, agricultural waste presents cost-effective and sustainable advantages over traditional methods, particularly in heavy metal contamination remediation.

15.3.2 Economic Benefits

Utilizing agricultural waste for environmental decontamination offers cost-effective solutions due to its ready availability and minimal processing costs compared to traditional methods (Karić et al., 2022). With straightforward application techniques and long-term benefits like soil fertility improvement, agricultural waste-based remediation minimizes ongoing maintenance expenses. Moreover, it promotes environmental sustainability by diverting waste from landfills and reducing synthetic chemical usage. Community involvement and social benefits further enhance its appeal, fostering local engagement and sustainable farming practices. In essence, agricultural waste-based remediation presents a holistic and economically advantageous approach to addressing environmental pollution and promoting sustainable development. Further, utilizing agricultural waste for environmental decontamination creates jobs in waste collection, processing, and application, including tasks like sorting, composting, and soil amendment. Research and development efforts drive innovation in remediation techniques, while community engagement initiatives raise awareness and educate about sustainability. Additionally, it fosters entrepreneurship, leading to the establishment of small businesses focused on waste management and environmental consultancy (Mihai et al., 2023).

Several research studies have underscored the economic evaluation of agro waste solutions for metal decontamination regarding their effectiveness, adaptability, and environmental sustainability (Galinato et al., 2011; Roberts et al., 2010). The economic efficiency of these solutions varies based on factors such as the type of raw materials used, the technology employed for conversion, and the incorporation of subsidies or carbon credits for greenhouse gas mitigation (Cleary and Caspersen, 2015). Studies indicate that the adoption of smart agricultural practices not only enhances crop yields and overall production but also improves the economic well-being of farmers while mitigating the adverse impacts of climate change. Biochar derived from agro waste demonstrates long-term potential for agricultural enhancement compared to inorganic fertilizers (Khan et al., 2021). Typically, biochar derived from different feedstock has been examined in laboratory settings or small-scale field trials, resulting in limited insights into its sustainable profitability (Figure 15.3) (Fytili and Zabaniotou, 2018).

Government subsidies play a pivotal role in bolstering investment in the bio-based economy derived from agricultural waste. These subsidies can be allocated to stakeholders either during the

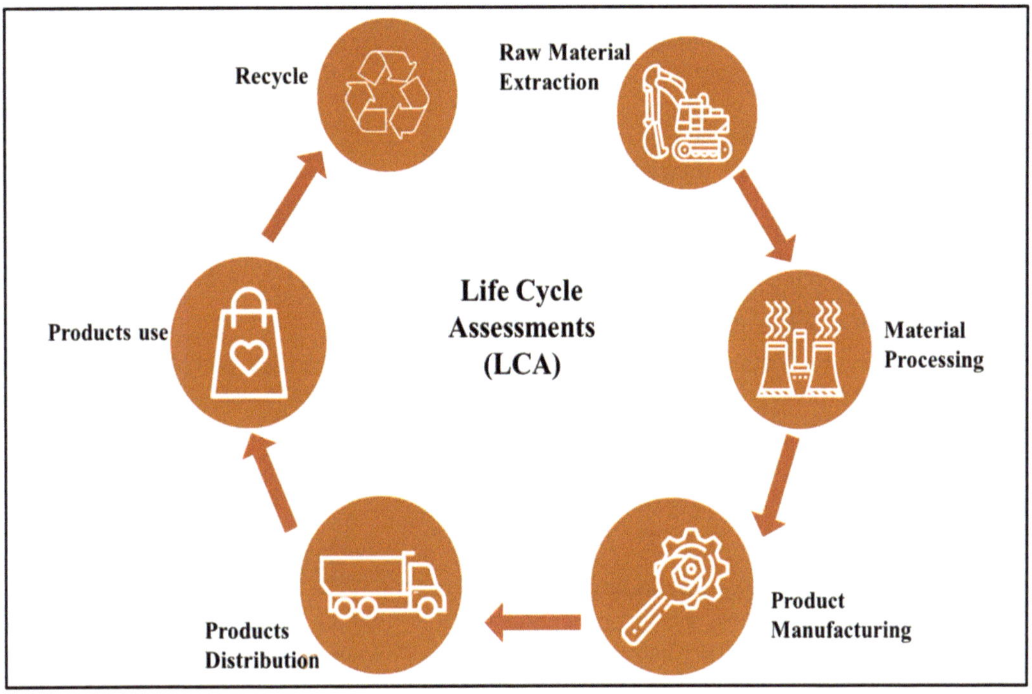

FIGURE 15.3 Stages and components of life cycle assessment: A scientific perspective. (Modified from Rebitzer et al., 2004.)

procurement of technological equipment or upon the sale of end products. Additionally, in developing countries, low-interest-rate banking facilities or government loans tailored for small- and medium-scale investors further facilitate investment in this sector (Khan et al., 2021).

15.4 LCA OF AGRICULTURAL WASTE REMEDIATION TECHNIQUES

LCA gives a comprehensive perspective on the influence on the environment by evaluating a number of environmental indicators, including emissions of greenhouse gases, energy consumption, water use, land occupancy, and the potential for toxicity (Jolliet et al., 2018). There is comprehensive knowledge of the environmental consequences connected with a certain product or process as a result of the assessment, which takes into consideration the resources used, the emissions emitted, and the waste created at each step (Loiseau et al., 2012). LCA is a method utilized in scientific inquiry to ascertain the environmental impact of a product or service (Jacquemin et al., 2012). It takes into consideration everything that occurs with the product or service, beginning at the moment it is manufactured and ending with the moment it is discarded (Figure 15.3). One way to evaluate the influence that various goods or services have on the environment is to utilize LCA (Corominas et al., 2020). Additionally, it may be used to determine whether or not there are strategies to enhance the environmental performance of a product or service.

15.4.1 Defining the Stages and Components of LCA: An Analytical Approach

15.4.1.1 Goal and Scope Definition

At this juncture, it is imperative to establish the objectives and parameters of the LCA research. This entails identifying the intended audience, the purpose of the assessment, and delineating the system boundaries by defining the processes and inputs to be considered. Additionally, determining the functional unit, or the basis for comparison, and designating the evaluator(s) are essential components of this phase (Sala et al., 2015).

15.4.1.2 Inventory Analysis

This step involves collecting information on the resources utilized within agriculture, such as fertilizers, pesticides, energy, water, agricultural products, alongside the by-products, emissions, and waste generated at each stage of the cycle. A life cycle inventory (LCI) entails compiling quantified input and output data (Huysveld et al., 2015).

15.4.1.3 Impact Assessment

The possible environmental implications of the inventory data are then assessed. At this point, we're going to classify the impacts according to several criteria, such as biodiversity loss, energy consumption, land usage, water pollution, and climate change. Analyzing the inventory data allows for the calculation of possible effects using mathematical models or specialized software tools (Finch et al., 2023).

15.4.1.4 Interpretation

Sharing and comprehending the findings of the LCA study represents the ultimate phase. This includes analyzing the results, deriving conclusions, and pinpointing areas for enhancement. To validate the findings, sensitivity or uncertainty analyses may be imperative. LCA proves invaluable in assessing and regulating the environmental impact of agricultural value chains, given agriculture's substantial role in pollution, biodiversity decline, and greenhouse gas emissions (Nabavi-Pelesaraei et al., 2021). Some important reasons which make LCA crucial in agricultural value chains are as follows: environmental footprint analysis, identifying hotspots and environmental impacts, sustainability evaluation, supply chain management, policy development, consumer information.

The environmental impact of farming may be thoroughly evaluated with the use of LCA. By identifying the most environmentally damaging links in the value chain, we can implement measures that will reduce down on waste, pollution, and emissions. Applying LCA to agricultural value chains may help pinpoint problem areas and provide insight into the environmental implications of various processing, manufacturing, and distribution steps (Gava et al., 2020). Through LCA, we can identify the most impactful parts of a project in terms of energy consumption, emissions, and resource loss. Stakeholders may then use this information to set priorities and execute specific activities to lessen these effects. LCA offers a method for calculating how well agricultural methods, goods, and systems do in terms of sustainability (Sieverding et al., 2020). It allows stakeholders to compare alternatives based on their environmental implications, which helps them make educated choices and choose more sustainable solutions. LCA, supply chain management choices may be informed by supplier environmental performance evaluations (Seifert et al., 2021). With its help, businesses may find sustainable suppliers to work with, which improves the agricultural value chain overall. Agricultural officials may benefit greatly from the insights provided by LCA when crafting successful rules and strategies. It is useful for creating eco-labeling schemes, creating environmental goals, and creating incentives to encourage sustainable activities all along the value chain. Consumers may learn about the ecological footprint of agricultural goods via LCA findings (McAuliffe et al., 2020). Environmental product declarations or eco-labels allow shoppers to learn more about a product's impact on the environment before purchasing it.

Major challenges of applying LCA in agriculture are the availability and quality of data, system boundaries and scope considerations, complexity and interconnectedness of agricultural systems, uncertainties and interpretation, evolving standards and methodologies. The whole life cycle, from extracting raw materials to disposing of them, is based on accurate and reliable data in LCA. On the other hand, collecting data from the agriculture sector may be challenging due to the wide variety of farming methods, geographical differences, and associated systems. Selecting the right inputs and processes to include or omit, as well as establishing suitable time and functional units, may be a challenging task. The completeness and comparability of LCA findings might be affected by competing viewpoints among stakeholders over which value chain stages should be included (Sevigné-Itoiz et al., 2021). Ensuring relevant and accurate evaluations requires careful deliberation and clear disclosure of the study's scope. Soil, water, energy, ecosystems, animals, crops, and agriculture all interact in complex manners. The indirect effects of these interdependencies might be hard to measure and give labeled to. Weather changes, land use shifts, and market dynamics are some of the other aspects that make LCA in agriculture difficult.

15.4.2 Environmental Impact Assessment

Environmental impact assessments (EIAs) help decision-makers find out what the environmental effects of a project may be, how different options would fare, and how to put in place the right monitoring, management, preventative, and mitigation strategies (Nita et al., 2022). One of the most important parts of an EIA is determining the project's social effect, as the two go hand in hand. An environmental management plan (EMP) describing the steps to be taken at different points in the project's lifespan to lessen the impact of potential hazards and bring them down to an acceptable level is required as part of the EIA. The production of agricultural waste is a problem that affects the whole globe (Sehker and Beukering, 1998). The amount and complex composition of agriculture waste have increased due to urbanization and the improvement of living standards in developing country cities (Glawe et al., 2005). Agricultural waste disposal facility is one project that might have devastating environmental consequences in the absence of an EIA. The modification of natural ecosystems to support agriculture production involves many practices that can impact the environment, including, but not limited to, tillage, use of fertilizers, pesticides and herbicides, water management practices (irrigation, drainage), and land use conversion (forest, wetland conversion to agriculture). The main objective of EIA projects in agricultural waste is to understand the current situation of agricultural and, given the current situation, appropriate enforcement strategies should be presented to improve the quality of the environment and reduce pollution caused by fertilizers.

Sustainable agriculture has the ability to contribute to the creation of a more resilient and sustainable future for the whole world. A rising number of people are coming to the realization that recycling and circular economy methods have the potential to convert trash into useful resources (Kurniawan et al., 2022). To assess the possibilities and consequences of recycling agricultural waste, a fresh and all-encompassing viewpoint is needed, one that takes into account the whole cycle of these recycling processes. Life cycle thinking is a methodology that sheds light on sustainable practices and the far-reaching environmental impacts of recycling agricultural waste (Dahiya et al., 2020; Zeug et al., 2023). The screening phase is the initial step in the EIA procedure, in order to establish if an environmental assessment is necessary for a project. A comprehensive EIA, a limited EIA (including preliminary evaluation or more analysis), or no EIA at all is typically the outcome (Noble, 2023). According to FAO (2011), agriculture projects are defined as farming activities that alter the species composition (and biodiversity) of the surface vegetation over the long term or alter the way land is farmed to be more intensively used.

Scoping is the first step in doing an EIA and it restricts the amount of data that needs to be gathered and evaluated (Noble, 2023). An EIA's scoping is an important first step. Its goal is to catalogue the proposed project's most significant problems and effects. The three main components of a project's scope are the activity itself, the existing environmental conditions, and the alternatives to the current plan. Proposed activity for this step provides the project's location, as well as its spatial and temporal bounds, and a physical description of the site. In a physical description, one can usually find details on the soil, the terrain, the drainage, the flora, the animals, the weather, the land use, the biodiversity, the roads, the utilities, the historic sites, and any adjoining properties that may be negatively impacted. Researchers utilized pre-existing resource and land-cover maps to establish geographical and temporal limits in a study that examined the impact of farming on climate change in terms of area (Merem et al., 2019).

The baseline assessment involves the selection of value component to be included in the assessment, and the assessment of the condition and changes in the value component. If the goal of an agricultural project is to install an irrigation system, then soil salinization would be an important value component indicator as soil quality is a major value component. In order to assess potential future effects, it is helpful to keep tabs on the present state of an area's environment through the collection and analysis of baseline data. Examining potential changes brought about by actions taken in previous years is another possible application of baseline data. Predicting and evaluating the possible environmental repercussions that may arise from the proposed project is what impact assessment is all about (Noble, 2023). An analysis of large GIS analytics in agriculture conducted by Sharma et al. (2020) showed a rise in the utilization of prescriptive analytics based on multi-criteria decision analysis (MCDA) across all areas of agriculture. According to Sharma et al. (2020), MCDA approaches show promise for impact assessment analysis, while its most prevalent use is in land use appropriateness analyses.

The goal of impact management is to lessen the negative effects of a project's proposed implementation by planning for and executing appropriate responses to any problems (Noble, 2023). In order to identify locations that need impact management following the activity's implementation, buffer zones might be utilized. To avoid nonpoint-source contamination caused by farming, establish buffer zones surrounding the area (Narumalani et al., 1997). The researchers created buffer zones surrounding the different types of land cover using spatial approaches after identifying and characterizing the land cover using satellite imagery. Monitoring and follow-up include monitoring of the effects of the mitigation measures proposed and implemented (Noble, 2023). Any newly planned agricultural project can undergo an EIA using the four-stage framework. In addition to new agricultural practices (such as the construction of irrigation or drainage systems), the framework may be used to evaluate the effects of yearly activities on current farms.

15.5 INTEGRATION OF TECHNO-ECONOMIC AND LCA RESULTS

Integration of TEA and LCA provides a comprehensive framework for enhancing processes and products with a dual focus on economic viability and environmental sustainability (Figure 15.4). By concurrently assessing these dimensions, decision-makers can steer toward prudent choices, facilitating long-lasting value creation while minimizing adverse environmental effects (Ferdous et al., 2023). TEA emerges as a pivotal instrument for appraising the capital and operational outlays associated with diverse processes, predicated on their material balance and energy requisites across varying operational magnitudes. Furthermore, it elucidates the feasibility of industrial or commercial-scale implementation of agricultural waste treatment systems. However, the global commercialization of agricultural and bio-waste treatment encounters significant economic impediments (Awasthi et al., 2021). Within a comprehensive TEA framework, meticulous mass and energy balances are established for disparate processes, encompassing feedstock collection, handling, product formation, power consumption, waste management, and transportation, thus facilitating the estimation of capital and operational expenditures (Zimmermann et al., 2020) (Figure 15.4).

The fundamental aim of LCA is to provide a comprehensive understanding of environmental impacts and waste management strategies. However, specific impact categories continue to be inadequately represented within conventional LCA frameworks (Zhang et al., 2021). Notably, methodologies for assessing land use impacts, including biodiversity effects, and resource factors, such as freshwater consumption, present challenges that require refinement. Enhanced methodologies can be cultivated through increased collaboration with related disciplines. In addition to the predominant focus on outdoor pollutant exposure, significant efforts have been directed toward understanding the implications of agricultural waste and heavy metal contamination on human health. The evolution of indoor exposure models and the potential integration of various health dimensions into

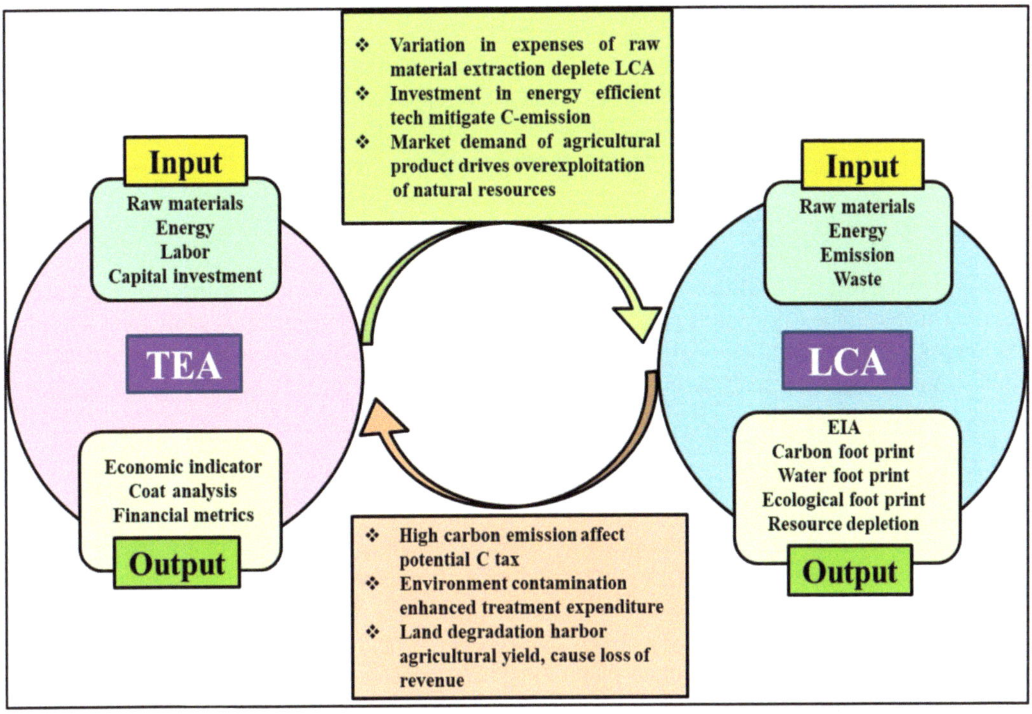

FIGURE 15.4 Integration of techno-economic analysis and life cycle assessment results for agricultural waste.

social LCA methodologies offer avenues for deeper insights (Rodríguez et al., 2018). TEA and LCA mutually enhance one another and represent their symbiotic connection as given below.

15.5.1 Evolutionary Comprehension

The integration of TEA and LCA results in a holistic assessment of a technology or product. TEA evaluates economic viability by analyzing variables like profitability, revenue generation, operational expenses, and capital costs, while LCA scrutinizes environmental effects across all stages of the life cycle, from raw material extraction to disposal (Hu et al., 2019).

15.5.2 Decision-Making

The combination of TEA and LCA data empowers decision-makers to comprehensively assess the trade-offs between economic advantages and environmental effects. This allows them to make informed decisions that prioritize both sustainability objectives and economic viability (Ferdous et al., 2023).

15.5.3 Identifying the Opportunity to Improvement

Through a comparative analysis of TEA and LCA outcomes, workers can identify avenues to maximize both economic and environmental performance. Integrating these evaluations facilitates the discovery of cost-effective methods for reducing environmental impacts (Escobar and Laibach, 2021).

15.5.4 Holistic Perspective

TEA and LCA collectively provide a holistic perspective on the assessment of technology or products. While LCA primarily focuses on environmental impacts, TEA typically emphasizes economic considerations. By integrating both, stakeholders can attain a more comprehensive understanding of the overall performance of the technology or product (Kumar and Verma, 2021).

15.5.5 Risk Assessment

The utilization of TEA facilitates the assessment of financial risks associated with technology, encompassing uncertainties in capital and operational costs as well as revenue predictions. Conversely, LCA aids in identifying potential environmental hazards and impacts throughout the product's life cycle. Integrating these analyses enhances the depth of risk evaluation (Langhorst et al., 2022).

15.5.6 Policy Making

Integrated TEA and LCA data offer policymakers' valuable tools to formulate sustainable policies that promote economic growth alongside environmental preservation. By comprehensively grasping the economic implications of environmental regulations and vice versa, legislators can develop more equitable and efficient policies (Sadhukhan et al., 2019). Ultimately, the complementary outcomes of TEA and LCA provide a comprehensive understanding of the environmental and economic characteristics of technologies or products. By integrating these assessments, stakeholders can pinpoint areas for improvement, make informed decisions, and create policies that facilitate sustainable development (Cheng et al., 2023).

The necessity for harmonizing LCA and TEA guidelines to facilitate comprehensive and practical assessments of early-stage systems has been discussed in the literature, particularly in discussions concerning the development of specific guidelines for assessing carbon capture systems (Davidson et al., 2021; Kim et al., 2020).

15.6 SYNERGIES BETWEEN ECONOMIC VIABILITY AND ENVIRONMENTAL SUSTAINABILITY

Synergies between economic viability and environmental sustainability in the context of heavy metal remediation can be effectively analyzed through the integration of LCA and TEA (Jones et al., 2017). Here's how these methodologies contribute to understanding and enhancing the interplay between economic viability and environmental sustainability. The synergies between economic viability and environmental sustainability in heavy metal remediation are crucial for ensuring effective and sustainable solutions. Here are some key points highlighting these synergies.

15.6.1 Cost-Effective Technologies

Increasing demand for heavy metal remediation solutions emphasizes the necessity for environmentally sustainable and economically viable options. Recent technological strides have paved the way for the development of cost-effective strategies like phytoremediation, bioremediation, and electrokinetic remediation. These methodologies offer proficient removal of heavy metals from soil, water, and air, alleviating financial burdens (Awasthi et al., 2021; Mahmud et al., 2021).

15.6.2 Resource Efficiency

Sustainable approaches to heavy metal remediation strive to enhance resource efficiency and diminish waste production. Through the adoption of methodologies that maximize the utilization of energy, water, and materials, financial expenses can be minimized alongside reductions in environmental repercussions linked to the extraction, consumption, and disposal of resources (Kumar et al., 2019).

15.6.3 Value Recovery

Innovative approaches to heavy metal remediation often include the concept of value recovery, where remediated materials are reused, recycled, or repurposed to extract value. For example, recovered heavy metals can be sold as secondary raw materials, generating revenue and offsetting remediation costs. This circular economy approach not only enhances economic viability but also reduces the environmental footprint of remediation activities by reducing the need for virgin materials and waste disposal (Espada et al., 2018).

15.6.4 Long-Term Benefits

Allocating resources toward sustainable heavy metal remediation holds promise for generating long-term economic benefits through the mitigation of risks stemming from environmental contamination. Through remedial interventions, it becomes feasible to forestall further environmental degradation, preserve essential ecosystem services, and safeguard human health, thereby reducing potential future outlays associated with cleanup endeavors, healthcare costs, and legal disputes. Additionally, the rehabilitation of contaminated sites can bolster property values and foster economic revitalization within impacted communities (Yang et al., 2024).

15.6.5 Regulatory Compliance

Compliance with environmental regulations is vital for enterprises operating in industries susceptible to heavy metal pollution. Embracing sustainable remediation technologies and practices not only facilitates compliance with regulatory mandates but also reduces the risk of fines, penalties, and reputational damage stemming from non-compliance. By incorporating environmental considerations into their business strategies, organizations can enhance their competitive advantage and tap into new markets where sustainability is a primary consideration (Chen and Ding, 2023).

15.6.6 Stakeholder Engagement

Collaboration with a spectrum of stakeholders, encompassing government agencies, local communities, environmental organizations, and industry partners, constitutes a fundamental aspect of crafting and executing efficacious heavy metal remediation initiatives. Involving stakeholders in decision-making processes facilitates consensus-building, enhances transparency, and uncovers synergies among economic, environmental, and social objectives. By addressing the diverse needs and concerns of stakeholders, sustainable remediation endeavors can garner broader support and yield enduring positive outcomes. The inherent alignment between economic viability and environmental sustainability underscores the significance of embracing integrated and holistic approaches that reconcile economic imperatives with environmental preservation and social accountability. Through the utilization of innovative technologies, optimization of resource utilization, valorization of remediated materials, and active stakeholder engagement, sustainable remediation strategies hold the potential to deliver tangible benefits for both the economy and the environment (Kim et al., 2020; Sadhukhan et al., 2019).

15.6.7 Challenges and Opportunities

Many challenges arise while considering green alternatives for heavy metal removal, especially the use of agricultural waste. One significant hurdle is the need for extensive TEA and LCA to determine the feasibility and sustainability of these technologies. Conducting such assessments requires knowledge, time, and resources, which are not always available. In addition, the efficacy and effectiveness of heavy metal remediation utilizing agricultural waste can be affected by several factors, including the type and concentration of contaminants, soil qualities, and environmental conditions. Developing standardized protocols and methodologies for analyzing the performance of these environmentally friendly alternatives in diverse contexts is essential for ensuring reliable outcomes and influencing decision-making processes. In addition, while agricultural waste has the potential to be a cost-effective and sustainable resource for heavy metal remediation, issues such as scalability, transportation, and supply chain management must be addressed.

By scrutinizing the TEA, we may create cost-effective strategies for utilizing agricultural waste in heavy metal remediation. This may involve assessing several treatment methods, such as phytoremediation or biochar application, to determine their efficacy as well as economic viability. Furthermore, TEA can identify possibilities to improve resource use, reduce operating expenses, and enhance overall profitability. Concurrently, LCA offers information about the environmental footprint of these green solutions. LCA identifies potential environmental hotspots and mitigation options by examining the full remediation process, from raw material extraction to waste disposal. By combining TEA and LCA, stakeholders may make informed decisions to progress with green heavy metal remediation approaches, realizing their full potential for sustainable development.

15.7 FUTURE DIRECTIONS AND RECOMMENDATIONS

There is a need for continued exploration of novel agricultural waste sources and their effectiveness in heavy metal remediation. This could include looking into lesser known agricultural by-products or residues, as well as optimizing the processing techniques to enhance adsorption capacities and selectivity for specific heavy metals. Additionally, incorporating advanced technologies like nanomaterials or biochar modification could considerably improve the efficacy of agricultural waste-based remediation approaches. Research efforts could be focused on understanding the mechanisms underlying these advancements, as well as ensuring the safety and environmental sustainability of such approaches. Furthermore, conducting comprehensive TEA and LCA will be crucial for evaluating the feasibility and sustainability of implementing agricultural waste-based remediation at scale. This involves not only assessing the costs and benefits of different treatment methods but also considering their long-term environmental impacts and comparing them with conventional remediation techniques. A significant socio-economic benefit of employing agricultural waste for heavy metal remediation is the creation of new economic opportunities in rural areas. Local farmers and agricultural communities can capitalize on the demand for their waste products, thereby diversifying income streams, enhancing economic resilience, and contributing to poverty alleviation and rural development. Implementation of such solutions may create jobs in waste collection, processing, and application, stimulating employment opportunities in rural areas. Utilizing green solutions not only encourages sustainability by minimizing the use of synthetic chemicals and greenhouse gas emissions associated with traditional remediation approaches but also benefits communities through improved soil, water, and air quality, promoting public health and well-being.

ACKNOWLEDGMENT

Authors are grateful to the Banaras Hindu University, Varanasi, for providing necessary support for this work.

REFERENCES

Abadi, M., Zamani, A., Khosravi, Y. and Homayouni, S., 2023. Comprehensive risk assessment of heavy metals in topsoil and their risks to human health and ecosystem in Zanjan Mineral Province, Western Iran. *Water, Air, & Soil Pollution*, 234(10), p. 627.

Abbar, B., Alem, A., Marcotte, S., Pantet, A., Ahfir, N.D., Bizet, L. and Duriatti, D., 2017. Experimental investigation on removal of heavy metals (Cu^{2+}, Pb^{2+}, and Zn^{2+}) from aqueous solution by flax fibres. *Process Safety and Environmental Protection*, 109, pp. 639–647.
Abdulrasaq, O.O. and Basiru, O.G., 2010. Removal of copper (II), iron (III) and lead (II) ions from mono-component simulated waste effluent by adsorption on coconut husk. *African Journal of Environmental Science and Technology*, 4(6).
Ahmad, T. and Danish, M., 2022. A review of avocado waste-derived adsorbents: Characterizations, adsorption characteristics, and surface mechanism. *Chemosphere*, 296, p. 134036.
Ajmal, M., Khan, A.H., Ahmad, S. and Ahmad, A., 1998. Role of sawdust in the removal of copper(II) from industrial wastes. *Water Research*, 32(10), pp. 3085–3091.
Ajmal, M., Rao, R.A.K., Ahmad, R. and Ahmad, J., 2000. Adsorption studies on *Citrus reticulata* (fruit peel of orange): Removal and recovery of Ni(II) from electroplating wastewater. *Journal of Hazardous Materials*, 79(1–2), pp. 117–131.
Akolgo, G.A., Awafo, E.A., Essandoh, E.O., Owusu, P.A., Uba, F. and Adu-Poku, K.A., 2021. Assessment of the potential of charred briquettes of sawdust, rice and coconut husks: Using water boiling and user acceptability tests. *Scientific African*, 12, p. e00789.
Akram, R., Amin, A., Hashmi, M.Z., Wahid, A., Mubeen, M., Hammad, H.M., Fahad, S. and Nasim, W., 2017. Fate of antibiotics in soil. *Antibiotics and Antibiotics Resistance Genes in Soils: Monitoring, Toxicity, Risk Assessment and Management*, pp. 207–220.
Alalwan, H.A., Kadhom, M.A. and Alminshid, A.H., 2020. Removal of heavy metals from wastewater using agricultural byproducts. *Journal of Water Supply: Research and Technology—AQUA*, 69(2), pp. 99–112.
Ali, A., Guo, D., Mahar, A., Wang, P., Ma, F., Shen, F., Li, R. and Zhang, Z., 2017. Phytoextraction of toxic trace elements by Sorghum bicolor inoculated with *Streptomyces pactum* (Act12) in contaminated soils. *Ecotoxicology and Environmental Safety*, 139, pp. 202–209.
Al-Wabel, M.I., Usman, A.R., El-Naggar, A.H., Aly, A.A., Ibrahim, H.M., Elmaghraby, S. and Al-Omran, A., 2015. Conocarpus biochar as a soil amendment for reducing heavy metal availability and uptake by maize plants. *Saudi Journal of Biological Sciences*, 22(4), pp. 503–511.
Aman, T., Kazi, A.A., Sabri, M.U. and Bano, Q., 2008. Potato peels as solid waste for the removal of heavy metal copper(II) from waste water/industrial effluent. *Colloids and Surfaces B: Biointerfaces*, 63(1), pp. 116–121.
Argun, M.E. and Dursun, S., 2007. Activation of pine bark surface with NaOH for lead removal. *Journal of International Environmental Application and Science*, 2, pp. 5–10.
Awasthi, M.K., Sarsaiya, S., Wainaina, S., Rajendran, K., Awasthi, S.K., Liu, T., Duan, Y., Jain, A., Sindhu, R., Binod, P. and Pandey, A., 2021. Techno-economics and life-cycle assessment of biological and thermochemical treatment of bio-waste. *Renewable and Sustainable Energy Reviews*, 144, p. 110837.
Ayob, S., Othman, N., Altowayti, W.A.H., Khalid, F.S., Bakar, N.A., Tahir, M. and Soedjono, E.S., 2021. A review on adsorption of heavy metals from wood-industrial wastewater by oil palm waste. *Journal of Ecological Engineering*, 22(3).
Baby Shaikh, R., Saifullah, B. and Rehman, F.U., 2018. Greener method for the removal of toxic metal ions from the wastewater by application of agricultural waste as an adsorbent. *Water*, 10(10), p. 1316.
Bansal, M., Singh, D. and Garg, V.K., 2009. A comparative study for the removal of hexavalent chromium from aqueous solution by agriculture wastes' carbons. *Journal of Hazardous Materials*, 171(1–3), pp. 83–92.
Belete, G., 2017. A review on elimination of heavy metals from wastewater using agricultural wastes as adsorbents. *Science Journal of Analytical Chemistry*, 5(5), pp. 72–75.
Boruah, P., Sarma, A. and Bhattacharyya, K.G., 2015. Removal of Ni (II) ions from aqueous solution by using low cost biosorbent prepared from jackfruit (*Artocarpus heterophyllus*) leaf powder.
Boumediene, M., Benaïssa, H., George, B., Molina, S. and Merlin, A., 2018. Effects of pH and ionic strength on methylene blue removal from synthetic aqueous solutions by sorption onto orange peel and desorption study. *Journal of Material and Environmental Science*, 9(6), pp. 1700–1711.
Bulut, Y., 2007. Removal of heavy metals from aqueous solution by sawdust adsorption. *Journal of Environmental Sciences*, 19(2), pp. 160–166.
Bustillo-Lecompte, C.F. and Mehrvar, M., 2015. Slaughterhouse wastewater characteristics, treatment, and management in the meat processing industry: A review on trends and advances. *Journal of Environmental Management*, 161, pp. 287–302.
Chamarthy, S., Seo, C.W. and Marshall, W.E., 2001. Adsorption of selected toxic metals by modified peanut shells. *Journal of Chemical Technology & Biotechnology*, 76(6), pp. 593–597.
Chen, S. and Ding, Y., 2023. Tackling heavy metal pollution: Evaluating governance models and frameworks. *Sustainability*, 15(22), p. 15863.

Cheng, J., Singh, H.S.M., Zhang, Y.C. and Wang, S.Y., 2023. The impact of business intelligence, big data analytics capability, and green knowledge management on sustainability performance. *Journal of Cleaner Production*, 429, p. 139410.

Cherdchoo, W., Nithettham, S. and Charoenpanich, J., 2019. Removal of Cr(VI) from synthetic wastewater by adsorption onto coffee ground and mixed waste tea. *Chemosphere*, 221, pp. 758–767.

Chinyelu, E., Ngozi, U., Kate, O. and Sixtus, O., 2015. Use of unmodified orange peel for the adsorption of Cd (II), Pb (II) and Hg (II) ions in aqueous solutions. *American Journal of Physical Chemistry*, 4(3), pp. 21–29.

Chong, Z.T., Soh, L.S. and Yong, W.F., 2023. Valorization of agriculture wastes as biosorbents for adsorption of emerging pollutants: Modification, remediation and industry application. *Results in Engineering*, 17, p. 100960.

Choudhary, M., Muduli, M. and Ray, S., 2022. A comprehensive review on nitrate pollution and its remediation: Conventional and recent approaches. *Sustainable Water Resources Management*, 8(4), p. 113.

Chowdhury, Z., Zain, S.M. and Rashid, A.K., 2011. Equilibrium isotherm modeling, kinetics and thermodynamics study for removal of lead from waste water. *Journal of Chemistry*, 8(1), pp. 333–339.

Chu, K. and Hashim, M., 2002. Adsorption characteristics of trivalent chromium on palm oil fuel ash. *Clean Technologies and Environmental Policy*, 4, pp. 8–15.

Cleary, J. and Caspersen, J.P., 2015. Comparing the life cycle impacts of using harvest residue as feedstock for small-and large-scale bioenergy systems (part I). *Energy*, 88, pp. 917–926.

Corominas, L., Byrne, D.M., Guest, J.S., Hospido, A., Roux, P., Shaw, A. and Short, M.D., 2020. The application of life cycle assessment (LCA) to wastewater treatment: A best practice guide and critical review. *Water Research*, 184, p. 116058.

Cui, H., Fan, Y., Fang, G., Zhang, H., Su, B. and Zhou, J., 2016. Leachability, availability and bioaccessibility of Cu and Cd in a contaminated soil treated with apatite, lime and charcoal: A five-year field experiment. *Ecotoxicology and Environmental Safety*, 134, pp. 148–155.

Dahiya, S., Katakojwala, R., Ramakrishna, S. and Mohan, S.V., 2020. Biobased products and life cycle assessment in the context of circular economy and sustainability. *Materials Circular Economy*, 2, pp. 1–28.

Dai, Y., Sun, Q., Wang, W., Lu, L., Liu, M., Li, J., Yang, S., Sun, Y., Zhang, K., Xu, J. and Zheng, W., 2018. Utilizations of agricultural waste as adsorbent for the removal of contaminants: A review. *Chemosphere*, 211, pp. 235–253.

Daifullah, A.A.M., Girgis, B.S. and Gad, H.M.H., 2003. Utilization of agro-residues (rice husk) in small waste water treatment plans. *Materials Letters*, 57(11), pp. 1723–1731.

Datta, S., Singh, S., Ramamurthy, P.C., Angurana, R., Dhanjal, D.S., Katoch, V., Kumar, A., Kapoor, D., Singh, J. and Singh, J., 2023. Treatment approaches for bio-contaminants in organic wastes. In K. Huang, S. Ahmad Bhat and G. Cui (Eds.), *Fate of Biological Contaminants During Recycling of Organic Wastes* (pp. 123–141). Elsevier.

Davidson, M.G., Elgie, S., Parsons, S. and Young, T.J., 2021. Production of HMF, FDCA and their derived products: A review of life cycle assessment (LCA) and techno-economic analysis (TEA) studies. *Green Chemistry*, 23(9), pp. 3154–3171.

Duque-Acevedo, M., Belmonte-Ureña, L.J., Cortés-García, F.J. and Camacho-Ferre, F., 2020. Agricultural waste: Review of the evolution, approaches and perspectives on alternative uses. *Global Ecology and Conservation*, 22, p. e00902.

Escobar, N. and Laibach, N., 2021. Sustainability check for bio-based technologies: A review of process-based and life cycle approaches. *Renewable and Sustainable Energy Reviews*, 135, p. 110213.

FAO, 2011. Aquaculture Development. 5. Use of Wild Fish as Feeding Aquaculture. FAO Technical Guidelines for Responsible Fisheries. No. 5, Suppl. 5. Rome: FAO. 2011. 79 p.

Farooq, U., Khan, M.A., Athar, M. and Kozinski, J.A., 2011. Effect of modification of environmentally friendly biosorbent wheat (*Triticum aestivum*) on the biosorptive removal of cadmium (II) ions from aqueous solution. *Chemical Engineering Journal*, 171(2), pp. 400–410.

Feng, N.C. and Guo, X.Y., 2012. Characterization of adsorptive capacity and mechanisms on adsorption of copper, lead and zinc by modified orange peel. *Transactions of Nonferrous Metals Society of China*, 22(5), pp. 1224–1231.

Feng, N., Guo, X., Liang, S., Zhu, Y. and Liu, J., 2011. Biosorption of heavy metals from aqueous solutions by chemically modified orange peel. *Journal of Hazardous Materials*, 185(1), pp. 49–54.

Ferdous, J., Bensebaa, F. and Pelletier, N., 2023. Integration of LCA, TEA, process simulation and optimization: A systematic review of current practices and scope to propose a framework for pulse processing pathways. *Journal of Cleaner Production*, 402, p. 136804.

Finch, T., Bradbury, R.B., Bradfer-Lawrence, T., Buchanan, G.M., Copping, J.P., Massimino, D., Smith, P., Peach, W.J. and Field, R.H., 2023. Spatially targeted nature-based solutions can mitigate climate change and nature loss but require a systems approach. *One Earth*, 6(10), pp. 1350–1374.
Food and Agriculture Organization of the United Nations (FAO), 2017b. Strategic Work of FAO for Sustainable Food and Agriculture.
Food and Agriculture Organization of the United Nations (FAO), 2017c. The State of Food and Agriculture 2017. Leveraging Food Systems for Inclusive Rural Transformation.
Food and Agriculture Organization of the United Nations (FAO), 2017a. The Future of Food and Agriculture: Trends and Challenges.
Fytili, D. and Zabaniotou, A., 2018. Circular economy synergistic opportunities of decentralized thermochemical systems for bioenergy and biochar production fueled with agro-industrial wastes with environmental sustainability and social acceptance: A review. *Current Sustainable/Renewable Energy Reports*, 5, pp. 150–155.
Galinato, S.P., Yoder, J.K. and Granatstein, D., 2011. The economic value of biochar in crop production and carbon sequestration. *Energy Policy*, 39(10), pp. 6344–6350.
Garg, U., Kaur, M.P., Jawa, G.K., Sud, D. and Garg, V.K., 2008. Removal of cadmium (II) from aqueous solutions by adsorption on agricultural waste biomass. *Journal of Hazardous Materials*, 154(1–3), pp. 1149–1157.
Gava, O., Bartolini, F., Venturi, F., Brunori, G. and Pardossi, A., 2020. Improving policy evidence base for agricultural sustainability and food security: A content analysis of life cycle assessment research. *Sustainability*, 12(3), p. 1033.
Glawe, U., Visvanathan, C. and Alamgir, M., 2005, July. Solid waste management in least developed Asian countries – A comparative analysis. In *International Conference on Integrated Solid Waste Management in Southeast Asian Cities* (pp. 5–7).
Gönen, F. and Serin, D.S., 2012. Adsorption study on orange peel: Removal of Ni (II) ions from aqueous solution. *African Journal of Biotechnology*, 11(5), pp. 1250–1258.
Guechi, E.K. and Hamdaoui, O., 2016. Evaluation of potato peel as a novel adsorbent for the removal of Cu(II) from aqueous solutions: Equilibrium, kinetic, and thermodynamic studies. *Desalination and Water Treatment*, 57(23), pp. 10677–10688.
Habineza, A., Zhai, J., Ntakirutimana, T., Qiu, F.P., Li, X. and Wang, Q., 2017. Heavy metal removal from wastewaters by agricultural waste low-cost adsorbents: Hindrances of adsorption technology to the large scale industrial application—A review. *Desalination and Water Treatment*, 78, pp. 192–214.
Hegazi, H.A., 2013. Removal of heavy metals from wastewater using agricultural and industrial wastes as adsorbents. *HBRC Journal*, 9(3), pp. 276–282.
Hu, L., Pan, H., Zhou, Y. and Zhang, M., 2011. Methods to improve lignin's reactivity as a phenol substitute and as replacement for other phenolic compounds: A brief review. *BioResources*, 6(3), pp. 3515–3525.
Hu, A.H., Chen, C.H., Huang, L.H., Chung, M.H., Lan, Y.C. and Chen, Z., 2019. Environmental impact and carbon footprint assessment of Taiwanese agricultural products: A case study on Taiwanese Dongshan Tea. *Energies*, 12(1), p. 138.
Hu, Y., Wang, J., Yang, Y., Li, S., Wu, Q., Nepovimova, E., Zhang, X. and Kuca, K., 2024. Revolutionizing soil heavy metal remediation: Cutting-edge innovations in plant disposal technology. *Science of the Total Environment*, 918, p. 170577.
Huang, K., Xiu, Y. and Zhu, H., 2013. Selective removal of Cr (VI) from aqueous solution by adsorption on mangosteen peel. *Environmental Science and Pollution Research*, 20, pp. 5930–5938.
Hussain, S., Anjali, K.P., Hassan, S.T. and Dwivedi, P.B., 2018. Waste tea as a novel adsorbent: A review. *Applied Water Science*, 8, pp. 1–16.
Huysveld, S., De Meester, S., Peiren, N., Muylle, H., Lauwers, L. and Dewulf, J., 2015. Resource use assessment of an agricultural system from a life cycle perspective – A dairy farm as case study. *Agricultural Systems*, 135, pp. 77–89.
Ibrahim, H., Abid, M.K. and Zain, H.H.M., 2020. Heavy Metal Ions Adsorption from Aqueous Solution by Jackfruit Peel as Activated Biochar Low-Cost Adsorbent. *Journal of Advanced Research in Fluid Mechanics and Thermal Sciences*, 67(2), pp. 154–165.
Israel, U. and Eduok, U.M., 2012. Biosorption of zinc from aqueous solution using coconut (*Cocos nucifera* L) coir dust. *Archives of Applied Science Research*, 4(2), pp. 809–819.
Jacquemin, L., Pontalier, P.Y. and Sablayrolles, C., 2012. Life cycle assessment (LCA) applied to the process industry: A review. *The International Journal of Life Cycle Assessment*, 17, pp. 1028–1041.

Jolliet, O., Antón, A., Boulay, A.M., Cherubini, F., Fantke, P., Levasseur, A., McKone, T.E., Michelsen, O., Milà i Canals, L., Motoshita, M. and Pfister, S., 2018. Global guidance on environmental life cycle impact assessment indicators: Impacts of climate change, fine particulate matter formation, water consumption and land use. *The International Journal of Life Cycle Assessment*, 23, pp. 2189–2207.

Jones, C., Gilbert, P., Raugei, M., Mander, S. and Leccisi, E., 2017. An approach to prospective consequential life cycle assessment and net energy analysis of distributed electricity generation. *Energy Policy*, 100, pp. 350–358.

Johns, M.M., Marshall, W.E. and Toles, C.A., 1998. Agricultural by-products as granular activated carbons for adsorbing dissolved metals and organics. *Journal of Chemical Technology & Biotechnology: International Research in Process, Environmental AND Clean Technology*, 71(2), pp. 131–140.

Kadirvelu, K., Thamaraiselvi, K. and Namasivayam, C., 2001. Removal of heavy metals from industrial wastewaters by adsorption onto activated carbon prepared from an agricultural solid waste. *Bioresource Technology*, 76(1), pp. 63–65.

Kannan, N. and Sundaram, M.M., 2001. Kinetics and mechanism of removal of methylene blue by adsorption on various carbons—A comparative study. *Dyes and Pigments*, 51(1), pp. 25–40.

Karić, N., Maia, A.S., Teodorović, A., Atanasova, N., Langergraber, G., Crini, G., Ribeiro, A.R. and Đolić, M., 2022. Bio-waste valorisation: Agricultural wastes as biosorbents for removal of (in)organic pollutants in wastewater treatment. *Chemical Engineering Journal Advances*, 9, p. 100239.

Karthikeyan, T., Rajgopal, S. and Miranda, L.R., 2005. Chromium (VI) adsorption from aqueous solution by *Hevea brasilinesis* sawdust activated carbon. *Journal of Hazardous Materials*, 124(1–3), pp. 192–199.

Khan, N., Chowdhary, P., Gnansounou, E. and Chaturvedi, P., 2021. Biochar and environmental sustainability: Emerging trends and techno-economic perspectives. *Bioresource Technology*, 332, p. 125102.

Khan, N.A., Ali, S.I. and Ayub, S., 2001. Effect of pH on the removal of chromium (Cr)(VI) by sugar cane bagasse. *Sultan Qaboos University Journal for Science* [SQUJS], 6(2), pp. 13–19.

Kılıc, M., Kırbıyık, Ç., Çepelioğullar, Ö. and Pütün, A.E., 2013. Adsorption of heavy metal ions from aqueous solutions by bio-char, a by-product of pyrolysis. *Applied Surface Science*, 283, pp. 856–862.

Kim, H., Jang, G. and Yoon, Y., 2020. Specific heavy metal/metalloid sensors: Current state and perspectives. *Applied Microbiology and Biotechnology*, 104(3), pp. 907–914.

Koul, B., Yakoob, M. and Shah, M.P., 2022. Agricultural waste management strategies for environmental sustainability. *Environmental Research*, 206, p. 112285.

Krishnan, K.A., Sreejalekshmi, K.G., Vimexen, V. and Dev, V.V., 2016. Evaluation of adsorption properties of sulphurised activated carbon for the effective and economically viable removal of Zn (II) from aqueous solutions. *Ecotoxicology and Environmental Safety*, 124, pp. 418–425.

Kumar, B. and Verma, P., 2021. Life cycle assessment: Blazing a trail for bioresources management. *Energy Conversion and Management: X*, 10, p. 100063.

Kumar, S., Prasad, S., Yadav, K.K., Shrivastava, M., Gupta, N., Nagar, S., Bach, Q.V., Kamyab, H., Khan, S.A., Yadav, S. and Malav, L.C., 2019. Hazardous heavy metals contamination of vegetables and food chain: Role of sustainable remediation approaches – A review. *Environmental Research*, 179, p. 108792.

Kurniawan, T.A., Othman, M.H.D., Hwang, G.H. and Gikas, P., 2022. Unlocking digital technologies for waste recycling in Industry 4.0 era: A transformation towards a digitalization-based circular economy in Indonesia. *Journal of Cleaner Production*, 357, p. 131911.

Langhorst, T., McCord, S., Zimmermann, A., Müller, L., Cremonese, L., Strunge, T., Wang, Y., Zaragoza, A.V., Wunderlich, J., Marxen, A. and Armstrong, K., 2022. Techno-Economic Assessment & Life Cycle Assessment Guidelines for CO_2 Utilization (Version 2.0). Global CO_2 Initiative. https://dx.doi.org/10.7302/4190

Lesmana, S.O., Febriana, N., Soetaredjo, F.E., Sunarso, J. and Ismadji, S., 2009. Studies on potential applications of biomass for the separation of heavy metals from water and wastewater. *Biochemical Engineering Journal*, 44(1), pp. 19–41.

Loiseau, E., Junqua, G., Roux, P. and Bellon-Maurel, V., 2012. Environmental assessment of a territory: An overview of existing tools and methods. *Journal of Environmental Management*, 112, pp. 213–225.

Mahale, K.K., Mokhasi, H.R. and Ashoka, H., 2016. Biosorption of nickel (II) from aqueous solutions using potato peel. *Research Journal of Chemical and Environmental Science* 4, pp. 96–101.

Mahmud, R., Moni, S.M., High, K. and Carbajales-Dale, M., 2021. Integration of techno-economic analysis and life cycle assessment for sustainable process design – A review. *Journal of Cleaner Production*, 317, p. 128247.

Malik, R., Ramteke, D.S. and Wate, S.R., 2007. Adsorption of malachite green on groundnut shell waste based powdered activated carbon. *Waste Management*, 27(9), pp. 1129–1138.

Malik, R. and Dahiya, S., 2017. An experimental and quantum chemical study of removal of utmostly quantified heavy metals in wastewater using coconut husk: A novel approach to mechanism. *International Journal of Biological Macromolecules*, 98, pp. 139–149.

Malkoc, E. and Nuhoglu, Y., 2006. Removal of Ni(II) ions from aqueous solutions using waste of tea factory: Adsorption on a fixed-bed column. *Journal of Hazardous Materials*, 135(1–3), pp. 328–336.

Mallampati, R., Xuanjun, L., Adin, A. and Valiyaveettil, S., 2015. Fruit peels as efficient renewable adsorbents for removal of dissolved heavy metals and dyes from water. *ACS Sustainable Chemistry & Engineering*, 3(6), pp. 1117–1124.

McAuliffe, G.A., Takahashi, T. and Lee, M.R., 2020. Applications of nutritional functional units in commodity-level life cycle assessment (LCA) of agri-food systems. *The International Journal of Life Cycle Assessment*, 25, pp. 208–221.

Meng, L., Sun, T., Li, M., Saleem, M., Zhang, Q. and Wang, C., 2019. Soil-applied biochar increases microbial diversity and wheat plant performance under herbicide fomesafen stress. *Ecotoxicology and Environmental Safety*, 171, pp. 75–83.

Memon, S.Q., Memon, N., Shah, S.W., Khuhawar, M.Y. and Bhanger, M.I., 2007. Sawdust—A green and economical sorbent for the removal of cadmium (II) ions. *Journal of Hazardous Materials*, 139(1), pp. 116–121.

Merem, E.C., Twumasi, Y., Wesley, J., Alsarari, M., Fageir, S., Crisler, M., Romorno, C., Olagbegi, D., Hines, A., Ochai, G.S. and Nwagboso, E., 2019. Regional assessment of climate change hazards in Southern Nigeria with GIS. *Journal of Safety Engineering*, 8(1), pp. 9–27.

Mihai, F.C., Plana, R., Arizmendiarrieta, J.S. and Irigoyen, I., 2023. Business models of composting for a circular economy. In S.P. Singh, A. Pandey, G. Du, and S. Kumar (Eds.), *Current Developments in Biotechnology and Bioengineering* (pp. 319–345). Elsevier.

Mohan, D. and Singh, K.P., 2002. Single-and multi-component adsorption of cadmium and zinc using activated carbon derived from bagasse—An agricultural waste. *Water Research*, 36(9), pp. 2304–2318.

Montanher, S.F., Oliveira, E.A. and Rollemberg, M.C., 2005. Removal of metal ions from aqueous solutions by sorption onto rice bran. *Journal of Hazardous Materials*, 117(2–3), pp. 207–211.

Moubarik, A. and Grimi, N., 2015. Valorization of olive stone and sugar cane bagasse by-products as biosorbents for the removal of cadmium from aqueous solution. *Food Research International*, 73, pp. 169–175.

Nabavi-Pelesaraei, A., Saber, Z., Mostashari-Rad, F., Ghasemi-Mobtaker, H. and Chau, K.W., 2021. Coupled life cycle assessment and data envelopment analysis to optimize energy consumption and mitigate environmental impacts in agricultural production. In J. Ren (Ed.), *Methods in Sustainability Science* (pp. 227–264). Elsevier.

Narayanasamy, S., Sundaram, V., Sundaram, T. and Vo, D.V.N., 2022. Biosorptive ascendency of plant based biosorbents in removing hexavalent chromium from aqueous solutions–Insights into isotherm and kinetic studies. *Environmental Research*, 210, p. 112902.

Narumalani, S., Zhou, Y. and Jensen, J.R., 1997. Application of remote sensing and geographic information systems to the delineation and analysis of riparian buffer zones. *Aquatic Botany*, 58(3–4), pp. 393–409.

Nhapi, I., Banadda, N., Murenzi, R., Sekomo, C.B. and Wali, U.G., 2011. Removal of heavy metals from industrial wastewater using rice husks. *The Open Environmental Engineering Journal*, 4(1), pp. 170–180.

Nita, A., Fineran, S. and Rozylowicz, L., 2022. Researchers' perspective on the main strengths and weaknesses of environmental impact assessment (EIA) procedures. *Environmental Impact Assessment Review*, 92, p. 106690.

Noble, B.F., 2023. EA simplification: Canadian processes and challenges. *Impact Assessment and Project Appraisal*, 41(3), pp. 212–214.

Okafor, P.C., Okon, P.U., Daniel, E.F. and Ebenso, E.E., 2012. Adsorption capacity of coconut (*Cocos nucifera* L.) shell for lead, copper, cadmium and arsenic from aqueous solutions. *International Journal of Electrochemical Science*, 7(12), pp. 12354–12369.

Olayebi, O.O. and Adebayo, A.T., 2017. Removal of Heavy Metals from Petroleum Refinery Effluent Using Coconut Shell Effluent Using Coconut Shell-Based Activated Based Activated Carbon. *CARD International Journal of Engineering and Emerging Scientific Discovery*, 2, pp. 102–117.

Oliveira, E.A., Montanher, S.F., Andrade, A.D., Nobrega, J.A. and Rollemberg, M.C., 2005. Equilibrium studies for the sorption of chromium and nickel from aqueous solutions using raw rice bran. *Process Biochemistry*, 40(11), pp. 3485–3490.

Organization for Economic Co-operation and Development (OECD), 2019. Background Notes on Sustainable, Productive and Resilient Agro-Food Systems: Value Chains, Human Capital, and the 2030 Agenda. A Report to the G20 Agriculture Deputies July 2019.

Pérez, S., Ulloa, M., Flórez, E., Acelas, N., Ocampo-Pérez, R., Padilla-Ortega, E. and Forgionny, A., 2023. Valorization of lemon peels wastes into a potential adsorbent for simultaneous removal of copper ion (Cu^{2+}) and Congo red from wastewater. *Environmental Nanotechnology, Monitoring & Management*, 20, p. 100795.

Rajamohan, N., 2014. Biosorption of mercury onto protonated pistachio hull wastes-effect of variables and kinetic experiments. *International Journal of Chemical Engineering and Applications*, 5(5), p. 415.

Rebitzer, G., Ekvall, T., Frischknecht, R., Hunkeler, D., Norris, G., Rydberg, T., Schmidt, W.P., Suh, S., Weidema, B.P. and Pennington, D.W., 2004. Life cycle assessment: Part 1: Framework, goal and scope definition, inventory analysis, and applications. *Environment International*, 30(5), pp. 701–720.

Roberts, K.G., Gloy, B.A., Joseph, S., Scott, N.R., Lehmann, J., 2010. Life cycle assessment of biochar systems: Estimating the energetic, economic, and climate change potential. *Environmental Science & Technology*, 44, pp. 827–833.

Rodríguez, R., Espada, J.J., Gallardo, M., Molina, R. and López-Muñoz, M.J., 2018. Life cycle assessment and techno-economic evaluation of alternatives for the treatment of wastewater in a chrome-plating industry. *Journal of Cleaner Production*, 172, pp. 2351–2362.

Saba, B., Christy, A.D. and Shah, A., 2024. Hydrochar for pollution remediation: Effect of process parameters, adsorption modeling, life cycle assessment and techno-economic evaluation. *Resources, Conservation and Recycling*, 202, pp. 107–359.

Sadhukhan, J., Gadkari, S., Martinez-Hernandez, E., Ng, K.S., Shemfe, M., Torres-Garcia, E. and Lynch, J., 2019. Novel macroalgae (seaweed) biorefinery systems for integrated chemical, protein, salt, nutrient and mineral extractions and environmental protection by green synthesis and life cycle sustainability assessments. *Green Chemistry*, 21(10), pp. 2635–2655.

Sala, S., Ciuffo, B. and Nijkamp, P., 2015. A systemic framework for sustainability assessment. *Ecological Economics*, 119, pp. 314–325.

Sallau, A.B., Aliyu, S. and Ukuwa, S., 2012. Biosorption of chromium (VI) from aqueous solution by corn cob powder. *International Journal of Environment and Bioenergy*, 4(3), pp. 131–140.

Saravanan, A., Jeevanantham, S. and Karishma, S., 2022. Biochar as sustainable strategy for remediation and regeneration of heavy metal–Contaminated soil. In R. Thapar Kapoor and M.P. Shah (Eds.), *BioChar: Applications for Bioremediation of Contaminated Systems* (pp. 417–430). De Gruyter.

Sciban, M., Klasnja, M. and Skrbic, B., 2006. Modified hardwood sawdust as adsorbent of heavy metal ions from water. *Wood Science and Technology*, 40(3), pp. 217–227.

Sehker, M. and Beukering, P.V., 1998. Integrated Solid Waste Management: A Perspective on Bangalore (India) (No. 24, pp. 277–295).

Seidavi, A.R., Zaker-Esteghamati, H. and Scanes, C.G., 2019. Present and potential impacts of waste from poultry production on the environment. *World's Poultry Science Journal*, 75(1), pp. 29–42.

Seifert, C., Koep, L., Wolf, P. and Guenther, E., 2021. Life cycle assessment as decision support tool for environmental management in hospitals: A literature review. *Health Care Management Review*, 46(1), pp. 12–24.

Sevigné-Itoiz, E., Mwabonje, O., Panoutsou, C. and Woods, J., 2021. Life cycle assessment (LCA): Informing the development of a sustainable circular bioeconomy. *Philosophical Transactions of the Royal Society A*, 379(2206), p. 20200352.

Shabani, N., Akhtari, S. and Sowlati, T., 2013. Value chain optimization of forest biomass for bioenergy production: A review. *Renewable and Sustainable Energy Reviews*, 23, pp. 299–311.

Shaheen, S.M., Niazi, N.K., Hassan, N.E., Bibi, I., Wang, H., Tsang, D.C., Ok, Y.S., Bolan, N. and Rinklebe, J., 2019. Wood-based biochar for the removal of potentially toxic elements in water and wastewater: A critical review. *International Materials Reviews*, 64(4), pp. 216–247.

Shakya, A. and Agarwal, T., 2019. Removal of Cr (VI) from water using pineapple peel derived biochars: Adsorption potential and re-usability assessment. *Journal of Molecular Liquids*, 293, p. 111497.

Shang, H., Li, Y., Liu, J., Wan, Y., Feng, Y. and Yu, Y., 2020. Preparation of nitrogen doped magnesium oxide modified biochar and its sorption efficiency of lead ions in aqueous solution. *Bioresource Technology*, 314, p. 123708.

Sharma, R., Parhi, S. and Shishodia, A., 2020. Industry 4.0 applications in agriculture: Cyber-physical agricultural systems (CPASs). In *Advances in Mechanical Engineering: Select Proceedings of ICAME 2020*, pp. 807–813.

Sharma, R.K., Agrawal, M. and Marshall, F., 2007. Heavy metal contamination of soil and vegetables in suburban areas of Varanasi, India. *Ecotoxicology and Environmental Safety*, 66(2), pp. 258–266.

Sharma, S., Tiwari, S., Hasan, A., Saxena, V. and Pandey, L.M., 2018. Recent advances in conventional and contemporary methods for remediation of heavy metal-contaminated soils. *3 Biotech*, 8, pp. 1–18.

Shen, Z., Som, A.M., Wang, F., Jin, F., McMillan, O. and Al-Tabbaa, A., 2016. Long-term impact of biochar on the immobilisation of nickel(II) and zinc(II) and the revegetation of a contaminated site. *Science of the Total Environment*, 542, pp. 771–776.

Shukla, S.R. and Pai, R.S., 2005. Removal of Pb (II) from solution using cellulose-containing materials. *Journal of Chemical Technology & Biotechnology: International Research in Process, Environmental & Clean Technology*, 80(2), pp. 176–183.

Sieverding, H., Kebreab, E., Johnson, J.M., Xu, H., Wang, M., Grosso, S.J.D., Bruggeman, S., Stewart, C.E., Westhoff, S., Ristau, J. and Kumar, S., 2020. A life cycle analysis (LCA) primer for the agricultural community. *Agronomy Journal*, 112(5), pp. 3788–3807.

Singh, P.K. and Sharma, R.K., 2021. Wastewater reuse in peri-urban agriculture ecosystem: Current scenario, consequences, and control measures. *Water Pollution and Management Practices*, pp. 121–146.

Singh, P.K., Yadav, J.S., Kumar, I., Kumar, U. and Sharma, R.K., 2022. Carpet industry irrigational sources risk assessment: Heavy metal contaminated vegetables and cereal crops in northern India. *Toxicology Reports*, 9, pp. 1906–1919.

Sulyman, M., Namiesnik, J. and Gierak, A., 2017. Low-cost adsorbents derived from agricultural by-products/wastes for enhancing contaminant uptakes from wastewater: A review. *Polish Journal of Environmental Studies*, 26(3).

Tang, Y., Zhao, W., Gao, L., Zhu, G., Jiang, Y., Rui, Y. and Zhang, P., 2023. Harnessing synergy: Integrating agricultural waste and nanomaterials for enhanced sustainability. *Environmental Pollution*, 341, p. 123023.

Taty-Costodes, V.C., Fauduet, H., Porte, C. and Delacroix, A., 2003. Removal of Cd (II) and Pb (II) ions, from aqueous solutions, by adsorption onto sawdust of *Pinus sylvestris*. *Journal of Hazardous Materials*, 105(1–3), pp. 121–142.

Tee, G.T., Gok, X.Y. and Yong, W.F., 2022. Adsorption of pollutants in wastewater via biosorbents, nanoparticles and magnetic biosorbents: A review. *Environmental Research*, 212, p. 113248.

Thangaraju, S., Pulivarthi, M.K. and Natarajan, V., 2020. Waste from oil-seed industry: A sustainable approach. In M. Thakur, V.K. Modi, R. Khedkar and K. Singh (Eds.), *Sustainable Food Waste Management: Concepts and Innovations* (pp. 177–190). Springer.

Tripathi, N., Hills, C.D., Singh, R.S. and Atkinson, C.J., 2019. Biomass waste utilisation in low-carbon products: Harnessing a major potential resource. *NPJ Climate and Atmospheric Science*, 2(1), p. 35.

Tushar, Q., Salehi, S., Santos, J., Zhang, G., Bhuiyan, M.A., Arashpour, M. and Giustozzi, F., 2023. Application of recycled crushed glass in road pavements and pipeline bedding: An integrated environmental evaluation using LCA. *Science of the Total Environment*, 881, p. 163488.

Wasewar, K.L., Atif, M., Prasad, B. and Mishra, I.M., 2009. Batch adsorption of zinc on tea factory waste. *Desalination*, 244(1–3), pp. 66–71.

Wilson, K., Yang, H., Seo, C.W. and Marshall, W.E., 2006. Select metal adsorption by activated carbon made from peanut shells. *Bioresource Technology*, 97(18), pp. 2266–2270.

Wong, K.K., Lee, C.K., Low, K.S. and Haron, M.J., 2003. Removal of Cu and Pb by tartaric acid modified rice husk from aqueous solutions. *Chemosphere*, 50(1), pp. 23–28.

Yadav, G., Yadav, N., Sultana, M. and Ahmaruzzaman, M., 2023. A comprehensive review on low-cost waste-derived catalysts for environmental remediation. *Materials Research Bulletin*, 164, p. 112261.

Yang, H., Feng, Q., Zhu, J., Liu, G., Dai, Y., Zhou, Q., Xia, S., Wu, Z. and Zhang, Y., 2024. Towards sustainable futures: A review of sediment remediation and resource valorization techniques. *Journal of Cleaner Production*, 435(2), p. 140529.

Yang, M., Chen, L., Wang, J., Msigwa, G., Osman, A.I., Fawzy, S., Rooney, D.W. and Yap, P.S., 2023. Circular economy strategies for combating climate change and other environmental issues. *Environmental Chemistry Letters*, 21(1), pp. 55–80.

Yang, X., Lu, K., McGrouther, K., Che, L., Hu, G., Wang, Q., Liu, X., Shen, L., Huang, H., Ye, Z. and Wang, H., 2017. Bioavailability of Cd and Zn in soils treated with biochars derived from tobacco stalk and dead pigs. *Journal of Soils and Sediments*, 17, pp. 751–762.

Younas, F., Younas, S., Bibi, I., Farooqi, Z.U.R., Hameed, M.A., Mohy-Ud-Din, W., Shehzad, M.T., Hussain, M.M., Shakil, Q., Shahid, M. and Niazi, N.K., 2024. A critical review on the separation of heavy metal(loid)s from the contaminated water using various agricultural wastes. *International Journal of Phytoremediation*, 26(3), pp. 349–368.

Yousef, R., Qiblawey, H. and El-Naas, M.H., 2020. Adsorption as a process for produced water treatment: A review. *Processes*, 8(12), p. 1657.

Yu, B., Zhang, Y., Shukla, A., Shukla, S.S. and Dorris, K.L., 2000. The removal of heavy metal from aqueous solutions by sawdust adsorption—removal of copper. *Journal of Hazardous Materials*, 80(1–3), pp. 33–42.

Zakir, M., Botahala, L., Ramang, M., Fauziah, S. and Abdussamad, B., 2013. Elektrodeposisi Logam Mn pada Permukaan Karbon Aktif Sekam Padi dengan Iradiasi Ultrasonik. *Indonesia Chimica Acta*, 6(2), pp. 9–18.
Zein, R., Suhaili, R., Earnestly, F. and Munaf, E., 2010. Removal of Pb (II), Cd (II) and Co (II) from aqueous solution using *Garcinia mangostana* L. fruit shell. *Journal of Hazardous Materials*, 181(1–3), pp. 52–56.
Zeug, W., Bezama, A. and Thrän, D., 2023. Life cycle sustainability assessment for sustainable bioeconomy, societal-ecological transformation and beyond. In F. Hesser, et al. (Eds.), *Progress in Life Cycle Assessment 2021* (pp. 131–159). Springer.
Zhang, J., Qin, Q., Li, G. and Tseng, C.H., 2021. Sustainable municipal waste management strategies through life cycle assessment method: A review. *Journal of Environmental Management*, 287, p. 112238.
Zhao, F., Zou, G., Shan, Y., Ding, Z., Dai, M. and He, Z., 2019. Coconut shell derived biochar to enhance water spinach (*Ipomoea aquatica* Forsk) growth and decrease nitrogen loss under tropical conditions. *Scientific Reports*, 9(1), p. 20291.
Zhao, Y., Qamar, S.A., Qamar, M., Bilal, M. and Iqbal, H.M., 2021. Sustainable remediation of hazardous environmental pollutants using biochar-based nanohybrid materials. *Journal of Environmental Management*, 300, p. 113762.
Zhou, N., Chen, H., Xi, J., Yao, D., Zhou, Z., Tian, Y. and Lu, X., 2017. Biochars with excellent Pb (II) adsorption property produced from fresh and dehydrated banana peels via hydrothermal carbonization. *Bioresource Technology*, 232, pp. 204–210.
Zimmermann, A.W., Wunderlich, J., Müller, L., Buchner, G.A., Marxen, A., Michailos, S., Armstrong, K., Naims, H., McCord, S., Styring, P. and Sick, V., 2020. Techno-economic assessment guidelines for CO_2 utilization. *Frontiers in Energy Research*, 8, p. 5.

Index

Note: Page numbers in **bold** and *italics* refer to tables and figures, respectively.

D

E

I

V

W

X

Z